MEI structured mathematics

Pure
Mathematics 6

TERRY HEARD
DAVID MARTIN

Series Editor: Roger Porkess

MEI Structured Mathematics is supported by industry:
BNFL, Casio, Esso, GEC, Intercity, JCB, Lucas, The National Grid Company,
Sharp, Texas Instruments, Thorn EMI

Acknowledgements

The authors and publishers would like to thank the following companies, institutions and individuals who have given permission to reproduce copyright materials. The publishers will be happy to make arrangments with any copyright holders whom it has not been possible to contact.

Helicon Publishing Ltd; *The World Weather Guide*, Pearce & Smith 1993 (p. 106).

The illustrations were drawn by Jeff Edwards and Hugh Neill.

Photographs:
(Page 98) Science Photo Library; (page 177) British Library Reproduction from *A History of Mathematics* by Carl Boyer (Wiley International Edition, 1968).

British Library Cataloguing in Publication Data

Heard, Terry
   Pure mathematics 6. – (MEI structured mathematics)
   1.Mathematics 2.Mathematics – Problems, exercises, etc.
   I.Title II.Martin, David III.Mathematics in Education and Industry
   510

ISBN 0 340 688017

First published 1998
Impression number   10 9 8 7 6 5 4 3 2 1
Year   2004 2003 2002 2001 2000 1999 1998

Typeset by the Alden Group, Oxford.
Printed in Great Britain for Hodder & Stoughton Educational, a division of Hodder Headline PLC, 338 Euston Road, London NW1 3BH by Scotprint Ltd, Musselburgh, Scotland.

MEI Structured Mathematics

Mathematics is not only a beautiful and exciting subject in its own right but also one that underpins many other branches of learning. It is consequently fundamental to the success of a modern economy.

MEI Structured Mathematics is designed to increase substantially the number of people taking the subject post-GCSE, by making it accessible, interesting and relevant to a wide range of students.

It is a credit accumulation scheme based on 45 hour components which may be taken individually or aggregated to give:

3 Components   AS Mathematics
6 Components   A Level Mathematics
9 Components   A Level Mathematics + AS Further Mathematics
12 Components   A Level Mathematics + A Level Further Mathematics

Components may alternatively be combined to give other A or AS certifications (in Statistics, for example) or they may be used to obtain credit towards other types of qualification.

The course is examined by OCR (previously OCEAC), with examinations held in January and June each year.

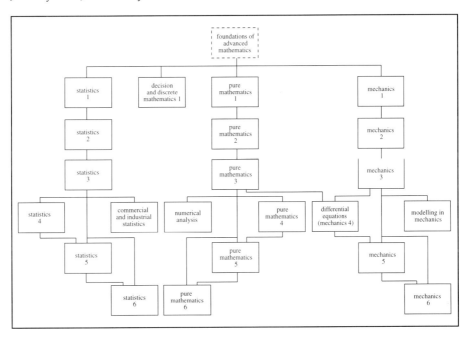

This is one of the series of books written to support the course. Its position within the whole scheme can be seen in the diagram above.

Mathematics in Education and Industry is a curriculum development body which aims to promote the links between Education and Industry in Mathematics, and to produce relevant examination and teaching syllabuses and support material. Since its foundation in the 1960s, MEI has provided syllabuses for GCSE (or O Level), Additional Mathematics and A Level.

For more information about MEI Structured Mathematics or other syllabuses and materials, write to MEI Office, 11 Market Street, Bradford-on-Avon BA15 1LL.

Introduction

This book completes the series covering the Pure Mathematics components of the MEI Structured Mathematics course by exploring five main topics. Since you need to prepare just three of these for the assessment of the course, the five chapters are almost completely independent and may be tackled in any order. The exceptions to this are that the work on envelopes (chapter 4) uses partial differentiation (chapter 3), and knowledge of eigenvectors from chapter 1 will help you at the end of chapter 5. Of course, we hope that you will be tempted to read the whole book!

The text is intended to be accessible to students who have to work on their own for much of the time. The frequent activities are a means of getting you involved in your learning (replacing to some extent the interaction you might have with a teacher), and should not be skipped; answers or hints to activities and all exercises are provided.

The topics here overlap considerably with some university work, so we hope that this book may help to smooth the transition from secondary to tertiary education. For this reason (and for mathematical reasons too) we have sometimes gone beyond what is strictly needed for examination purposes (as detailed in the current syllabus), either in the extent of the coverage or in the level of rigour used. We hope too that the material presented here may be useful to students on some other further mathematics or higher education courses.

Many people have helped us with this book. Nick Lord, Gerry Leversha and Randal Cousins all contributed valuable ideas, as did the Series Editor, Roger Porkess, who took a close interest in the development of the whole book. Mike Jones read the various drafts and checked the answers; little escapes his eagle eye, and his detailed comments and suggestions have led to many improvements. The editorial staff at Hodder and Stoughton have been unfailingly helpful, and our colleagues and families, especially Heather Martin, have spurred us on when energy was flagging. We gratefully acknowledge the substantial debt we owe to all these friends.

Terry Heard and David Martin

Contents

Vectors and matrices

If you cannot solve the proposed problem try to solve first some related problem.

George Polya (1887–1985)

Building works at an airport require the use of a crane near the end of the runway.

How far is it from the top of the crane to the flight path of the plane?

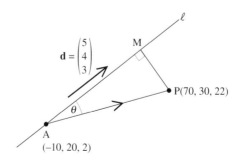

plane
taking off

runway

crane

Calculating distances

Distance of a point from a line

To answer the question above you need to know the flight path and the position of the top of the crane. Working in metres, suppose planes take off along the line ℓ: $\dfrac{x+10}{5} = \dfrac{y-20}{4} = \dfrac{z-2}{3}$ and that the top of the crane is at P(70, 30, 22), as illustrated in figure 1.1.

$$\mathbf{d} = \begin{pmatrix} 5 \\ 4 \\ 3 \end{pmatrix}$$

M

P(70, 30, 22)

θ

A
(−10, 20, 2)

Figure 1.1

The shortest distance from a point P to a straight line ℓ is measured along the line which is perpendicular to ℓ. This distance, PM in figure 1.1, is the distance from the point P to the line ℓ. The vector product (see *Pure Mathematics 4*) provides a convenient way of calculating such distances.

Since PM is perpendicular to ℓ

$$PM = AP \sin PAM$$

$$= |\overrightarrow{AP} \times \hat{\mathbf{d}}|$$

$\mathbf{a} \times \mathbf{b} = |\mathbf{a}||\mathbf{b}| \sin \theta \, \hat{\mathbf{n}},$
where $\hat{\mathbf{n}}$ is a unit vector perpendicular to $\mathbf{a}$ and $\mathbf{b}$.

where $\hat{\mathbf{d}}$ is the unit vector parallel to $\mathbf{d}$.

Using the convention that $\mathbf{a}$ and $\mathbf{p}$ represent the position vectors of A and P respectively, we may write:

$$PM = |(\mathbf{p} - \mathbf{a}) \times \hat{\mathbf{d}}|$$

Since $\mathbf{d} = |\mathbf{d}|\hat{\mathbf{d}}$, we may also write:

$$PM = \frac{|(\mathbf{p} - \mathbf{a}) \times \mathbf{d}|}{|\mathbf{d}|}$$

a result which is harder to remember but easier to use.

Returning to calculating the distance from the top of the crane to the flight path, we have:

$$\mathbf{p} = \begin{pmatrix} 70 \\ 30 \\ 22 \end{pmatrix}, \mathbf{a} = \begin{pmatrix} -10 \\ 20 \\ 2 \end{pmatrix} \text{ and } \mathbf{d} = \begin{pmatrix} 5 \\ 4 \\ 3 \end{pmatrix} \text{ so that:}$$

$$\overrightarrow{AP} = \mathbf{p} - \mathbf{a} = \begin{pmatrix} 70 \\ 30 \\ 22 \end{pmatrix} - \begin{pmatrix} -10 \\ 20 \\ 2 \end{pmatrix} = \begin{pmatrix} 80 \\ 10 \\ 20 \end{pmatrix} = 10 \begin{pmatrix} 8 \\ 1 \\ 2 \end{pmatrix} \text{ and}$$

$$\begin{pmatrix} a_1 \\ a_2 \\ a_3 \end{pmatrix} \times \begin{pmatrix} b_1 \\ b_2 \\ b_3 \end{pmatrix} = \begin{pmatrix} a_2 b_3 - a_3 b_2 \\ a_3 b_1 - a_1 b_3 \\ a_1 b_2 - a_2 b_1 \end{pmatrix}$$

$$\overrightarrow{AP} \times \mathbf{d} = 10 \begin{pmatrix} 8 \\ 1 \\ 2 \end{pmatrix} \times \begin{pmatrix} 5 \\ 4 \\ 3 \end{pmatrix} = 10 \begin{pmatrix} 3 - 8 \\ 10 - 24 \\ 32 - 5 \end{pmatrix} = 10 \begin{pmatrix} -5 \\ -14 \\ 27 \end{pmatrix}.$$

Therefore:

$$|\overrightarrow{AP} \times \mathbf{d}| = 10\sqrt{(-5)^2 + (-14)^2 + 27^2} = 10\sqrt{950} = 10\sqrt{25 \times 38} = 50\sqrt{38}$$

while $|\mathbf{d}| = \sqrt{5^2 + 4^2 + 3^2} = \sqrt{50} = 5\sqrt{2}$.

Therefore the (shortest) distance from P to ℓ is:

$$PM = \frac{50\sqrt{38}}{5\sqrt{2}} = 10\sqrt{19} \approx 43.6 \text{ metres.}$$

The technique just illustrated does not tell you which point on line ℓ is closest to P. In such situations the scalar product provides a way of finding the position of M, the foot of the perpendicular from P to ℓ. In figure 1.2, A, with position vector $\mathbf{a}$, is any (known) point on ℓ, and $\mathbf{d}$ is any vector parallel to ℓ. The position vector of M is:

$$m = a + \overrightarrow{AM}$$

$$= a + \lambda d$$

where λ is a measure of the length of AM.

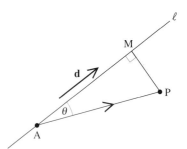

Figure 1.2

Now the length of AM is $AP \cos PAM = |\overrightarrow{AP} \cdot \hat{d}|$. The modulus function is needed here because $\overrightarrow{AP} \cdot \hat{d}$ is only positive when $\overrightarrow{AM}$ and d have the same sense (i.e. point in the same direction, as illustrated). When $\overrightarrow{AM}$ and d have opposite senses $\overrightarrow{AP} \cdot \hat{d}$ is negative. In both cases, $\overrightarrow{AM} = (\overrightarrow{AP} \cdot \hat{d})\hat{d}$, which

may also be written as $\dfrac{\overrightarrow{AP} \cdot d}{|d|^2} d$ since $d = |d|\hat{d}$. Thus the foot of the

perpendicular from the point P to the line $r = a + \lambda d$ is $a + (\overrightarrow{AP} \cdot \hat{d})\hat{d}$. You may prefer to write this as:

$$a + ((p - a) \cdot \hat{d})\hat{d} \quad \text{or} \quad a + \frac{\overrightarrow{AP} \cdot d}{|d|^2} d \quad \text{or} \quad a + \frac{(p - a) \cdot d}{|d|^2} d.$$

EXAMPLE

Find the coordinates of M, the foot of the perpendicular from the top of the crane, $P(70, 30, 22)$ to the line ℓ:

$$\frac{x + 10}{5} = \frac{y - 20}{4} = \frac{z - 2}{3}$$

representing the flight path.

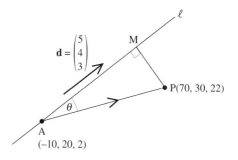

Solution

From the Cartesian equations of ℓ you may, as before, take A as the point

$(-10, 20, 2)$ and the direction vector for ℓ as $d = \begin{pmatrix} 5 \\ 4 \\ 3 \end{pmatrix}$. Then $\overrightarrow{AP} = p - a = 10 \begin{pmatrix} 8 \\ 1 \\ 2 \end{pmatrix}$

and $|d| = \sqrt{50}$.

Now $\overrightarrow{AP} \cdot \mathbf{d} = 10 \begin{pmatrix} 8 \\ 1 \\ 2 \end{pmatrix} \cdot \begin{pmatrix} 5 \\ 4 \\ 3 \end{pmatrix} = 10(40 + 4 + 6) = 500$ so that:

$$\frac{\overrightarrow{AP} \cdot \mathbf{d}}{|\mathbf{d}|^2} = \frac{500}{(\sqrt{50})^2} = 10 \quad \text{and} \quad \overrightarrow{AM} = \frac{\overrightarrow{AP} \cdot \mathbf{d}}{|\mathbf{d}|^2} \mathbf{d} = 10\mathbf{d} = \begin{pmatrix} 50 \\ 40 \\ 30 \end{pmatrix}.$$

So $\mathbf{m} = \mathbf{a} + \overrightarrow{AM} = \begin{pmatrix} -10 \\ 20 \\ 2 \end{pmatrix} + \begin{pmatrix} 50 \\ 40 \\ 30 \end{pmatrix} = \begin{pmatrix} 40 \\ 60 \\ 32 \end{pmatrix}$ and M is $(40, 60, 32)$.

Knowing the coordinates of both M and P allows us to use an alternative method of finding the length of PM:

$$PM = \sqrt{(40 - 70)^2 + (60 - 30)^2 + (32 - 22)^2} = \sqrt{1900} \approx 43.6 \text{ metres.}$$

Alternative solution

M lies on $\ell \Rightarrow$ M is of the form $(-10 + 5\lambda, \ 20 + 4\lambda, \ 2 + 3\lambda)$

$$\Rightarrow \overrightarrow{PM} = \begin{pmatrix} -80 + 5\lambda \\ -10 + 4\lambda \\ -20 + 3\lambda \end{pmatrix};$$

$\overrightarrow{PM}$ is perpendicular to $\ell \Rightarrow \begin{pmatrix} -80 + 5\lambda \\ -10 + 4\lambda \\ -20 + 3\lambda \end{pmatrix} \cdot \begin{pmatrix} 5 \\ 4 \\ 3 \end{pmatrix} = 0$

$$\Rightarrow -400 + 25\lambda - 40 + 16\lambda - 60 + 9\lambda = 0$$

$$\Rightarrow \lambda = 10 \Rightarrow \text{M is } (40, 60, 32).$$

At the start of this chapter you saw that the distance from the point P, with position vector $\mathbf{p}$, to the line $\mathbf{r} = \mathbf{a} + \lambda \mathbf{d}$ may be written in any of the following ways:

$$|\overrightarrow{AP} \times \hat{\mathbf{d}}| = |(\mathbf{p} - \mathbf{a}) \times \hat{\mathbf{d}}| = \frac{|(\mathbf{p} - \mathbf{a}) \times \mathbf{d}|}{|\mathbf{d}|}.$$

As the vector product of vectors $\mathbf{a}$ and $\mathbf{b}$ is a vector perpendicular to both $\mathbf{a}$ and $\mathbf{b}$ these results presuppose that you are working in three dimensions. When you are working in two dimensions (in which case vectors have only two components) you can either quote the formula obtained in part (iii) of the activity below or adapt the process described in that activity.

Activity

Let points in the plane $z = 0$ correspond to points of two-dimensional space, so that the point $R'(x, y, 0)$ in three-dimensional space corresponds to the point $R(x, y)$ in two-dimensional space. Use the following steps to find the distance from $P(x_1, y_1)$ to the line $ax + by + c = 0$.

(i) The point A' in three-dimensional space corresponds to $A\left(0, -\dfrac{c}{b}\right)$. Write down the coordinates of A'.

(ii) Find p, q, r so that $p\mathbf{i} + q\mathbf{j} + r\mathbf{k}$ is parallel to the line in three-dimensional space which corresponds to the line $ax + by + c = 0$ in two-dimensional space.

(iii) Use the formula $\dfrac{|(\mathbf{p} - \mathbf{a}) \times \mathbf{d}|}{|\mathbf{d}|}$ to show that the distance from $P(x_1, y_1)$ to the line $ax + by + c = 0$ is $\left|\dfrac{ax_1 + by_1 + c}{\sqrt{a^2 + b^2}}\right|$.

(iv) Show that the expression $\dfrac{ax_1 + by_1 + c}{\sqrt{a^2 + b^2}}$ changes sign if the point (x_1, y_1) is moved to the opposite side of the line $ax + by + c = 0$.

The shortest distance between two skew lines

The shortest distance between two skew lines, ℓ_1 and ℓ_2, is measured along a line which is perpendicular to both ℓ_1 and ℓ_2. Figure 1.3 shows lines ℓ_1 and ℓ_2. Let π be the plane containing ℓ_2 and parallel to ℓ_1. Drop perpendiculars from points on ℓ_1 to π to form ℓ_1', the projection of ℓ_1 on π. The common perpendicular of ℓ_1 and ℓ_2 is the perpendicular from ℓ_1 that passes through Q, the point of intersection of ℓ_2 and ℓ_1', shown as PQ in the diagram.

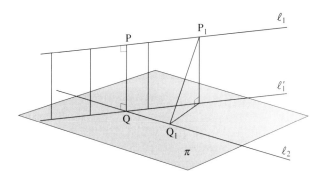

Figure 1.3

Activity

Explain why PQ is shorter than any other line (such as P_1Q_1) joining lines ℓ_1 and ℓ_2.

If ℓ_1 and ℓ_2 have equations $\mathbf{r} = \mathbf{a} + t\mathbf{d}$ and $\mathbf{r} = \mathbf{b} + t\mathbf{e}$ respectively, and you want to find the length of the common perpendicular PQ, it is tempting to start with the known result $PQ = |\overrightarrow{AP} \times \hat{\mathbf{d}}|$, but this time you do not know the position of P. However, you do know a point on each line, A and B, and a vector parallel to each line, $\mathbf{d}$ and $\mathbf{e}$ respectively.

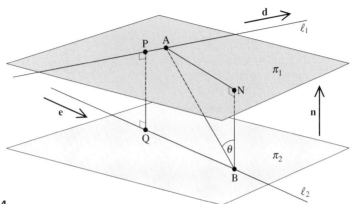

Figure 1.4

Figure 1.4 shows lines ℓ_1 and ℓ_2 and two parallel planes:

π_1 contains ℓ_1 and is parallel to ℓ_2

π_2 contains ℓ_2 and is parallel to ℓ_1.

Then PQ, the common perpendicular of ℓ_1 and ℓ_2, has the same length as any other perpendicular between the two planes, such as BN, where B is a point on ℓ_2. If A is any point on ℓ_1 and angle $ABN = \theta$:

$$PQ = BN = BA\cos\theta = |\overrightarrow{BA} \cdot \hat{\mathbf{n}}|$$

where $\hat{\mathbf{n}}$ is a unit vector parallel to BN, i.e. perpendicular to both planes. The modulus function is used to ensure a positive answer: the vector $\hat{\mathbf{n}}$ may be directed from π_1 towards π_2, making $\overrightarrow{BA} \cdot \hat{\mathbf{n}}$ negative.

Since π_1 and π_2 are parallel to ℓ_1 and ℓ_2, which are parallel to $\mathbf{d}$ and $\mathbf{e}$ respectively, you may take $\mathbf{d} \times \mathbf{e}$ as $\mathbf{n}$, then $\hat{\mathbf{n}} = \dfrac{(\mathbf{d} \times \mathbf{e})}{|\mathbf{d} \times \mathbf{e}|}$. Then:

$$PQ = BN = |\overrightarrow{BA} \cdot \hat{\mathbf{n}}| = \left| \frac{\overrightarrow{BA} \cdot (\mathbf{d} \times \mathbf{e})}{|\mathbf{d} \times \mathbf{e}|} \right| = \left| \frac{(\mathbf{a} - \mathbf{b}) \cdot (\mathbf{d} \times \mathbf{e})}{|\mathbf{d} \times \mathbf{e}|} \right|.$$

Our final result is just a complicated way of expressing $BA\cos\theta$!

EXAMPLE

Find the shortest distance between the lines $\ell_1 \colon \dfrac{x-8}{1} = \dfrac{y-9}{2} = \dfrac{z+2}{-3}$

and $\ell_2 \colon \dfrac{x-6}{1} = \dfrac{y}{-1} = \dfrac{z+2}{-2}$.

Solution

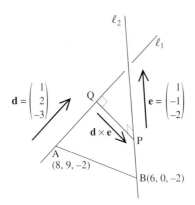

As shown on the diagram line ℓ_1 contains the point $A(8, 9, -2)$ and is parallel to $\mathbf{d} = \mathbf{i} + 2\mathbf{j} - 3\mathbf{k}$; line ℓ_2 contains the point $B(6, 0, -2)$ and is parallel to $\mathbf{e} = \mathbf{i} - \mathbf{j} - 2\mathbf{k}$.

Now
$$\mathbf{a} - \mathbf{b} = \begin{pmatrix} 8 \\ 9 \\ -2 \end{pmatrix} - \begin{pmatrix} 6 \\ 0 \\ -2 \end{pmatrix} = \begin{pmatrix} 2 \\ 9 \\ 0 \end{pmatrix}$$

and
$$\mathbf{d} \times \mathbf{e} = \begin{pmatrix} 1 \\ 2 \\ -3 \end{pmatrix} \times \begin{pmatrix} 1 \\ -1 \\ -2 \end{pmatrix} = \begin{pmatrix} -4 - 3 \\ -3 + 2 \\ -1 - 2 \end{pmatrix} = \begin{pmatrix} -7 \\ -1 \\ -3 \end{pmatrix}$$

so that
$$(\mathbf{a} - \mathbf{b}) . (\mathbf{d} \times \mathbf{e}) = \begin{pmatrix} 2 \\ 9 \\ 0 \end{pmatrix} . \begin{pmatrix} -7 \\ -1 \\ -3 \end{pmatrix} = -14 - 9 = -23$$

and $|\mathbf{d} \times \mathbf{e}| = \sqrt{7^2 + 1^2 + 3^2} = \sqrt{59}$.

Therefore the shortest (i.e. perpendicular) distance between ℓ_1 and ℓ_2 is:
$$\left| \frac{(\mathbf{a} - \mathbf{b}) . (\mathbf{d} \times \mathbf{e})}{|\mathbf{d} \times \mathbf{e}|} \right| = \frac{23}{\sqrt{59}} \approx 2.99 \, \text{units}.$$

As previously, this technique for finding the shortest distance between two skew lines does not tell you the positions of the ends of the shortest line segment joining the two lines. One way of doing that involves the following steps.

1. Find the equation of the plane π which contains ℓ_1 and is perpendicular to ℓ_2.

2. Find P, the point where line ℓ_2 meets plane π.

3. Find $\mathbf{d} \times \mathbf{e}$.

4. Find a general point R on the line through P parallel to $\mathbf{d} \times \mathbf{e}$.

5. By equating R to the general point on ℓ_1 find Q, the point where the line through P parallel to $\mathbf{d} \times \mathbf{e}$ meets ℓ_1.

Distance of a point from a plane

The distance from the point $P(x_1, y_1, z_1)$ to the plane π with equation $ax + by + cz + d = 0$ is PM where M is the foot of the perpendicular from P to the plane. Notice that PM is parallel to $\mathbf{n} = a\mathbf{i} + b\mathbf{j} + c\mathbf{k}$, the normal to plane π. Let R (with position vector $\mathbf{r}$) be any point on plane π.

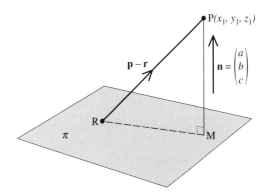

Figure 1.5

If the angle between vectors $\mathbf{p} - \mathbf{r}$ and $\mathbf{n}$ is acute (as illustrated in figure 1.5):

$$PM = RP \cos RPM = \overrightarrow{RP} \cdot \hat{\mathbf{n}} = (\mathbf{p} - \mathbf{r}) \cdot \hat{\mathbf{n}}$$

If the angle between $\mathbf{p} - \mathbf{r}$ and $\mathbf{n}$ is obtuse, $\cos RPM$ is negative and $PM = -(\mathbf{p} - \mathbf{r}) \cdot \hat{\mathbf{n}}$. As R is any point on π you may use $\mathbf{r} = -\dfrac{d}{c}\mathbf{k}$ which means that:

$$(\mathbf{p} - \mathbf{r}) \cdot \mathbf{n} = \begin{pmatrix} x_1 \\ y_1 \\ z_1 + \dfrac{d}{c} \end{pmatrix} \cdot \begin{pmatrix} a \\ b \\ c \end{pmatrix} = ax_1 + by_1 + cz_1 + d$$

and $PM = |(\mathbf{p} - \mathbf{r}) \cdot \hat{\mathbf{n}}| = \left| \dfrac{(\mathbf{p} - \mathbf{r}) \cdot \mathbf{n}}{|\mathbf{n}|} \right| = \left| \dfrac{ax_1 + by_1 + cz_1 + d}{\sqrt{a^2 + b^2 + c^2}} \right|.$

Notice how the formula for the distance of a point from a plane (in three-dimensional space) resembles the formula given on page 6 for the distance of a point from a line (in two-dimensional space).

Activity

It is noticed that $ax_1 + by_1 + cz_1 + d$ and $ax_2 + by_2 + cz_2 + d$ have opposite signs. Explain the significance of this observation.

1. Calculate the distance from P to the line ℓ.

(i) $\mathrm{P}(1,-2,3)$; ℓ: $\begin{cases} x = 2t+1 \\ y = 2t+5 \\ z = -t-1 \end{cases}$

(ii) $\mathrm{P}(2,3,-5)$; ℓ: $\mathbf{r} = \begin{pmatrix} 4 \\ 3 \\ 4 \end{pmatrix} + t \begin{pmatrix} 6 \\ -7 \\ 6 \end{pmatrix}$

(iii) $\mathrm{P}(8,9,1)$; ℓ: $\dfrac{x-6}{12} = \dfrac{y-5}{-9} = \dfrac{z-11}{-8}$

2. Find the distance from P to the line ℓ.

(i) $\mathrm{P}(8,9)$; ℓ: $3x+4y+5 = 0$

(ii) $\mathrm{P}(5,-4)$; ℓ: $6x-3y+3 = 0$

(iii) $\mathrm{P}(4,-4)$; ℓ: $8x+15y+11 = 0$

3. Find the distance from P to plane π.

(i) $\mathrm{P}(5,4,0)$; π: $6x+6y+7z+1 = 0$

(ii) $\mathrm{P}(7,2,-2)$; π: $12x-9y-8z+3 = 0$

(iii) $\mathrm{P}(-4,-5,3)$; π: $8x+5y-3z-4 = 0$

4. Are P and Q on the same side of the plane π? Justify your answer.

(i) $\mathrm{P}(1,1,3)$; $\mathrm{Q}(1,1,2)$; π: $5x-7y+2z-3 = 0$

(ii) $\mathrm{P}(0,3,5)$; $\mathrm{Q}(2,-1,1)$; π: $11x+5y+z+2 = 0$

(iii) $\mathrm{P}(1,2,2)$; $\mathrm{Q}(5,-2,3)$; π: $4x+6y-3z+1 = 0$

5. Find the shortest distance between lines ℓ and ℓ'.

(i) ℓ: $x-2 = \dfrac{y-3}{2} = \dfrac{z-4}{2}$;

ℓ': $\dfrac{x-2}{2} = \dfrac{y-9}{-2} = z-1$

(ii) ℓ: $\dfrac{x-8}{4} = \dfrac{y+2}{3} = \dfrac{z-7}{5}$;

ℓ': $\dfrac{x-2}{2} = \dfrac{y+6}{-6} = \dfrac{z-1}{-9}$

(iii) ℓ: $\dfrac{x+5}{8} = \dfrac{y-6}{6} = \dfrac{z-1}{3}$;

ℓ': $\dfrac{x-5}{5} = y-8 = z-3$

6. Find the coordinates of the foot of the perpendicular from P to the line ℓ.

(i) $\mathrm{P}(0,0,0)$; ℓ: $x-14 = -\tfrac{1}{3}(y+4) = z-18$

(ii) $\mathrm{P}(10,3,-14)$; ℓ: $x-11 = 5-y = z+1$

(iii) $\mathrm{P}(13,4,2)$;
ℓ: $\mathbf{r} = 2\mathbf{i} - 8\mathbf{j} + 21\mathbf{k} + t(\mathbf{i} - 2\mathbf{j} + 3\mathbf{k})$

7. Find the coordinates of A on ℓ and B on ℓ' such that AB is as short as possible.

(i) ℓ: $\begin{cases} x = 5t-6 \\ y = 4t \\ z = 3t-3 \end{cases}$ ℓ': $\begin{cases} x = t+6 \\ y = 2t-8 \\ z = 3t-1 \end{cases}$

(ii) ℓ: $\dfrac{x-5}{2} = \dfrac{y+2}{-3} = \dfrac{z-3}{-4}$

ℓ': $\dfrac{x-10}{3} = \dfrac{y-1}{-1} = \dfrac{z+7}{-6}$

(iii) ℓ: $\mathbf{r} = \begin{pmatrix} -5 \\ 10 \\ 12 \end{pmatrix} + t \begin{pmatrix} 3 \\ -1 \\ 4 \end{pmatrix}$

ℓ': $\mathbf{r} = \begin{pmatrix} 21 \\ -7 \\ 3 \end{pmatrix} + t \begin{pmatrix} 2 \\ 1 \\ -1 \end{pmatrix}$

8. Two straight lines are said to be *skew* if they do not intersect and are not parallel. Distinct points with position vectors **a** and **b** lie on the first of two such lines and distinct points with position vectors **c** and **d** lie on the second. Find an expression for the shortest distance between the two lines in terms of **a**, **b**, **c** and **d**.

Find the shortest distance between the two skew lines defined by the points
$\mathbf{a} = \mathbf{i} - 2\mathbf{j} + 3\mathbf{k}$, $\mathbf{b} = 2\mathbf{i} + 5\mathbf{k}$, $\mathbf{c} = 4\mathbf{i} + \mathbf{j} - \mathbf{k}$, $\mathbf{d} = -2\mathbf{i} + 3\mathbf{j} + 4\mathbf{k}$.

[MEI]

9.

The plane π has equation $\mathbf{n} \cdot \mathbf{r} + d = 0$.

Reflection in π maps the point P to P'.

Show that $\mathbf{p}' = \mathbf{p} - \dfrac{2}{|\mathbf{n}|^2}(\mathbf{n} \cdot \mathbf{p} + d)\mathbf{n}$.

10. The diagram shows skew lines ℓ_1 and ℓ_2. Plane π contains ℓ_1 and is perpendicular to ℓ_2. Line ℓ_2 intersects π at Q, and lines are drawn from Q to $P_1, P_2, P_3, \ldots$ on ℓ_1. Use this as the basis of an explanation as to why the shortest distance between two skew lines is along the common perpendicular.

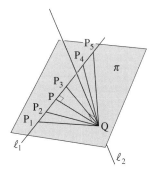

11. (i) Lines ℓ and ℓ' are skew, and P is a point on neither ℓ nor ℓ'. By considering the plane containing ℓ and P, or otherwise, prove that there is at most one line through P which intersects both ℓ and ℓ'.

    (ii) Lines ℓ and ℓ' have equations
    $x - 5 = -y = z + 1$ and
    $\dfrac{x-5}{2} = \dfrac{y-5}{-2} = \dfrac{z-4}{-3}$.

    Find equations for the line through $P(2,5,4)$ which intersects both ℓ and ℓ'.

12. The point P has coordinates $(4, k, 5)$, where k is a constant.

    The line L has equation $\mathbf{r} = \begin{pmatrix} 1 \\ 0 \\ -4 \end{pmatrix} + t\begin{pmatrix} 1 \\ 2 \\ -2 \end{pmatrix}$.

    The line M has equation $\mathbf{r} = \begin{pmatrix} 4 \\ k \\ 5 \end{pmatrix} + t\begin{pmatrix} 7 \\ 3 \\ -4 \end{pmatrix}$.

    (i) Show that the shortest distance from the point P to the line L is
    $\frac{1}{3}\sqrt{5(k^2 + 12k + 117)}$.

    (ii) Find (in terms of k) the shortest distance between the lines L and M.

(iii) Find the value of k for which the lines L and M intersect.

(iv) When $k = 12$, show that the distances in parts (i) and (ii) are equal. In this case, find the equation of the line which is perpendicular to, and intersects, both L and M. [MEI]

13. Lines ℓ_1 and ℓ_2 have direction vectors $\mathbf{d}$ and $\mathbf{e}$ respectively; A is any point on ℓ_1 and B is any point on ℓ_2; $\hat{\mathbf{n}}$ is a unit vector parallel to the common perpendicular PQ.

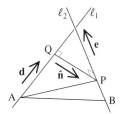

(i) Explain why $PQ = |\overrightarrow{PA} \cdot \hat{\mathbf{n}}|$.

(ii) Use the fact that $\overrightarrow{PA} = \overrightarrow{PB} + \overrightarrow{BA}$ to show that $\overrightarrow{PA} \cdot \hat{\mathbf{n}} = \overrightarrow{BA} \cdot \hat{\mathbf{n}}$ and deduce that
$$PQ = \left| \frac{\overrightarrow{BA} \cdot (\mathbf{d} \times \mathbf{e})}{|\mathbf{d} \times \mathbf{e}|} \right|.$$

14. Two ships are moving with constant, but different, velocities $\mathbf{v}_1$ and $\mathbf{v}_2$. At noon the ships have position vectors $\mathbf{a}_1$ and $\mathbf{a}_2$ respectively with respect to a fixed origin. The distance between the ships is decreasing. Show that $(\mathbf{v}_1 - \mathbf{v}_2) \cdot (\mathbf{a}_1 - \mathbf{a}_2) < 0$.

The ship with velocity $\mathbf{v}_1$ will sight the second ship if the distance between the ships is, at any time, less than or equal to d, where $d < |\mathbf{a}_1 - \mathbf{a}_2|$. Show that sighting will occur provided
$$d^2 \geqslant |\mathbf{a}_1 - \mathbf{a}_2|^2 - \frac{((\mathbf{v}_1 - \mathbf{v}_2) \cdot (\mathbf{a}_1 - \mathbf{a}_2))^2}{|\mathbf{v}_1 - \mathbf{v}_2|^2}.$$

If this condition is just satisfied, find the time after noon when sighting occurs. [MEI]

15. A and B have position vectors $\mathbf{a}$ and $\mathbf{b}$, and $\hat{\mathbf{n}}$ is a unit vector. Prove that:

(i) $|(\mathbf{a} - \mathbf{b}) \cdot \hat{\mathbf{n}}|$ is the shortest distance from A to π, the plane through B perpendicular to $\hat{\mathbf{n}}$;

(ii) $|(\mathbf{a} - \mathbf{b}) \times \hat{\mathbf{n}}|$ is the shortest distance from A to ℓ, the line through B parallel to $\hat{\mathbf{n}}$.

Scalar triple product

In the last section we developed the formula $\left| \dfrac{(\mathbf{a} - \mathbf{b}) \cdot (\mathbf{d} \times \mathbf{e})}{|\mathbf{d} \times \mathbf{e}|} \right|$ for the shortest distance between two skew lines. This contains the expression $(\mathbf{a} - \mathbf{b}) \cdot (\mathbf{d} \times \mathbf{e})$, an example of a *scalar triple product*. The simplest form of a scalar triple product is $\mathbf{a} \cdot (\mathbf{b} \times \mathbf{c})$.

For Discussion

Does it matter if the brackets are omitted from $\mathbf{a} \cdot (\mathbf{b} \times \mathbf{c})$?

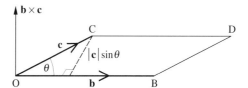

Figure 1.6

The magnitude of the vector product $\mathbf{b} \times \mathbf{c}$ is $|\mathbf{b} \times \mathbf{c}| = |\mathbf{b}||\mathbf{c}| \sin \theta$ which is the area of the parallelogram OBDC with sides OB and OC, illustrated in figure 1.6.

A parallelepiped is a polyhedron with six faces, each of which is a parallelogram. The volume of any parallelepiped is the product of its height and the area of its base.

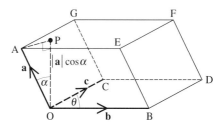

Figure 1.7

Figure 1.7 shows parallelepiped OBDCAEFG, with parallelogram OBDC as base. The area of the base is $|\mathbf{b} \times \mathbf{c}|$. The height of the parallelepiped is OP, where P is the foot of the perpendicular from O to plane AEFG. Using α to denote angle AOP, the length of OP is $|\mathbf{a}| \cos \alpha$, and the volume of parallelepiped OBDCAEFG is

$$V = |\mathbf{a}||\mathbf{b} \times \mathbf{c}| \cos \alpha.$$

If, as illustrated in figure 1.7, $\mathbf{a}, \mathbf{b}, \mathbf{c}$ is a right-handed set of vectors, $\mathbf{b} \times \mathbf{c}$ has the same direction and sense as $\overrightarrow{OP}$, so that $|\mathbf{a}||\mathbf{b} \times \mathbf{c}| \cos \alpha = \mathbf{a} \cdot (\mathbf{b} \times \mathbf{c})$ and we have $V = \mathbf{a} \cdot (\mathbf{b} \times \mathbf{c})$.

You will recall that $\mathbf{b} \times \mathbf{c}$ is defined as $|\mathbf{b}||\mathbf{c}| \sin \theta \hat{\mathbf{n}}$ where $\hat{\mathbf{n}}$ is a unit vector with the same direction and sense as $\overrightarrow{OP}$; this means that:

$$\mathbf{a} \cdot (\mathbf{b} \times \mathbf{c}) = (\mathbf{a} \cdot \hat{\mathbf{n}})|\mathbf{b}||\mathbf{c}| \sin \theta = (\mathbf{a} \cdot \hat{\mathbf{n}})|\mathbf{b} \times \mathbf{c}|.$$

However, if $\mathbf{a}$, $\mathbf{b}$, $\mathbf{c}$ is a left-handed set of vectors, $\mathbf{b} \times \mathbf{c}$ and $\overrightarrow{OP}$ are parallel but with opposite sense; this means that the angle between $\mathbf{a}$ and $\mathbf{b} \times \mathbf{c}$ is $\pi - \alpha$, an obtuse angle, so that $V = |\mathbf{a}||\mathbf{b} \times \mathbf{c}| \cos \alpha = -\mathbf{a} \cdot (\mathbf{b} \times \mathbf{c})$.

Therefore the scalar triple product $\mathbf{a} \cdot (\mathbf{b} \times \mathbf{c})$ is the signed volume of the parallelepiped with edges represented by $\mathbf{a}$, $\mathbf{b}$ and $\mathbf{c}$.

As the volume of the parallelepiped does not depend on which parallelogram you take as the base:

$$\mathbf{a} \cdot (\mathbf{b} \times \mathbf{c}) = \mathbf{b} \cdot (\mathbf{c} \times \mathbf{a}) = \mathbf{c} \cdot (\mathbf{a} \times \mathbf{b})$$
$$= -\mathbf{a} \cdot (\mathbf{c} \times \mathbf{b}) = -\mathbf{b} \cdot (\mathbf{a} \times \mathbf{c}) = -\mathbf{c} \cdot (\mathbf{b} \times \mathbf{a}).$$

That is, cyclic interchange of the vectors does not affect the value of the scalar triple product, but non-cyclic interchange of the vectors multiplies the product by -1.

Activity

What (if any) difference does interchanging the $\cdot$ and the $\times$ make?

Activity

Show that:

(i) the volume of the tetrahedron OABC is $\frac{1}{6}|\mathbf{a} \cdot (\mathbf{b} \times \mathbf{c})|$;

(ii) the volume of the tetrahedron PQRS is $\frac{1}{6}|(\mathbf{q} - \mathbf{p}) \cdot ((\mathbf{r} - \mathbf{p}) \times (\mathbf{s} - \mathbf{p}))|$.

Evaluating scalar triple products

We now show you one way of evaluating a scalar triple product.

If $\mathbf{a} = \begin{pmatrix} a_1 \\ a_2 \\ a_3 \end{pmatrix}$, $\mathbf{b} = \begin{pmatrix} b_1 \\ b_2 \\ b_3 \end{pmatrix}$ and $\mathbf{c} = \begin{pmatrix} c_1 \\ c_2 \\ c_3 \end{pmatrix}$ then

$$\mathbf{a} \cdot (\mathbf{b} \times \mathbf{c}) = \begin{pmatrix} a_1 \\ a_2 \\ a_3 \end{pmatrix} \cdot \left(\begin{pmatrix} b_1 \\ b_2 \\ b_3 \end{pmatrix} \times \begin{pmatrix} c_1 \\ c_2 \\ c_3 \end{pmatrix} \right)$$

$$= \begin{pmatrix} a_1 \\ a_2 \\ a_3 \end{pmatrix} \cdot \begin{pmatrix} b_2 c_3 - b_3 c_2 \\ b_3 c_1 - b_1 c_3 \\ b_1 c_2 - b_2 c_1 \end{pmatrix}$$

These three components may be written as 2×2 determinants, e.g. $b_2 c_3 - b_3 c_2 = \begin{vmatrix} b_2 & c_2 \\ b_3 & c_3 \end{vmatrix}$. See *Pure Mathematics 4*, page 99.

$$= \begin{pmatrix} a_1 \\ a_2 \\ a_3 \end{pmatrix} \cdot \left(\begin{vmatrix} b_2 & c_2 \\ b_3 & c_3 \end{vmatrix} \mathbf{i} - \begin{vmatrix} b_1 & c_1 \\ b_3 & c_3 \end{vmatrix} \mathbf{j} + \begin{vmatrix} b_1 & c_1 \\ b_2 & c_2 \end{vmatrix} \mathbf{k} \right)$$

Note these minus signs.

$$= a_1 \begin{vmatrix} b_2 & c_2 \\ b_3 & c_3 \end{vmatrix} - a_2 \begin{vmatrix} b_1 & c_1 \\ b_3 & c_3 \end{vmatrix} + a_3 \begin{vmatrix} b_1 & c_1 \\ b_2 & c_2 \end{vmatrix}$$

In the next section you will see that this final result may be written as the

3×3 *determinant* $\begin{vmatrix} a_1 & b_1 & c_1 \\ a_2 & b_2 & c_2 \\ a_3 & b_3 & c_3 \end{vmatrix}$ (see page 20).

EXAMPLE

Evaluate $\mathbf{a} \cdot (\mathbf{b} \times \mathbf{c})$ when $\mathbf{a} = 3\mathbf{i} + 2\mathbf{j} + 5\mathbf{k}$, $\mathbf{b} = 4\mathbf{i} - 3\mathbf{j} + 6\mathbf{k}$, and $\mathbf{c} = 2\mathbf{i} + 3\mathbf{j} + 5\mathbf{k}$, and interpret the result.

Solution

Note the minus sign.

$$\mathbf{a} \cdot (\mathbf{b} \times \mathbf{c}) = \begin{vmatrix} 3 & 4 & 2 \\ 2 & -3 & 3 \\ 5 & 6 & 5 \end{vmatrix} = 3 \begin{vmatrix} -3 & 3 \\ 6 & 5 \end{vmatrix} - 2 \begin{vmatrix} 4 & 2 \\ 6 & 5 \end{vmatrix} + 5 \begin{vmatrix} 4 & 2 \\ -3 & 3 \end{vmatrix}$$

$$= 3 \times (-33) - 2 \times 8 + 5 \times 18 = -25.$$

The volume of the parallelepiped which has edges given by $\mathbf{a}$, $\mathbf{b}$ and $\mathbf{c}$ is 25 units. The negative sign indicates that $\mathbf{a}$, $\mathbf{b}$ and $\mathbf{c}$, in that order, form a left-handed set of vectors.

Geometrical applications of the scalar triple product

To test whether two lines meet

Our discussion of the scalar triple product arose from our construction of the formula $\left| \dfrac{(\mathbf{a} - \mathbf{b}) \cdot (\mathbf{d} \times \mathbf{e})}{|\mathbf{d} \times \mathbf{e}|} \right|$ for the shortest distance between two skew lines ℓ_1 and ℓ_2, with equations $\mathbf{r} = \mathbf{a} + t\mathbf{d}$ and $\mathbf{r} = \mathbf{b} + t\mathbf{e}$ respectively.

Activity

What happens if ℓ_1 and ℓ_2 are parallel?

Clearly, if ℓ_1 and ℓ_2 intersect, the shortest distance between them is zero, so that $\dfrac{(\mathbf{a} - \mathbf{b}) \cdot (\mathbf{d} \times \mathbf{e})}{|\mathbf{d} \times \mathbf{e}|} = 0 \Leftrightarrow (\mathbf{a} - \mathbf{b}) \cdot (\mathbf{d} \times \mathbf{e}) = 0$, providing an easily applied test as to whether the two lines $\mathbf{r} = \mathbf{a} + t\mathbf{d}$ and $\mathbf{r} = \mathbf{b} + t\mathbf{e}$ intersect or not.

EXAMPLE

Decide whether the line $\ell_1: \mathbf{r} = -4\mathbf{i} + 2\mathbf{j} + \mathbf{k} + t(2\mathbf{i} + 3\mathbf{j} + 2\mathbf{k})$ meets the line $\ell_2: \dfrac{x - 2}{3} = \dfrac{y - 5}{3} = z + 1.$

Solution

The points $A(-4, 2, 1)$ and $B(2, 5, -1)$ are on ℓ_1 and ℓ_2 respectively, so:

$$\mathbf{a} - \mathbf{b} = \begin{pmatrix} -4 \\ 2 \\ 1 \end{pmatrix} - \begin{pmatrix} 2 \\ 5 \\ -1 \end{pmatrix} = \begin{pmatrix} -6 \\ -3 \\ 2 \end{pmatrix}.$$

The respective direction vectors are $\mathbf{d} = \begin{pmatrix} 2 \\ 3 \\ 2 \end{pmatrix}$ and $\mathbf{e} = \begin{pmatrix} 3 \\ 3 \\ 1 \end{pmatrix}$.

Then $(\mathbf{a} - \mathbf{b}) \cdot (\mathbf{d} \times \mathbf{e}) = \begin{pmatrix} -6 \\ -3 \\ 2 \end{pmatrix} \cdot \left(\begin{pmatrix} 2 \\ 3 \\ 2 \end{pmatrix} \times \begin{pmatrix} 3 \\ 3 \\ 1 \end{pmatrix} \right)$

$$= \begin{pmatrix} -6 \\ -3 \\ 2 \end{pmatrix} \cdot \begin{pmatrix} -3 \\ 4 \\ -3 \end{pmatrix} = 18 - 12 - 6 = 0$$

proving that ℓ_1 and ℓ_2 meet.

To test whether four points are coplanar

If O, A, B and C are in the same plane the volume of tetrahedron OABC is zero so that $\mathbf{a} \cdot (\mathbf{b} \times \mathbf{c}) = 0$. Conversely, if $\mathbf{a} \cdot (\mathbf{b} \times \mathbf{c}) = 0$, the volume of tetrahedron OABC is zero, the tetrahedron has collapsed, and the four points O, A, B and C are contained in a single plane. (They may, of course, all fall on one straight line, a special case of being coplanar.)

If you want to test whether the four points P, Q, R and S are coplanar, translate all four points so that one vertex, P say, is moved to the origin. Then the (signed) volume of the tetrahedron PQRS is equal to $\frac{1}{6}(\mathbf{q} - \mathbf{p}) \cdot ((\mathbf{r} - \mathbf{p}) \times (\mathbf{s} - \mathbf{p}))$; the points P, Q, R and S are coplanar if and only if this scalar triple product is 0.

EXAMPLE

Show that the points $P(-1, 1, 2)$, $Q(2, 3, -4)$, $R(2, 2, 6)$ and $S(4, 3, 5)$ are not all contained in a single plane.

Solution

$$\overrightarrow{PQ} = \begin{pmatrix} 2 \\ 3 \\ -4 \end{pmatrix} - \begin{pmatrix} -1 \\ 1 \\ 2 \end{pmatrix} = \begin{pmatrix} 3 \\ 2 \\ -6 \end{pmatrix}, \qquad \overrightarrow{PR} = \begin{pmatrix} 2 \\ 2 \\ 6 \end{pmatrix} - \begin{pmatrix} -1 \\ 1 \\ 2 \end{pmatrix} = \begin{pmatrix} 3 \\ 1 \\ 4 \end{pmatrix},$$

$$\overrightarrow{PS} = \begin{pmatrix} 4 \\ 3 \\ 5 \end{pmatrix} - \begin{pmatrix} -1 \\ 1 \\ 2 \end{pmatrix} = \begin{pmatrix} 5 \\ 2 \\ 3 \end{pmatrix}.$$

Therefore $\overrightarrow{PQ} \cdot (\overrightarrow{PR} \times \overrightarrow{PS}) = \begin{pmatrix} 3 \\ 2 \\ -6 \end{pmatrix} \cdot \left(\begin{pmatrix} 3 \\ 1 \\ 4 \end{pmatrix} \times \begin{pmatrix} 5 \\ 2 \\ 3 \end{pmatrix} \right)$

$$= \begin{pmatrix} 3 \\ 2 \\ -6 \end{pmatrix} \cdot \begin{pmatrix} -5 \\ 11 \\ 1 \end{pmatrix} = -15 + 22 - 6 = 1.$$

The scalar triple product is not zero so P, Q, R and S are not all in the same plane.

(Notice that the volume of tetrahedron PQRS is $\frac{1}{6} \times 1 = \frac{1}{6}$.)

To test whether three vectors form a right- or left-handed set

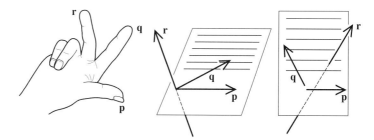

Figure 1.8 Vectors p, q and r form a right-handed set

Assuming you are not double-jointed, if the thumb of your right hand points in the direction of vector **p**, your index finger in the direction of **q**, and your second finger in the direction of **r** then the vectors **p**, **q** and **r** (in that order) form a right-handed set. Figure 1.8 illustrates this and shows other right-handed sets of vectors: in these **p** is drawn to the right across a piece of paper, **q** is drawn 'up' the paper (not necessarily 'straight up') and **r** rises 'up out of' the paper (not necessarily at right angles to the paper).

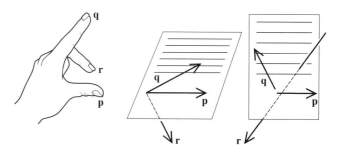

Figure 1.9 Vectors p, q and r form a left-handed set

Figure 1.9 illustrates left-handed sets of vectors, where, as before, **p** points to the right across the paper, **q** points 'up' the paper but now **r** points 'down', again not necessarily at right angles to the paper.

Activity

The vectors **p**, **q** and **r** (in that order) form a right-handed set. Show that **q**, **r**, **p** (in that order) is also a right-handed set but that **p**, **r**, **q** (in that order) is a left-handed set.

(This means that cyclical interchange does *not* change the 'handedness' of a set of three vectors, but non-cyclical interchange does change the 'handedness'.)

The vector $\mathbf{a} \times \mathbf{b}$ is defined as being in the direction $\hat{\mathbf{n}}$ such that **a**, **b** and $\hat{\mathbf{n}}$ (in that order) form a right-handed set. When **a**, **b** and **c** (in that order) form a right-handed set the angle between **c** and $\mathbf{a} \times \mathbf{b}$ is acute so that $\mathbf{c} \cdot (\mathbf{a} \times \mathbf{b})$ is positive. When **a**, **b** and **c** (in that order) form a left-handed set the angle between **c** and $\mathbf{a} \times \mathbf{b}$ is obtuse and $\mathbf{c} \cdot (\mathbf{a} \times \mathbf{b})$ is negative. Thus the scalar triple product provides an efficient test of 'handedness'. (As cyclical interchange of the vectors has no effect you can test using $\mathbf{a} \cdot (\mathbf{b} \times \mathbf{c})$ instead of $\mathbf{c} \cdot (\mathbf{a} \times \mathbf{b})$.)

Exercise 1B

1. Find $\mathbf{a} \cdot (\mathbf{b} \times \mathbf{c})$ in the following cases.

(i) $\mathbf{a} = \begin{pmatrix} 3 \\ 4 \\ 6 \end{pmatrix}, \mathbf{b} = \begin{pmatrix} 5 \\ 8 \\ -3 \end{pmatrix}, \mathbf{c} = \begin{pmatrix} 1 \\ 2 \\ 3 \end{pmatrix}$

(ii) $\mathbf{a} = \begin{pmatrix} 5 \\ -2 \\ 2 \end{pmatrix}, \mathbf{b} = \begin{pmatrix} 3 \\ 1 \\ 5 \end{pmatrix}, \mathbf{c} = \begin{pmatrix} 3 \\ 1 \\ -2 \end{pmatrix}$

(iii) $\mathbf{a} = \begin{pmatrix} -2 \\ -5 \\ 1 \end{pmatrix}, \mathbf{b} = \begin{pmatrix} 4 \\ 3 \\ 2 \end{pmatrix}, \mathbf{c} = \begin{pmatrix} 2 \\ 3 \\ 2 \end{pmatrix}$

2. Does line ℓ meet line ℓ'?

(i) $\ell: \mathbf{r} = \begin{pmatrix} 5 \\ 0 \\ 1 \end{pmatrix} + t \begin{pmatrix} 1 \\ 3 \\ 1 \end{pmatrix}$;

$\ell': \mathbf{r} = \begin{pmatrix} 5 \\ 9 \\ 3 \end{pmatrix} + t \begin{pmatrix} 4 \\ 3 \\ 2 \end{pmatrix}$

(ii) $\ell: \begin{cases} x = 2t - 2 \\ y = 4 - 3t; \\ z = 2t - 3 \end{cases}$ $\ell': \begin{cases} x = 3t - 1 \\ y = 3 - 5t \\ z = 4 - t \end{cases}$

(iii) $\ell: \dfrac{x+1}{5} = y + 3 = \dfrac{z-1}{7}$;

$\ell': \dfrac{x+2}{4} = \dfrac{y-4}{-5} = \dfrac{z-3}{3}$

3. Are A, B, C and D coplanar?

(i) $A(1, 2, 3)$; $B(6, 5, 4)$; $C(7, 8, 9)$; $D(12, 11, 10)$

(ii) $A(-2, -3, 3)$; $B(1, -2, -1)$; $C(4, -1, -5)$; $D(7, 3, 6)$

(iii) $A(-2, -2, 1)$; $B(-3, -1, 11)$; $C(1, 2, 6)$; $D(3, 2, 5)$

4. Decide whether a, b and c (in that order) form a right- or left-handed set.

(i) $\mathbf{a} = -3\mathbf{i} + 2\mathbf{j} - 3\mathbf{k}$, $\mathbf{b} = 3\mathbf{i} - \mathbf{j} + 3\mathbf{k}$, $\mathbf{c} = 2\mathbf{i} + 3\mathbf{j} + 5\mathbf{k}$

(ii) $\mathbf{a} = 2\mathbf{i} + 2\mathbf{j} + 6\mathbf{k}$, $\mathbf{b} = \mathbf{i} + \mathbf{j} + 2\mathbf{k}$, $\mathbf{c} = 3\mathbf{i} + 3\mathbf{j} + 4\mathbf{k}$

(iii) $\mathbf{a} = \begin{pmatrix} 2 \\ 4 \\ -5 \end{pmatrix}, \mathbf{b} = \begin{pmatrix} -5 \\ 3 \\ 2 \end{pmatrix}, \mathbf{c} = \begin{pmatrix} 3 \\ -1 \\ 3 \end{pmatrix}$

5. Using the given coordinates find the volume of parallelepiped $PQRSP'Q'R'S'$.

(i) $P(0, 0, 0)$, $Q(1, 2, 5)$, $S(7, 3, 4)$, $P'(2, 3, 4)$

(ii) $P(3, -2, 4)$, $Q(1, 4, 5)$, $S(6, 2, 7)$, $P'(2, 3, 5)$

(iii) $P(1, 2, 3)$, $Q(6, 5, 4)$, $S(7, 8, 9)$, $P'(12, 11, 10)$

(iv) $P(-3, -1, 1)$, $Q(3, 2, 4)$, $S(7, -2, 6)$, $P'(3, 5, 2)$

6. Using the given coordinates find the volume of tetrahedron ABCD.

   (i)   A$(0,0,0)$, B$(1,0,0)$, C$(1,2,0)$, D$(1,2,3)$

   (ii) A$(2,3,1)$, B$(5,4,-2)$, C$(3,6,5)$, D$(3,2,-1)$

   (iii) A$(-5,1,-6)$, B$(7,-3,5)$, C$(6,5,7)$,
         D$(3,-1,0)$

7. Points A, B and C have position vectors **a**, **b** and **c**. Prove that $(\mathbf{a} \times \mathbf{b}) \cdot \mathbf{a} = 0$ and show that $\mathbf{b} \times \mathbf{c} + \mathbf{c} \times \mathbf{a} + \mathbf{a} \times \mathbf{b}$ is perpendicular to the plane ABC.

   [Hint: Use the distributive property of vector product over addition,
   $$\mathbf{p} \times (\mathbf{q} + \mathbf{r}) = (\mathbf{p} \times \mathbf{q}) + (\mathbf{p} \times \mathbf{r});$$
   see *Pure Mathematics 4*.]

8. Show that $(\mathbf{a} \times \mathbf{b}) \cdot (\mathbf{c} \times \mathbf{d}) = \begin{vmatrix} \mathbf{a} \cdot \mathbf{c} & \mathbf{a} \cdot \mathbf{d} \\ \mathbf{b} \cdot \mathbf{c} & \mathbf{b} \cdot \mathbf{d} \end{vmatrix}$.

9. Two lines have equations $\mathbf{r} = \mathbf{a} + \lambda\mathbf{d}$, $\mathbf{r} = \mathbf{b} + \mu\mathbf{e}$.

   (i) Show that the lines intersect if $\mathbf{a} \cdot (\mathbf{d} \times \mathbf{e}) = \mathbf{b} \cdot (\mathbf{d} \times \mathbf{e})$.

   (ii) Suppose the lines intersect at C. Explain why λ and μ can be found such that $\mathbf{a} + \lambda\mathbf{d} = \mathbf{b} + \mu\mathbf{e}$. By forming the scalar product of both sides with $\mathbf{a} \times \mathbf{e}$ show that
   $$\mathbf{c} = \mathbf{a} + \frac{\mathbf{a} \cdot (\mathbf{b} \times \mathbf{e})}{\mathbf{a} \cdot (\mathbf{d} \times \mathbf{e})}\mathbf{d}.$$
   Find a similar expression for **c** in the form $\mathbf{b} + \mu\mathbf{e}$.

10. The diagram shows the plan of five white and four coloured spheres pushed together on a horizontal table. The radius of all the spheres is 1 unit. The centres of the spheres are marked, and the x- and y-axes are shown, with the origin at O, the centre of a coloured sphere.

    A second layer is formed by placing four additional coloured spheres, also of unit radius, on top of the first layer in the positions marked U. Each of these spheres touches three spheres in the first layer: C is the centre of the sphere that touches the spheres with centres O, A and B. The second diagram shows both layers.

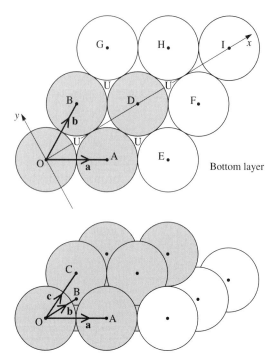

Bottom layer

(i)   Show that the centres of the eight coloured spheres form the vertices of a parallelepiped with edges given by vectors **a**, **b** and **c**.

(ii) Show that $\mathbf{a} = \sqrt{3}\mathbf{i} - \mathbf{j}$ and express **b** and **c** in terms of **i**, **j** and **k**.

(iii) Find the volume of the parallelepiped with edges given by **a**, **b** and **c**.

(iv) The parallelepiped with edges given by **a**, **b** and **c** contains parts of eight spheres; explain why these parts sum to one whole sphere, and show that the volume of this sphere is just over 74% of the volume of the parallelepiped.

(This stacking arrangement, known as *hexagonal close packing*, is important in crystallography: elements in which the atoms are packed this way include magnesium, titanium, zinc and cobalt.)

Determinant and inverse of a 3 × 3 matrix

Determinants

We start this section by considering a particular way of solving the matrix equation $\mathbf{Mr} = \mathbf{d}$, where $\mathbf{M}$ is the 3 × 3 matrix $\begin{pmatrix} a_1 & b_1 & c_1 \\ a_2 & b_2 & c_2 \\ a_3 & b_3 & c_3 \end{pmatrix}$, $\mathbf{r} = \begin{pmatrix} x \\ y \\ z \end{pmatrix}$ and

$\mathbf{d} = \begin{pmatrix} d_1 \\ d_2 \\ d_3 \end{pmatrix}$. Finding $\mathbf{r}$ by solving the matrix equation $\mathbf{Mr} = \mathbf{d}$ is equivalent to

finding the three unknowns x, y and z by solving the three simultaneous equations:

$$a_1 x + b_1 y + c_1 z = d_1$$

$$a_2 x + b_2 y + c_2 z = d_2$$

$$a_3 x + b_3 y + c_3 z = d_3.$$

These three equations can be written as the single vector equation:

$$x \begin{pmatrix} a_1 \\ a_2 \\ a_3 \end{pmatrix} + y \begin{pmatrix} b_1 \\ b_2 \\ b_3 \end{pmatrix} + z \begin{pmatrix} c_1 \\ c_2 \\ c_3 \end{pmatrix} = \begin{pmatrix} d_1 \\ d_2 \\ d_3 \end{pmatrix}$$

which can be written more compactly as:

$$x\mathbf{a} + y\mathbf{b} + z\mathbf{c} = \mathbf{d} \qquad \qquad ①$$

with obvious meaning for the vectors $\mathbf{a}$, $\mathbf{b}$, $\mathbf{c}$ and $\mathbf{d}$.

The vector $\mathbf{b} \times \mathbf{c}$ is perpendicular to both $\mathbf{b}$ and $\mathbf{c}$ so
$\mathbf{b} \cdot (\mathbf{b} \times \mathbf{c}) = \mathbf{c} \cdot (\mathbf{b} \times \mathbf{c}) = 0$. This means you can eliminate y and z from ① by forming the scalar product of both sides with $\mathbf{b} \times \mathbf{c}$:

$$x\mathbf{a} + y\mathbf{b} + z\mathbf{c} = \mathbf{d}$$

$$\Rightarrow \quad x\mathbf{a} \cdot (\mathbf{b} \times \mathbf{c}) + y\mathbf{b} \cdot (\mathbf{b} \times \mathbf{c}) + z\mathbf{c} \cdot (\mathbf{b} \times \mathbf{c}) = \mathbf{d} \cdot (\mathbf{b} \times \mathbf{c})$$

$$\Rightarrow \quad x\mathbf{a} \cdot (\mathbf{b} \times \mathbf{c}) = \mathbf{d} \cdot (\mathbf{b} \times \mathbf{c})$$

$$\Rightarrow \quad x = \frac{\mathbf{d} \cdot (\mathbf{b} \times \mathbf{c})}{\mathbf{a} \cdot (\mathbf{b} \times \mathbf{c})}$$

> You can divide by $\mathbf{a} \cdot (\mathbf{b} \times \mathbf{c})$ as it is a scalar.

provided $\mathbf{a} \cdot (\mathbf{b} \times \mathbf{c}) \neq 0$.

Similarly $y = \dfrac{\mathbf{d} \cdot (\mathbf{c} \times \mathbf{a})}{\mathbf{b} \cdot (\mathbf{c} \times \mathbf{a})}$ provided $\mathbf{b} \cdot (\mathbf{c} \times \mathbf{a}) \neq 0$

and $\quad z = \dfrac{\mathbf{d} \cdot (\mathbf{a} \times \mathbf{b})}{\mathbf{c} \cdot (\mathbf{a} \times \mathbf{b})}$ provided $\mathbf{c} \cdot (\mathbf{a} \times \mathbf{b}) \neq 0$.

The cyclic properties of scalar triple products show that these three conditions are identical, and the results can be re-written as:

$$x = \frac{\mathbf{d} \cdot (\mathbf{b} \times \mathbf{c})}{\mathbf{a} \cdot (\mathbf{b} \times \mathbf{c})}, \quad y = \frac{\mathbf{a} \cdot (\mathbf{d} \times \mathbf{c})}{\mathbf{a} \cdot (\mathbf{b} \times \mathbf{c})}, \quad z = \frac{\mathbf{a} \cdot (\mathbf{b} \times \mathbf{d})}{\mathbf{a} \cdot (\mathbf{b} \times \mathbf{c})} \text{ provided } \mathbf{a} \cdot (\mathbf{b} \times \mathbf{c}) \neq 0.$$

(This result is known as *Cramer's rule*, after Gabriel Cramer who published it in 1750, although Maclaurin seems to have known the result as early as 1729. It is usual to express scalar triple products as determinants – see below.)

The scalar triple product $\mathbf{a} \cdot (\mathbf{b} \times \mathbf{c})$ is known as the *determinant* of $\mathbf{M}$ and is variously denoted by Δ, $\det\mathbf{M}$, $\begin{vmatrix} a_1 & b_1 & c_1 \\ a_2 & b_2 & c_2 \\ a_3 & b_3 & c_3 \end{vmatrix}$ or $|\mathbf{a}\ \mathbf{b}\ \mathbf{c}|$.

Evaluating a 3 × 3 determinant

On page 13 you saw that:

$$\begin{vmatrix} b_2 & c_2 \\ b_3 & c_3 \end{vmatrix} = b_2 c_3 - b_3 c_2$$

$$\det\mathbf{M} = \mathbf{a} \cdot (\mathbf{b} \times \mathbf{c}) = \begin{vmatrix} a_1 & b_1 & c_1 \\ a_2 & b_2 & c_2 \\ a_3 & b_3 & c_3 \end{vmatrix} = a_1 \begin{vmatrix} b_2 & c_2 \\ b_3 & c_3 \end{vmatrix} - a_2 \begin{vmatrix} b_1 & c_1 \\ b_3 & c_3 \end{vmatrix} + a_3 \begin{vmatrix} b_1 & c_1 \\ b_2 & c_2 \end{vmatrix}.$$

This is known as *expanding* $\det\mathbf{M}$ *by the first column*. Notice that a_1 is multiplied by what is known as its *minor*, the 2 × 2 determinant $\begin{vmatrix} b_2 & c_2 \\ b_3 & c_3 \end{vmatrix}$

obtained by deleting the row and the column containing a_1: $\begin{vmatrix} a_1 & b_1 & c_1 \\ a_2 & b_2 & c_2 \\ a_3 & b_3 & c_3 \end{vmatrix}$.

Other minors are obtained similarly; the minor of a_2 is:

$$\begin{vmatrix} a_1 & b_1 & c_1 \\ a_2 & b_2 & c_2 \\ a_3 & b_3 & c_3 \end{vmatrix} = \begin{vmatrix} b_1 & c_1 \\ b_3 & c_3 \end{vmatrix} = b_1 c_3 - b_3 c_1.$$

Alternatively you may expand $\det\mathbf{M}$ by the second or third column, obtaining the $\mathbf{b} \cdot (\mathbf{c} \times \mathbf{a})$ version of $\det\mathbf{M}$:

$$-b_1 \begin{vmatrix} a_2 & c_2 \\ a_3 & c_3 \end{vmatrix} + b_2 \begin{vmatrix} a_1 & c_1 \\ a_3 & c_3 \end{vmatrix} - b_3 \begin{vmatrix} a_1 & c_1 \\ a_2 & c_2 \end{vmatrix}$$

or the $\mathbf{c} \cdot (\mathbf{a} \times \mathbf{b})$ version of $\det\mathbf{M}$:

$$c_1 \begin{vmatrix} a_2 & b_2 \\ a_3 & b_3 \end{vmatrix} - c_2 \begin{vmatrix} a_1 & b_1 \\ a_3 & b_3 \end{vmatrix} + c_3 \begin{vmatrix} a_1 & b_1 \\ a_2 & b_2 \end{vmatrix}.$$

The signs attached to the minors alternate as shown: $\begin{vmatrix} + & - & + \\ - & + & - \\ + & - & + \end{vmatrix}$.

A minor together with its correct sign is known as a *cofactor* and is denoted by the corresponding capital letter, e.g. the cofactor of a_3 is A_3. Thus

$$A_3 = + \begin{vmatrix} b_1 & c_1 \\ b_2 & c_2 \end{vmatrix}, \quad C_2 = - \begin{vmatrix} a_1 & b_1 \\ a_3 & b_3 \end{vmatrix}, \text{ and so on. Using this cofactor notation:}$$

$$\det\mathbf{M} = a_1 A_1 + a_2 A_2 + a_3 A_3$$
$$= b_1 B_1 + b_2 B_2 + b_3 B_3$$
$$= c_1 C_1 + c_2 C_2 + c_3 C_3.$$

Activity

(i) Prove that interchanging any two columns of a 3×3 determinant changes the sign of the determinant.

(ii) Deduce that a determinant with two identical columns is zero, and interpret this result geometrically.

(iii) Prove that cyclic interchange of the columns of a 3×3 determinant leaves the value of the determinant unchanged.

Other properties of determinants are developed in Exercise 1C, particularly in Questions 4, 6 and 9.

HISTORICAL NOTE

Although the early work on determinants was done by Vandermonde around 1776, the word 'determinant' was coined by Cauchy, in 1815, and the vertical line notation was introduced in 1841 by Arthur Cayley, a pioneer in the study of matrices.

EXAMPLE

Factorise the determinant $\Delta = \begin{vmatrix} 1 & 1 & 1 \\ x & y & z \\ x^3 & y^3 & z^3 \end{vmatrix}$.

Solution

(The obvious approach is to expand the determinant and then to factorise the resulting expression, but this expression consists of six terms of the form yz^3 and factorising this is not easy, so an alternative approach is used here.)

Determinant Δ may be thought of as a polynomial in x (or in y or in z as appropriate).

If y takes the same value as x the first two columns are identical and then $\Delta - 0$. Therefore $(x - y)$ is a factor of Δ by the factor theorem.

Similarly $(y - z)$ and $(z - x)$ are also factors of Δ.

As cyclic interchange of x, y and z leaves both Δ and $(x - y)(y - z)(z - x)$ unchanged, while non-cyclic interchange reverses the signs of both expressions, any further factors of Δ will be symmetrical in x, y and z. Since Δ is of degree 4 in x, y and z, the remaining factor is of degree 1, and so is of the form $k(x + y + z)$ where k is a number. Considering the coefficient of $x^3 y$ in the expansions of Δ and in $(x - y)(y - z)(z - x)k(x + y + z)$ shows that $k = -1$. Therefore $\Delta = -(x - y)(y - z)(z - x)(x + y + z)$.

1. Evaluate these determinants.

(i) $\begin{vmatrix} 1 & 3 & 2 \\ 5 & 1 & 4 \\ 6 & 2 & 3 \end{vmatrix}$
(ii) $\begin{vmatrix} 5 & -1 & -2 \\ -2 & 2 & 3 \\ 3 & 4 & 1 \end{vmatrix}$

(iii) $\begin{vmatrix} 5 & 6 & 4 \\ -4 & 0 & 1 \\ 7 & -3 & -5 \end{vmatrix}$
(iv) $\begin{vmatrix} 7 & -6 & -1 \\ -7 & 6 & 0 \\ 2 & -2 & 5 \end{vmatrix}$

2. Expand and simplify these determinants.

(i) $\begin{vmatrix} 1 & h & 2 \\ 5 & 10 & 10 \\ 2 & 4 & k \end{vmatrix}$
(ii) $\begin{vmatrix} x & y & z \\ y & z & x \\ x-y & y-z & z-x \end{vmatrix}$

(iii) $\begin{vmatrix} a & h & g \\ h & b & f \\ g & f & c \end{vmatrix}$

3. The following method of expanding

$$\det \mathbf{M} = \begin{vmatrix} a_1 & b_1 & c_1 \\ a_2 & b_2 & c_2 \\ a_3 & b_3 & c_3 \end{vmatrix} \text{ was devised by}$$

P.F. Sarrus, 1798–1861.

Copy the first and second rows below the third row; form diagonal products and sum them, as shown.

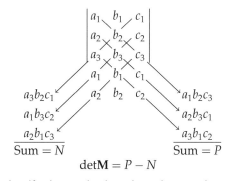

$$\det \mathbf{M} = P - N$$

Justify the method, and use it to evaluate the determinants in Question 1.

4. (i) Prove that $|k\mathbf{a} \quad \mathbf{b} \quad \mathbf{c}| = k|\mathbf{a} \quad \mathbf{b} \quad \mathbf{c}|$, where k is constant.

(ii) Explain in terms of volumes why multiplying all the elements in the first column by constant k multiplies the value of the determinant by k.

(iii) What happens if you multiply another column by k?

5. Given that $\begin{vmatrix} 1 & 2 & 3 \\ 6 & 4 & 5 \\ 7 & 5 & 1 \end{vmatrix} = 43$, use the property developed in Question 4 to evaluate the following without expanding the determinants.

(i) $\begin{vmatrix} 10 & 2 & 3 \\ 60 & 4 & 5 \\ 70 & 5 & 1 \end{vmatrix}$

(ii) $\begin{vmatrix} 4 & 10 & -21 \\ 24 & 20 & -35 \\ 28 & 25 & -7 \end{vmatrix}$

(iii) $\begin{vmatrix} x^4 & 1/x & 12y \\ 6x^4 & 2/x & 20y \\ 7x^4 & 5/2x & 4y \end{vmatrix}$

6. (i) Prove that $|\mathbf{a} + k\mathbf{b} \quad \mathbf{b} \quad \mathbf{c}| = |\mathbf{a} \quad \mathbf{b} \quad \mathbf{c}|$, where k is constant.

(ii) Interpret this geometrically in terms of volumes.

(iii) Prove that the value of a determinant is unchanged when you add any multiple of one column to any other column.

7. Use the property developed in Question 6 to evaluate the following.

(i) $\begin{vmatrix} 21 & 2 & 3 \\ 46 & 4 & 5 \\ 57 & 5 & 1 \end{vmatrix}$

(ii) $\begin{vmatrix} 19 & 14 & 20 \\ 23 & 27 & 25 \\ 15 & 26 & 17 \end{vmatrix}$

(iii) $\begin{vmatrix} 25 & 17 & 51 \\ 38 & 33 & 78 \\ 25 & 32 & 52 \end{vmatrix}$

8. In a Fibonacci sequence the third and subsequent terms satisfy $u_{r+2} = u_{r+1} + u_r$. Show that if $u_1, u_2, u_3, \ldots$ is a Fibonacci sequence then $\begin{vmatrix} u_1 & u_2 & u_3 \\ u_4 & u_5 & u_6 \\ u_7 & u_8 & u_9 \end{vmatrix} = 0$.

Exercise 1C continued

9. In this question A_1, A_2, A_3, etc. represent (as usual) the cofactors of a_1, a_2, a_3, etc.

(i) By considering expanding $|\mathbf{b} \quad \mathbf{b} \quad \mathbf{c}|$ by its first column show that $b_1 A_1 + b_2 A_2 + b_3 A_3 = 0$.

(ii) Prove that $a_1 C_1 + a_2 C_2 + a_3 C_3 = 0$.

(iii) Write down four other similar expressions which also evaluate to 0.

(Multiplying a column of $\det \mathbf{M}$ by cofactors belonging to a different column, and then adding, is known as *expanding by alien cofactors*. The result is always zero.)

10. You will need to use the results of Question 9 in this question.

(i) By multiplying out and simplifying

$$\begin{pmatrix} A_1 & A_2 & A_3 \\ B_1 & B_2 & B_3 \\ C_1 & C_2 & C_3 \end{pmatrix} \begin{pmatrix} a_1 & b_1 & c_1 \\ a_2 & b_2 & c_2 \\ a_3 & b_3 & c_3 \end{pmatrix}$$

find the inverse of $\begin{pmatrix} a_1 & b_1 & c_1 \\ a_2 & b_2 & c_2 \\ a_3 & b_3 & c_3 \end{pmatrix}$.

(ii) Will this method always produce the inverse of a 3×3 matrix?

11. Prove that the area of the triangle with vertices at $(x_1, y_1), (x_2, y_2), (x_3, y_3)$ is $\pm\frac{1}{2}\begin{vmatrix} x_1 & y_1 & 1 \\ x_2 & y_2 & 1 \\ x_3 & y_3 & 1 \end{vmatrix}$ and

interpret the equation $\begin{vmatrix} x & y & 1 \\ x_1 & y_1 & 1 \\ x_2 & y_2 & 1 \end{vmatrix} = 0$.

12. Explain why $\begin{vmatrix} x+1 & 4 & 4 \\ 5 & x+2 & 7 \\ 2 & 2 & x-3 \end{vmatrix} = 0$

can be described as a cubic equation.

Show that $x = 3$ is one root, and find the other two roots.

13. Prove that $\begin{vmatrix} a & b & c \\ c & a & b \\ b & c & a \end{vmatrix} \equiv (a+b+c)\begin{vmatrix} 1 & b & c \\ 1 & a & b \\ 1 & c & a \end{vmatrix}$

and hence deduce that:

$a^3 + b^3 + c^3 - 3abc$
$= (a + b + c)(a^2 + b^2 + c^2 - bc - ca - ab).$

14. Factorise these determinants.

(i) $\begin{vmatrix} 1 & a & bc \\ 1 & b & ca \\ 1 & c & ab \end{vmatrix}$
(ii) $\begin{vmatrix} 1 & 1 & 1 \\ x & y & z \\ x^2 & y^2 & z^2 \end{vmatrix}$

(iii) $\begin{vmatrix} 1 & 1 & 1 \\ x^2 & y^2 & z^2 \\ yz & zx & xy \end{vmatrix}$
(iv) $\begin{vmatrix} x & y & z \\ x^2 & y^2 & z^2 \\ yz & zx & xy \end{vmatrix}$

15. Show that $x, (x-1), (x+1)$ are factors of

$\Delta = \begin{vmatrix} 1 & 1 & 1 \\ 1 & x & x^2 \\ 1 & x^2 & x^4 \end{vmatrix}$ and factorise Δ completely.

Inverses

We can now find the inverse of the 3×3 matrix $\mathbf{M} = \begin{pmatrix} a_1 & b_1 & c_1 \\ a_2 & b_2 & c_2 \\ a_3 & b_3 & c_3 \end{pmatrix}$ by

applying Cramer's rule to three special cases, assuming that $\Delta = \det \mathbf{M} = \mathbf{a} \cdot (\mathbf{b} \times \mathbf{c}) \neq 0$.

Let $\mathbf{L}$ be a matrix such that $\mathbf{LM} = \mathbf{I}$.

Then $\mathbf{Mr} = \mathbf{d} \implies \mathbf{LMr} = \mathbf{Ld}$

$\qquad\qquad \implies \mathbf{Ir} = \mathbf{Ld}$

$\qquad\qquad \implies \mathbf{r} = \mathbf{Ld}$

so if $\mathbf{d} = \mathbf{i} = \begin{pmatrix} 1 \\ 0 \\ 0 \end{pmatrix}$, $\mathbf{r} = \mathbf{Li}$ which is just the first column of $\mathbf{L}$.

Replacing $\mathbf{d}$ by $\mathbf{i}$ in the expressions $x = \dfrac{\mathbf{d} \cdot (\mathbf{b} \times \mathbf{c})}{\mathbf{a} \cdot (\mathbf{b} \times \mathbf{c})}$, $y = \dfrac{\mathbf{d} \cdot (\mathbf{c} \times \mathbf{a})}{\mathbf{a} \cdot (\mathbf{b} \times \mathbf{c})}$,
$z = \dfrac{\mathbf{d} \cdot (\mathbf{a} \times \mathbf{b})}{\mathbf{a} \cdot (\mathbf{b} \times \mathbf{c})}$ gives:

$$x = \frac{\mathbf{i} \cdot (\mathbf{b} \times \mathbf{c})}{\Delta} = \frac{A_1}{\Delta}, \quad y = \frac{\mathbf{i} \cdot (\mathbf{c} \times \mathbf{a})}{\Delta} = \frac{B_1}{\Delta}, \quad z = \frac{\mathbf{i} \cdot (\mathbf{a} \times \mathbf{b})}{\Delta} = \frac{C_1}{\Delta}$$

so that the first column of $\mathbf{L}$ is $\begin{pmatrix} A_1/\Delta \\ B_1/\Delta \\ C_1/\Delta \end{pmatrix}$.

Similarly, by using $\mathbf{d} = \mathbf{j}$ and $\mathbf{d} = \mathbf{k}$, the second and third columns of $\mathbf{L}$ are
$\begin{pmatrix} A_2/\Delta \\ B_2/\Delta \\ C_2/\Delta \end{pmatrix}$ and $\begin{pmatrix} A_3/\Delta \\ B_3/\Delta \\ C_3/\Delta \end{pmatrix}$ respectively, so that $\mathbf{L} = \dfrac{1}{\Delta} \begin{pmatrix} A_1 & A_2 & A_3 \\ B_1 & B_2 & B_3 \\ C_1 & C_2 & C_3 \end{pmatrix}$.

(This result was obtained by a different method in Question 10 of Exercise 1C.)

The matrix $\begin{pmatrix} A_1 & A_2 & A_3 \\ B_1 & B_2 & B_3 \\ C_1 & C_2 & C_3 \end{pmatrix}$ is known as the *adjugate* (or *adjoint*) of $\mathbf{M}$,

denoted by $\text{adj}\mathbf{M}$. Note that $\text{adj}\mathbf{M}$ is formed by replacing each element of $\mathbf{M}$ by its cofactor, and then transposing, i.e. changing rows into columns and columns into rows. This process will be illustrated in the example on page 25.

Strictly, the inverse $\mathbf{L}$ just obtained is a *left-inverse* of $\mathbf{M}$ because $\mathbf{LM} = \mathbf{I}$. But if $\mathbf{R}$ is a right-inverse (i.e. $\mathbf{MR} = \mathbf{I}$):

$\mathbf{LM} = \mathbf{I} \implies (\mathbf{LM})\mathbf{R} = \mathbf{IR} = \mathbf{R} \implies \mathbf{L}(\mathbf{MR}) = \mathbf{R} \implies \mathbf{LI} = \mathbf{R} \implies \mathbf{L} = \mathbf{R}$

so any left-inverse of a matrix is also a right-inverse and vice-versa.

If $\mathbf{L}$ and $\mathbf{L}'$ are two left-inverses of $\mathbf{M}$, then $\mathbf{L}$ is also a right-inverse and:

$\mathbf{L}'\mathbf{M} = \mathbf{I} \implies (\mathbf{L}'\mathbf{M})\mathbf{L} = \mathbf{IL} \implies \mathbf{L}'(\mathbf{ML}) = \mathbf{L} \implies \mathbf{L}'\mathbf{I} = \mathbf{L} \implies \mathbf{L}' = \mathbf{L}$

so the inverse of $\mathbf{M}$, if it exists, is unique. So we are justified in saying $\mathbf{L}$ is the inverse of $\mathbf{M}$, usually denoted by $\mathbf{M}^{-1}$, where:

$$\mathbf{M}^{-1} = \frac{1}{\Delta} \text{adj}\mathbf{M} = \frac{1}{\Delta} \begin{pmatrix} A_1 & A_2 & A_3 \\ B_1 & B_2 & B_3 \\ C_1 & C_2 & C_3 \end{pmatrix}, \ \Delta \neq 0.$$

Activity

When we *transpose* a matrix $\mathbf{A}$ we form a new matrix, denoted by $\mathbf{A}^{\mathrm{T}}$, which has the same elements as $\mathbf{A}$ except that they are rearranged so that the element in the rth row, cth column of $\mathbf{A}$ becomes the element in the cth row, rth column of $\mathbf{A}^{\mathrm{T}}$: the first row of $\mathbf{A}$ becomes the first column of $\mathbf{A}^{\mathrm{T}}$, and so on.

(i) Let $\mathbf{A}$ be an $m \times n$ matrix and $\mathbf{B}$ be $n \times p$. Taking the rth row of $\mathbf{A}$ to be

$(r_1\ r_2\ r_3\ \ldots\ r_n)$ and the cth column of $\mathbf{B}$ to be $\begin{pmatrix} c_1 \\ c_2 \\ c_3 \\ \ldots \\ c_n \end{pmatrix}$, write down the

element in the rth row and cth column of $\mathbf{AB}$. Show that this is the same as the element in the cth row and rth column of the matrix product $\mathbf{B}^{\mathrm{T}}\mathbf{A}^{\mathrm{T}}$. Hence show that $(\mathbf{AB})^{\mathrm{T}} = \mathbf{B}^{\mathrm{T}}\mathbf{A}^{\mathrm{T}}$.
(Note that $\mathbf{A}$ and $\mathbf{B}$ do not need to be square matrices.)

(ii) By putting $\mathbf{A} = \mathbf{M}$ and $\mathbf{B} = \mathbf{M}^{-1}$, where $\mathbf{M}$ is a square matrix with $\det\mathbf{M} \neq 0$, use the fact that $\mathbf{I}^{\mathrm{T}} = \mathbf{I}$ to prove that $(\mathbf{M}^{-1})^{\mathrm{T}} = (\mathbf{M}^{\mathrm{T}})^{-1}$.

EXAMPLE

Find the inverse of $\mathbf{M} = \begin{pmatrix} 2 & 3 & 4 \\ 2 & -5 & 2 \\ -3 & 6 & -3 \end{pmatrix}$.

Solution

First evaluate the three cofactors of the elements in one column and hence find the determinant. We have chosen to work with the first column.

$$A_1 = \begin{vmatrix} -5 & 2 \\ 6 & -3 \end{vmatrix} = 15 - 12 = 3$$

$$A_2 = -\begin{vmatrix} 3 & 4 \\ 6 & -3 \end{vmatrix} = -(-9 - 24) = 33$$

$$A_3 = \begin{vmatrix} 3 & 4 \\ -5 & 2 \end{vmatrix} = 6 - (-20) = 26 \qquad \begin{aligned} \Delta &= 2A_1 + 2A_2 - 3A_3 \\ &= 2 \times 3 + 2 \times 33 - 3 \times 26 = -6 \end{aligned}$$

Since $\Delta \neq 0$, $\mathbf{M}^{-1}$ exists.

Next evaluate the cofactors of the elements in the other two columns.

$$B_1 = -\begin{vmatrix} 2 & 2 \\ -3 & -3 \end{vmatrix} = 0 \qquad\qquad C_1 = \begin{vmatrix} 2 & -5 \\ -3 & 6 \end{vmatrix} = 12 - 15 = -3$$

$$B_2 = \begin{vmatrix} 2 & 4 \\ -3 & -3 \end{vmatrix} = -6 - (-12) = 6 \qquad C_2 = -\begin{vmatrix} 2 & 3 \\ -3 & 6 \end{vmatrix} = -(12 - (-9)) = -21$$

$$B_3 = -\begin{vmatrix} 2 & 4 \\ 2 & 2 \end{vmatrix} = -(4 - 8) = 4 \qquad\qquad C_3 = \begin{vmatrix} 2 & 3 \\ 2 & -5 \end{vmatrix} = -10 - 6 = -16$$

Evaluate Δ by other expansions to check arithmetic.

$$\Delta = 3B_1 - 5B_2 + 6B_3 = 3 \times 0 - 5 \times 6 + 6 \times 4 = -6$$

$$\Delta = 4C_1 + 2C_2 - 3C_3 = 4(-3) + 2(-21) - 3(-16) = -6$$

$$\mathbf{M}^{-1} = \begin{pmatrix} 2 & 3 & 4 \\ 2 & -5 & 2 \\ -3 & 6 & -3 \end{pmatrix}^{-1} = \frac{1}{-6} \begin{pmatrix} 3 & 0 & -3 \\ 33 & 6 & -21 \\ 26 & 4 & -16 \end{pmatrix}^{T}$$

The matrix of cofactors …

… which is to be transposed …

$$= \frac{1}{-6} \begin{pmatrix} 3 & 33 & 26 \\ 0 & 6 & 4 \\ -3 & -21 & -16 \end{pmatrix}$$

… and multiplied by $\frac{1}{\Delta}$.

$$= \frac{1}{6} \begin{pmatrix} -3 & -33 & -26 \\ 0 & -6 & -4 \\ 3 & 21 & 16 \end{pmatrix}.$$

The final matrix may be written as $\begin{pmatrix} -\frac{1}{2} & -\frac{11}{2} & -\frac{13}{3} \\ 0 & -1 & -\frac{2}{3} \\ \frac{1}{2} & \frac{7}{2} & \frac{8}{3} \end{pmatrix}.$

The adjoint method is a reasonable way of finding the inverse of a 3×3 matrix, though it is important to check your arithmetic as it is very easy to make mistakes. But for larger matrices a routine known as the row operation method is used, as it requires far fewer arithmetic steps. For example, it takes about 10^8 steps to invert a 10×10 matrix by the adjoint method, but only about 3000 steps by row operations. The row operations method is easy to program for a computer – another major advantage – though care must be exercised to avoid rounding errors, and problems occur when the determinant is close to zero.

Determinant of a product

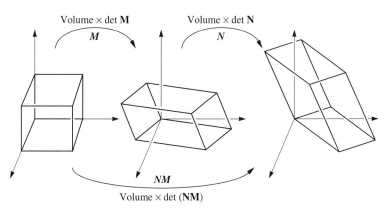

Figure 1.10

We now prove that if $\mathbf{M}$ and $\mathbf{N}$ are 3×3 matrices, then $\det(\mathbf{MN}) = \det\mathbf{M} \times \det\mathbf{N}$. An algebraic proof is difficult, but an alternative proof using geometry is quite straightforward. The transformation M with matrix $\mathbf{M}$ maps the unit cube with edges $\mathbf{i}$, $\mathbf{j}$, $\mathbf{k}$ to the parallelepiped with edges $\mathbf{a}$, $\mathbf{b}$, $\mathbf{c}$, with signed volume $\det\mathbf{M} = \mathbf{a} \cdot (\mathbf{b} \times \mathbf{c})$. Transformation M followed by transformation N, as illustrated in figure 3.10, is equivalent to the single transformation NM with matrix $\mathbf{NM}$ and volume scale factor $\det(\mathbf{NM})$. Clearly $\det(\mathbf{NM}) = \det\mathbf{N} \times \det\mathbf{M} = \det\mathbf{M} \times \det N = \det(\mathbf{MN})$.

Activity

Use the fact that $\det(\mathbf{NM}) = \det\mathbf{N} \times \det\mathbf{M}$ to prove that if $\det\mathbf{M} = 0$, matrix $\mathbf{M}$ does not have an inverse. (Such matrices are described as *singular*.)

Row properties

If $\det\mathbf{M} = \Delta \neq 0$, $\mathbf{M}^{-1}$ exists and is equal to $\dfrac{1}{\Delta}\text{adj}\mathbf{M}$. Therefore $\mathbf{M}\,\text{adj}\mathbf{M} = \Delta\mathbf{I}$. This may be rewritten as:

$$\begin{pmatrix} a_1 & b_1 & c_1 \\ a_2 & b_2 & c_2 \\ a_3 & b_3 & c_3 \end{pmatrix} \begin{pmatrix} A_1 & A_2 & A_3 \\ B_1 & B_2 & B_3 \\ C_1 & C_2 & C_3 \end{pmatrix} = \begin{pmatrix} \Delta & 0 & 0 \\ 0 & \Delta & 0 \\ 0 & 0 & \Delta \end{pmatrix}$$

Looking just at the top left-hand element of the product we have $a_1 A_1 + b_1 B_1 + c_1 C_1 = \Delta$. But $a_1 A_1 + b_1 B_1 + c_1 C_1$ is the expansion by the first column of $\det(\mathbf{M}^{\mathsf{T}})$, where $\mathbf{M}^{\mathsf{T}}$ is the transpose of $\mathbf{M}$. This means that $\det(\mathbf{M}^{\mathsf{T}}) = \det\mathbf{M}$. It also follows that a determinant may be evaluated by expanding by rows as well as by expanding by columns and that there are row properties corresponding to all the column properties established in Exercise 1C and the previous pages.

Activity

Some calculators handle matrices. Find out how to use such a calculator to find the inverse of a 3×3 matrix.

Exercise 1D

1. By finding the adjugate matrix find the inverses of the following, where possible.

(i) $\begin{pmatrix} 1 & 2 & 4 \\ 2 & 4 & 5 \\ 0 & 1 & 2 \end{pmatrix}$ (ii) $\begin{pmatrix} 3 & 2 & 6 \\ 5 & 3 & 11 \\ 7 & 4 & 16 \end{pmatrix}$

(iii) $\begin{pmatrix} 5 & 5 & -5 \\ -9 & 3 & -5 \\ -4 & -6 & 8 \end{pmatrix}$ (iv) $\begin{pmatrix} 6 & 5 & 6 \\ -5 & 2 & -4 \\ -4 & -6 & -5 \end{pmatrix}$

2. Given $\mathbf{M} = \begin{pmatrix} 1 & 2 & 2 \\ 3 & 2 & 3 \\ 4 & 1 & 1 \end{pmatrix}$ find:

(i) $\text{adj}\mathbf{M}$ (ii) $\mathbf{M}(\text{adj}\mathbf{M})$ (iii) $(\text{adj}\mathbf{M})\mathbf{M}$

(iv) $\det\mathbf{M}$ (v) $\det(\text{adj}\mathbf{M})$ (vi) $\text{adj}(\text{adj}\mathbf{M})$.

Comment on the answer to (vi).

3. Find the inverse of $\begin{pmatrix} 4 & -5 & 3 \\ 3 & 3 & -4 \\ 5 & 4 & -6 \end{pmatrix}$ and hence

solve $\begin{cases} 4x - 5y + 3z = 3 \\ 3x + 3y - 4z = 48 \\ 5x + 4y - 6z = 74 \end{cases}$.

4. Given $\mathbf{A} = \begin{pmatrix} 6 & 2 & -3 \\ 2 & 3 & 6 \\ 3 & -6 & 2 \end{pmatrix}$ evaluate $\mathbf{AA}^T$ and

hence, without doing further calculations, write down:

    (i) $\mathbf{A}^{-1}$ (ii) $\det\mathbf{A}$ (iii) $\mathbf{A}^T\mathbf{A}$.

5. Given $\mathbf{P} = \begin{pmatrix} -2 & 1 & 0 \\ 3 & 2 & 5 \\ 1 & 2 & 2 \end{pmatrix}$ and

$\mathbf{Q} = \begin{pmatrix} 2 & -1 & -2 \\ -1 & 2 & 2 \\ 2 & 3 & 3 \end{pmatrix}$ evaluate:

    (i) $\mathbf{PQ}$ and $\det(\mathbf{PQ})$
    (ii) $\mathbf{QP}$ and $\det(\mathbf{QP})$
    (iii) $\det\mathbf{P}$ (iv) $\det\mathbf{Q}$.

Verify that
$\det(\mathbf{PQ}) = \det(\mathbf{QP}) = (\det\mathbf{P}) \times (\det\mathbf{Q})$.

6. Given that the following matrices are singular find the values of x.

    (i) $\begin{pmatrix} 5 & 0 & 3 \\ 2 & x & 0 \\ 3 & 4 & 5 \end{pmatrix}$

    (ii) $\begin{pmatrix} 4 & 6 & -1 \\ -1 & 2 & -3 \\ 5 & x & 15 \end{pmatrix}$

    (iii) $\begin{pmatrix} 6 & 7 & -1 \\ 3 & x & 5 \\ 9 & 11 & x \end{pmatrix}$

    (iv) $\begin{pmatrix} 1-x & 1 & -2 \\ -1 & 2-x & 1 \\ 0 & 1 & -1-x \end{pmatrix}$

7. In this question $\mathbf{M}$ and $\mathbf{N}$ are square matrices of the same order.

    (i)   Prove that if $\mathbf{M}$ and $\mathbf{N}$ are non-singular, then $\mathbf{MN}$ is non-singular.

    (ii) By considering $(\mathbf{MN})(\mathbf{N}^{-1}\mathbf{M}^{-1})$, prove that if $\mathbf{MN}$ is non-singular then:
        (a) $(\mathbf{MN})^{-1} = \mathbf{N}^{-1}\mathbf{M}^{-1}$
        (b) $\text{adj}(\mathbf{MN}) = (\text{adj}\mathbf{N})(\text{adj}\mathbf{M})$.

8. The non-singular matrix $\mathbf{M}$ has the property that $\mathbf{MM}^T = \mathbf{M}^T\mathbf{M}$. Prove that:

    (i) $\mathbf{M}^T\mathbf{M}^{-1} = \mathbf{M}^{-1}\mathbf{M}^T$

    (ii) $\mathbf{N} = \mathbf{M}^{-1}\mathbf{M}^T \Rightarrow \mathbf{NN}^T = \mathbf{I}$.

9. Prove that if $\mathbf{M}$ is a non-singular 3×3 matrix then:

    (i) $\mathbf{M}(\text{adj}\mathbf{M}) = (\det\mathbf{M})\mathbf{I}$

    (ii) $\det(\text{adj}\mathbf{M}) = (\det\mathbf{M})^2$

    (iii) $\text{adj}(\text{adj}\mathbf{M}) = (\det\mathbf{M})\mathbf{M}$.

Do these results hold if $\mathbf{M}$ is singular? Justify your answer.

10. Prove that every matrix of the form $\begin{pmatrix} 1 & p & q \\ 0 & 1 & r \\ 0 & 0 & 1 \end{pmatrix}$
is non-singular, and that the inverse of such a matrix is of the same form.

11. A square matrix $\mathbf{M}$ is known as *orthogonal* if $\mathbf{M}^T\mathbf{M} = \mathbf{I}$. Prove that orthogonal matrices have the following properties.

    (i)   The vectors forming the columns of an orthogonal matrix:
        (a) have magnitude 1
        (b) are mutually perpendicular.
        ('Orthogonal' means 'perpendicular'.)

    (ii) The determinant of an orthogonal matrix is ± 1.

    (iii) Transformations represented by orthogonal matrices are *isometric*: i.e. they preserve length.

    Hints:
    1. Let $\mathbf{M}$ map P to P', where $\mathbf{p}' = \mathbf{Mp}$.
    2. Use $|\mathbf{q}' - \mathbf{p}'|^2 = (\mathbf{q}' - \mathbf{p}')^T(\mathbf{q}' - \mathbf{p}')$ to show that $(P'Q')^2 = (PQ)^2$.

Eigenvalues and eigenvectors

2 × 2 matrices

In *Pure Mathematics 4* you learnt how to identify lines which are invariant under a transformation. We now look at this idea in terms of matrices, but restrict our discussion to lines which pass through the origin.

As an example look at the effect of the transformation with matrix

$\mathbf{M} = \begin{pmatrix} 4 & 2 \\ 1 & 3 \end{pmatrix}$. Since $\begin{pmatrix} 4 & 2 \\ 1 & 3 \end{pmatrix}\begin{pmatrix} 1 \\ 1 \end{pmatrix} = \begin{pmatrix} 6 \\ 4 \end{pmatrix}$ the transformation defined by pre-multiplying position vectors by matrix $\mathbf{M}$ maps the vector $\begin{pmatrix} 1 \\ 1 \end{pmatrix}$ to $\begin{pmatrix} 6 \\ 4 \end{pmatrix}$.

Similarly the image of $\begin{pmatrix} k \\ k \end{pmatrix}$ is $\begin{pmatrix} 6k \\ 4k \end{pmatrix}$. Each point on the line $y = x$ can be represented by the position vector of the form $\begin{pmatrix} k \\ k \end{pmatrix}$ and the points with position vectors $\begin{pmatrix} 6k \\ 4k \end{pmatrix}$ form the line $y = \frac{2}{3}x$. This means that under the transformation represented by the matrix $\mathbf{M}$ the image of the line $y = x$ is the line $y = \frac{2}{3}x$. Similarly since $\begin{pmatrix} 4 & 2 \\ 1 & 3 \end{pmatrix}\begin{pmatrix} 1 \\ 2 \end{pmatrix} = \begin{pmatrix} 8 \\ 7 \end{pmatrix}$ the image of the line $y = 2x$ is the line $y = \frac{7}{8}x$.

Activity

(i) Find the images of the following position vectors under the transformation given by $\mathbf{M} = \begin{pmatrix} 4 & 2 \\ 1 & 3 \end{pmatrix}$.

(a) $\begin{pmatrix} 1 \\ 0 \end{pmatrix}$ (b) $\begin{pmatrix} 2 \\ 1 \end{pmatrix}$ (c) $\begin{pmatrix} 0 \\ 1 \end{pmatrix}$ (d) $\begin{pmatrix} -1 \\ 2 \end{pmatrix}$ (e) $\begin{pmatrix} -1 \\ 1 \end{pmatrix}$ (f) $\begin{pmatrix} -2 \\ 1 \end{pmatrix}$

(ii) Use your answers to (i) to find the equations of the images of the following lines.

(a) $y = 0$ (b) $y = \frac{1}{2}x$ (c) $x = 0$ (d) $y = -2x$ (e) $y = -x$ (f) $y = -\frac{1}{2}x$

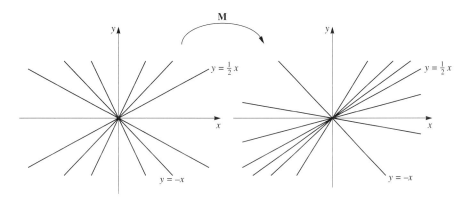

Figure 1.11

The information you have just gathered may be represented as in figure 1.11, where the object lines and their images are shown in separate diagrams. However we can show all the information on one diagram, as in figure 1.12, where (parts of) the object lines are shown at the centre of the diagram, and (parts of) their image lines are shown in the outer section of the diagram.

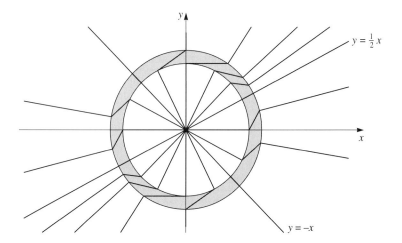

Figure 1.12

The shaded part of the diagram is not directly relevant but shows lines connecting each object line to its image. You will notice that there are two *invariant* lines, $y = \frac{1}{2}x$ and $y = -x$, which map to themselves under this transformation. The other lines appear to crowd towards $y = \frac{1}{2}x$, moving away from $y = -x$. This diagram prompts several questions.

- Are there other invariant lines?
- Why does $y = \frac{1}{2}x$ attract and $y = -x$ repel?
- Do all transformations behave like this?
- How can such lines be found efficiently?

Terminology

To answer these questions we need suitable terminology.

If **s** is a non-zero vector such that $\mathbf{Ms} = \lambda\mathbf{s}$, where **M** is a matrix and λ is a scalar we say that **s** is an *eigenvector* of **M**. The scalar λ is known as an *eigenvalue*.

Therefore, since
$$\begin{pmatrix} 4 & 2 \\ 1 & 3 \end{pmatrix}\begin{pmatrix} 2 \\ 1 \end{pmatrix} = \begin{pmatrix} 10 \\ 5 \end{pmatrix} = 5\begin{pmatrix} 2 \\ 1 \end{pmatrix}$$

and
$$\begin{pmatrix} 4 & 2 \\ 1 & 3 \end{pmatrix}\begin{pmatrix} -1 \\ 1 \end{pmatrix} = \begin{pmatrix} -2 \\ 2 \end{pmatrix} = 2\begin{pmatrix} -1 \\ 1 \end{pmatrix}$$

$\begin{pmatrix} 2 \\ 1 \end{pmatrix}$ and $\begin{pmatrix} -1 \\ 1 \end{pmatrix}$ are eigenvectors of the matrix $\mathbf{M} = \begin{pmatrix} 4 & 2 \\ 1 & 3 \end{pmatrix}$. The corresponding eigenvalues are 5 and 2 respectively. It will become evident later that these are the only two eigenvalues.

Properties of eigenvectors

Notice the following properties of eigenvectors.

1. All non-zero scalar multiples of $\begin{pmatrix} 2 \\ 1 \end{pmatrix}$ and $\begin{pmatrix} -1 \\ 1 \end{pmatrix}$ are also eigenvectors of $\mathbf{M}$, with (respectively) the same eigenvalues.

2. Under the transformation the eigenvector is enlarged by a scale factor equal to its eigenvalue.

3. The direction of an eigenvector is unchanged by the transformation. If the eigenvalue is negative the direction will be reversed – we regard this as a change of sense rather than a change of direction.

When finding eigenvectors you need to solve the equation $\mathbf{Ms} = \lambda\mathbf{s}$. Now:

$$\mathbf{Ms} = \lambda\mathbf{s}$$
$$\Leftrightarrow \quad \mathbf{Ms} - \lambda\mathbf{s} = \mathbf{0}$$
$$\Leftrightarrow \quad \mathbf{Ms} - \lambda\mathbf{Is} = \mathbf{0}$$
$$\Leftrightarrow \quad (\mathbf{M} - \lambda\mathbf{I})\mathbf{s} = \mathbf{0}$$

> $\mathbf{I}$ is the identity matrix; it is superfluous in this line.

> $\mathbf{I}$ is essential here.

Clearly $\mathbf{s} = \mathbf{0}$ is a solution, but you are seeking a non-zero solution for $\mathbf{s}$. For non-zero solutions you require $\det(\mathbf{M} - \lambda\mathbf{I}) = 0$. The equation $\det(\mathbf{M} - \lambda\mathbf{I}) = 0$ is known as the *characteristic equation* of $\mathbf{M}$. The left-hand side of the characteristic equation is a polynomial in λ; this polynomial is known as the *characteristic polynomial*.

(The German word for 'characteristic' is *eigen*: eigenvectors are also known as *characteristic* vectors; eigenvalues are also known as *characteristic* values.)

Finding eigenvectors

The following are the steps for finding eigenvectors, illustrated in the next example.

1. Form the characteristic equation: $\det(\mathbf{M} - \lambda\mathbf{I}) = 0$.

2. Solve the characteristic equation to find the eigenvalues, λ.

3. For each eigenvalue λ find a corresponding eigenvector $\mathbf{s}$ by solving $(\mathbf{M} - \lambda\mathbf{I})\mathbf{s} = \mathbf{0}$.

EXAMPLE

Find the eigenvectors of the matrix $\mathbf{M} = \begin{pmatrix} 4 & 2 \\ 1 & 3 \end{pmatrix}$.

Solution

1. Form the characteristic equation, $\det(\mathbf{M} - \lambda\mathbf{I}) = 0$.

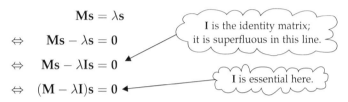

$$\mathbf{M} - \lambda\mathbf{I} = \begin{pmatrix} 4 & 2 \\ 1 & 3 \end{pmatrix} - \lambda\begin{pmatrix} 1 & 0 \\ 0 & 1 \end{pmatrix} = \begin{pmatrix} 4 - \lambda & 2 \\ 1 & 3 - \lambda \end{pmatrix}$$

so that $\quad \det(\mathbf{M} - \lambda\mathbf{I}) = 0 \quad \Leftrightarrow \quad \begin{vmatrix} 4 - \lambda & 2 \\ 1 & 3 - \lambda \end{vmatrix} = 0$

$$\Leftrightarrow \quad (4 - \lambda)(3 - \lambda) - 2 = 0$$

$$\Leftrightarrow \quad \lambda^2 - 7\lambda + 10 = 0.$$

2. Solve the characteristic equation to find the eigenvalues, λ.

$$\lambda^2 - 7\lambda + 10 = 0 \quad \Leftrightarrow \quad (\lambda - 2)(\lambda - 5) = 0$$

$$\Leftrightarrow \quad \lambda = 2 \text{ or } 5.$$

3. For each eigenvalue λ find a corresponding eigenvector $\mathbf{s}$ by solving $(\mathbf{M} - \lambda\mathbf{I})\mathbf{s} = \mathbf{0}$.

When $\lambda = 2$: $\quad (\mathbf{M} - \lambda\mathbf{I})\mathbf{s} = \mathbf{0} \quad \Leftrightarrow \quad \begin{pmatrix} 2 & 2 \\ 1 & 1 \end{pmatrix}\begin{pmatrix} x \\ y \end{pmatrix} = \begin{pmatrix} 0 \\ 0 \end{pmatrix}$, where $\mathbf{s} = \begin{pmatrix} x \\ y \end{pmatrix}$

$$\Leftrightarrow \quad x + y = 0 \quad \longleftarrow \quad \text{This tells you that if } y \text{ is any number, } k \text{ say, then } x \text{ is } -k.$$

$$\Leftrightarrow \quad \mathbf{s} = \begin{pmatrix} -k \\ k \end{pmatrix} = k\begin{pmatrix} -1 \\ 1 \end{pmatrix}.$$

When $\lambda = 5$: $\quad (\mathbf{M} - \lambda\mathbf{I})\mathbf{s} = \mathbf{0} \quad \Leftrightarrow \quad \begin{pmatrix} -1 & 2 \\ 1 & -2 \end{pmatrix}\begin{pmatrix} x \\ y \end{pmatrix} = \begin{pmatrix} 0 \\ 0 \end{pmatrix}$

$$\Leftrightarrow \quad -x + 2y = 0 \quad \longleftarrow \quad \text{If } y \text{ is any number, } k \text{ say, then } x \text{ is } 2k.$$

$$\Leftrightarrow \quad \mathbf{s} = \begin{pmatrix} 2k \\ k \end{pmatrix} = k\begin{pmatrix} 2 \\ 1 \end{pmatrix}.$$

Thus the eigenvectors are $\begin{pmatrix} -1 \\ 1 \end{pmatrix}$ and $\begin{pmatrix} 2 \\ 1 \end{pmatrix}$ or any scalar multiples of these vectors.

Expressing vectors in terms of the eigenvectors $\mathbf{s}_1 = \begin{pmatrix} -1 \\ 1 \end{pmatrix}$ and $\mathbf{s}_2 = \begin{pmatrix} 2 \\ 1 \end{pmatrix}$ explains why the line $y = \frac{1}{2}x$ attracts and the line $y = -x$ repels under the transformation with matrix $\mathbf{M} = \begin{pmatrix} 4 & 2 \\ 1 & 3 \end{pmatrix}$. Since $\mathbf{s}_1$ and $\mathbf{s}_2$ are non-zero and non-parallel you can express any position vector $\mathbf{p}$ as $\alpha\mathbf{s}_1 + \beta\mathbf{s}_2$. Then:

$$\mathbf{Mp} = \mathbf{M}(\alpha\mathbf{s}_1 + \beta\mathbf{s}_2)$$

$$= \alpha\mathbf{Ms}_1 + \beta\mathbf{Ms}_2$$

$$= 2\alpha\mathbf{s}_1 + 5\beta\mathbf{s}_2$$

showing that the image of $\mathbf{p}$ is attracted towards the eigenvector with the numerically larger eigenvalue.

Activity

Express three vectors in terms of $\mathbf{s}_1$ and $\mathbf{s}_2$. Illustrate the above property by drawing an accurate diagram showing your vectors, $\mathbf{s}_1$, $\mathbf{s}_2$, and their images under the transformation given by $\mathbf{M} = \begin{pmatrix} 4 & 2 \\ 1 & 3 \end{pmatrix}$.

3 × 3 matrices

The definitions of eigenvalue and eigenvector apply to all square matrices. The characteristic equation of matrix $\mathbf{M}$ is $\det(\mathbf{M} - \lambda\mathbf{I}) = 0$. When $\mathbf{M}$ is a 2×2 matrix the characteristic equation is quadratic, and may or may not have real roots. When $\mathbf{M}$ is a 3×3 matrix of real elements the characteristic equation is cubic, with real coefficients; this must have at least one real root. This proves that every real 3×3 matrix has at least one real eigenvector, and so every linear transformation of three-dimensional space has at least one invariant line. You use the same procedure as before for finding the eigenvalues and eigenvectors of a 3×3 matrix, though the work will generally be lengthy.

EXAMPLE

Find the eigenvectors of the matrix $\mathbf{M} = \begin{pmatrix} 3 & 1 & 1 \\ 1 & 3 & 1 \\ 1 & 1 & 3 \end{pmatrix}$.

Solution

$$\mathbf{M} - \lambda\mathbf{I} = \begin{pmatrix} 3 & 1 & 1 \\ 1 & 3 & 1 \\ 1 & 1 & 3 \end{pmatrix} - \lambda \begin{pmatrix} 1 & 0 & 0 \\ 0 & 1 & 0 \\ 0 & 0 & 1 \end{pmatrix} = \begin{pmatrix} 3-\lambda & 1 & 1 \\ 1 & 3-\lambda & 1 \\ 1 & 1 & 3-\lambda \end{pmatrix};$$

$$\det(\mathbf{M} - \lambda\mathbf{I}) = \begin{vmatrix} 3-\lambda & 1 & 1 \\ 1 & 3-\lambda & 1 \\ 1 & 1 & 3-\lambda \end{vmatrix}$$

$$= (3-\lambda)((3-\lambda)(3-\lambda) - 1) - ((3-\lambda) - 1) + (1 - (3-\lambda))$$

$$= -(\lambda^3 - 9\lambda^2 + 24\lambda - 20)$$

$$= -(\lambda - 5)(\lambda - 2)^2$$

so that $\det(\mathbf{M} - \lambda\mathbf{I}) = 0 \iff \lambda = 5$ or 2 (repeated root).

When $\lambda = 5$: $(\mathbf{M} - \lambda\mathbf{I})\mathbf{s} = \mathbf{0} \iff \begin{pmatrix} -2 & 1 & 1 \\ 1 & -2 & 1 \\ 1 & 1 & -2 \end{pmatrix} \begin{pmatrix} x \\ y \\ z \end{pmatrix} = \begin{pmatrix} 0 \\ 0 \\ 0 \end{pmatrix}$

$$\iff \left. \begin{array}{r} -2x + y + z = 0 \\ x - 2y + z = 0 \\ x + y - 2z = 0 \end{array} \right\} \iff x = y = z = k \text{ say,}$$

so that $\mathbf{s} = \begin{pmatrix} k \\ k \\ k \end{pmatrix} = k \begin{pmatrix} 1 \\ 1 \\ 1 \end{pmatrix}$ is an eigenvector, with eigenvalue 5.

When $\lambda = 2$, a repeated root:

$$(\mathbf{M} - \lambda\mathbf{I})\mathbf{s} = \mathbf{0} \quad \Leftrightarrow \quad \begin{pmatrix} 1 & 1 & 1 \\ 1 & 1 & 1 \\ 1 & 1 & 1 \end{pmatrix} \begin{pmatrix} x \\ y \\ z \end{pmatrix} = \begin{pmatrix} 0 \\ 0 \\ 0 \end{pmatrix}$$

$$\Leftrightarrow \quad x + y + z = 0$$

i.e. any vector in the plane $x + y + z = 0$ is an eigenvector, with eigenvalue 2.

A general vector in that plane is $\mathbf{s} = \begin{pmatrix} p \\ q \\ -p-q \end{pmatrix}$.

Thus the eigenvectors are $k\begin{pmatrix} 1 \\ 1 \\ 1 \end{pmatrix}$ and $\begin{pmatrix} p \\ q \\ -p-q \end{pmatrix}$ where p and q are not both zero.

The ideas above also apply to larger square matrices, but if $\mathbf{M}$ is $n \times n$, its characteristic equation is of degree n, and solving polynomial equations of higher degree is generally not straightforward – Evariste Galois (1811–32) proved that there is no general formula for solving polynomial equations of degree 5 or higher. In practice eigenvalues are not usually found by solving characteristic equations! Numerical methods will usually be applied to matrices, using a computer, with consequent problems caused by approximation and rounding errors.

Activity

The $n \times n$ matrix $\mathbf{M}$ has n eigenvalues $\lambda_1, \lambda_2, \ldots \lambda_n$, the roots of the polynomial equation $\det(\mathbf{M} - \lambda\mathbf{I}) = 0$.

(i) Imagine factorising the polynomial $\det(\mathbf{M} - \lambda\mathbf{I})$ into linear factors, and hence show that the product of the n eigenvalues is $\det\mathbf{M}$.

(ii) By considering the coefficient of the term in λ^{n-1} in the polynomial $\det(\mathbf{M} - \lambda\mathbf{I})$ show that the sum of the n eigenvalues is the sum of the elements on the leading diagonal of $\mathbf{M}$. This sum is known as the *trace* of matrix $\mathbf{M}$, $\text{tr}(\mathbf{M})$.

Exercise 1E

1. Find the eigenvalues and corresponding eigenvectors of these 2×2 matrices and check that the sum of the eigenvalues is the trace of the matrix.

(i) $\begin{pmatrix} 5 & 3 \\ 2 & 4 \end{pmatrix}$

(ii) $\begin{pmatrix} 7 & 2 \\ -12 & -4 \end{pmatrix}$

(iii) $\begin{pmatrix} 1 & 2 \\ 1 & 1 \end{pmatrix}$

(iv) $\begin{pmatrix} 1 & -1 \\ 1 & 3 \end{pmatrix}$

(v) $\begin{pmatrix} 1.1 & -0.4 \\ 0.2 & 0.2 \end{pmatrix}$

(vi) $\begin{pmatrix} p & 0 \\ 0 & q \end{pmatrix}, p \neq q$

2. Find the eigenvalues and corresponding eigenvectors of these 3×3 matrices and check that the sum of the eigenvalues is the trace of the matrix.

(i) $\begin{pmatrix} 3 & 0 & 0 \\ 0 & 2 & 1 \\ 0 & 0 & -1 \end{pmatrix}$

(ii) $\begin{pmatrix} 1 & 1 & 2 \\ 4 & 2 & -3 \\ 4 & 2 & 3 \end{pmatrix}$

(iii) $\begin{pmatrix} 1 & 1 & 2 \\ 5 & -2 & 1 \\ 1 & 1 & 2 \end{pmatrix}$

(iv) $\begin{pmatrix} 0 & 1 & -4 \\ -10 & 7 & -20 \\ -2 & 1 & -2 \end{pmatrix}$

(v) $\begin{pmatrix} 0 & 0 & 2 \\ 0 & 3 & 0 \\ 2 & 0 & 0 \end{pmatrix}$

(vi) $\begin{pmatrix} 1 & -3 & -3 \\ -8 & 6 & -3 \\ 8 & -2 & 7 \end{pmatrix}$

3. Matrix $\mathbf{M}$ is 2×2. Find the real eigenvalues of $\mathbf{M}$ and the corresponding eigenvectors when $\mathbf{M}$ represents:

(i) reflection in $y = x \tan \theta$

(ii) a rotation through angle θ about the origin.

4. Vector $\mathbf{s}$ is an eigenvector of matrix $\mathbf{A}$, with eigenvalue α, and also an eigenvector of matrix $\mathbf{B}$, with eigenvalue β. Prove that $\mathbf{s}$ is an eigenvector of

(i) $\mathbf{A} + \mathbf{B}$ (ii) $\mathbf{AB}$

and find the corresponding eigenvalues.

Hint: $\mathbf{s}$ is an eigenvector of $\mathbf{M} \Leftrightarrow \mathbf{Ms} = \lambda\mathbf{s}$, $\mathbf{s} \neq 0$.

5. Matrix $\mathbf{M}$ has eigenvalue λ with corresponding eigenvector $\mathbf{s}$; k is a non-zero scalar. Prove that the matrix $k\mathbf{M}$ has eigenvalue $k\lambda$ and that $\mathbf{s}$ is a corresponding eigenvector.

6. (i) Show that $\mathbf{r} = \begin{pmatrix} -2 \\ 0 \\ 1 \end{pmatrix}$ is an eigenvector

of $\mathbf{A} = \begin{pmatrix} 7 & 4 & -4 \\ 4 & 1 & 8 \\ -4 & 8 & 1 \end{pmatrix}$ and determine the

corresponding eigenvalue.

(ii) State two other eigenvectors of $\mathbf{A}$ which, together with $\mathbf{r}$, give three mutually perpendicular eigenvectors and state the corresponding eigenvalues.

(iii) What is the value of $\det\mathbf{A}$?   [O&C]

7. Matrix $\mathbf{M}$ is $n \times n$. For $n = 2$ and for $n = 3$ prove that if the sum of the elements in each row of $\mathbf{M}$ is 1 then 1 is an eigenvalue of $\mathbf{M}$. (This property holds for all values of n.)

8. Show that if λ is an eigenvalue of the square matrix $\mathbf{M}$ and the corresponding eigenvector is $\mathbf{s}$, then:

λ^2 is an eigenvalue of $\mathbf{M}^2$
λ^3 is an eigenvalue of $\mathbf{M}^3$
λ^n is an eigenvalue of $\mathbf{M}^n$

and even: λ^{-1} is an eigenvalue of $\mathbf{M}^{-1}$.

Show further that $\mathbf{s}$ is the corresponding eigenvector in all cases.

For (i) $\mathbf{M} = \begin{pmatrix} 4 & -1 \\ 2 & 1 \end{pmatrix}$ and

(ii) $\mathbf{M} = \begin{pmatrix} 3 & 2 & 2 \\ 1 & 4 & 1 \\ -2 & -4 & -1 \end{pmatrix}$

find the eigenvalues of:
(a) $\mathbf{M}$ (b) $\mathbf{M}^2$ (c) $\mathbf{M}^5$ (d) $\mathbf{M}^{-1}$.

9. The 2×2 matrix $\mathbf{M}$ has real eigenvalues λ_1, λ_2 and associated eigenvectors $\mathbf{s}_1, \mathbf{s}_2$, where $|\lambda_1| > |\lambda_2|$. By expressing any vector $\mathbf{v}$ in terms of $\mathbf{s}_1$ and $\mathbf{s}_2$, describe the behaviour of $\mathbf{M}^n\mathbf{v}$ as n increases when

(i) $|\lambda_1| < 1$ (ii) $|\lambda_1| = 1$ (iii) $|\lambda_1| > 1$.

10. Given $\mathbf{M} = -\mathbf{M}^\mathrm{T}$ prove that:
$\det(\mathbf{M} - k\mathbf{I}) = -\det(\mathbf{M} + k\mathbf{I})$.
Deduce that if λ is a non-zero eigenvalue of $\mathbf{M}$ then $-\lambda$ is also an eigenvalue of $\mathbf{M}$.
(Such matrices are called *skew-symmetric*.)

PM6

Exercise 1E continued

11. The self-drive camper-van hire firm DIY has depots at Calgary and Vancouver. The hire period commences on Saturday afternoon, and all vans are returned (to either depot) the following Saturday morning. Each week:

 • all DIY's vans are hired out

 • of the vans hired in Calgary, 50% are returned there, 50% to Vancouver

 • of the vans hired in Vancouver, 70% are returned there, 30% to Calgary.

 (i) One Saturday the Calgary depot has c vans and the Vancouver depot has v vans. Form matrix $\mathbf{M}$ so that the product $(c \quad v)\mathbf{M}$ gives the number of vans in each depot the following Saturday.

 (ii) At the start of the season each depot has 100 vans. Use matrix multiplication to find out how many vans will be at each depot two weeks later.

 (iii) Solve the equation $\mathbf{xM} = \mathbf{x}$ where $\mathbf{x} = (c \quad v)$ and explain the connection with eigenvalues.

 (iv) How many vans should DIY stock at Calgary and Vancouver if they want the number of vans available at those depots to remain constant?

 (The process described above is an example of a *Markov process*. The matrix governing it is a *transition* or *stochastic matrix*. It is common in such work to represent a state by a row vector, and to post-multiply by the transition matrix, as above. Each row of the transition matrix consists of non-negative elements with a sum of 1.)

12. At time t, the rabbit and wolf populations (r and w respectively) on a certain island are described by the differential equations:

$$\begin{cases} \dfrac{dr}{dt} = 5r - 3w \\ \dfrac{dw}{dt} = r + w \end{cases} \quad ①$$

 Throughout this question $\mathbf{p}$ represents $\begin{pmatrix} r \\ w \end{pmatrix}$ and $\mathbf{M}$ represents $\begin{pmatrix} 5 & -3 \\ 1 & 1 \end{pmatrix}$.

 (i) Show that the differential equations may be written as:

$$\dfrac{d\mathbf{p}}{dt} = \mathbf{Mp}. \quad ②$$

 (ii) Show that if $\mathbf{p} = \mathbf{p}_1(t)$ and $\mathbf{p} = \mathbf{p}_2(t)$ satisfy ② then $\mathbf{p} = a\mathbf{p}_1(t) + b\mathbf{p}_2(t)$ also satisfies ②, where a and b are constants.

 (iii) Show that if $\mathbf{p} = e^{\lambda t}\mathbf{k}$ satisfies ②, where $\mathbf{k}$ is constant, then $\mathbf{Mk} = \lambda \mathbf{k}$.

 (iv) Find the eigenvalues and eigenvectors of $\mathbf{M}$ and hence solve ① given that there are 1000 rabbits and 50 wolves at $t = 0$.

13. The number k is (numerically) the largest of the eigenvalues of 3×3 matrix $\mathbf{M}$, and $\mathbf{s}$ is a corresponding eigenvector; $\mathbf{v}_0$ is an arbitrary vector, and $\mathbf{v}_n = \mathbf{M}^n\mathbf{v}_0$ has components x_n, y_n and z_n.

 (i) Explain why, as n increases, $\mathbf{v}_n$ generally converges on a multiple of $\mathbf{s}$ and identify the occasions when this does not happen.

 (ii) What do you expect to notice about $\dfrac{x_n}{x_{n-1}}$, $\dfrac{y_n}{y_{n-1}}$ and $\dfrac{z_n}{z_{n-1}}$ as n increases?

Evaluating powers of square matrices

Pre-multiplying position vector $\mathbf{r}$ by a matrix $\mathbf{M}$ gives $\mathbf{r}'$, the position vector of R′, the image of R under the transformation represented by $\mathbf{M}$. If you apply the same transformation to R′ you get R″, with position vector $\mathbf{r}'' = \mathbf{M}(\mathbf{Mr}) = \mathbf{M}^2\mathbf{r}$. Higher powers of $\mathbf{M}$ arise if you continue to apply the same transformation. In this section you will learn to use eigenvalues and eigenvectors to evaluate powers of matrices.

The two statements $\begin{pmatrix} 4 & 2 \\ 1 & 3 \end{pmatrix}\begin{pmatrix} -1 \\ 1 \end{pmatrix} = \begin{pmatrix} -2 \\ 2 \end{pmatrix} = 2\begin{pmatrix} -1 \\ 1 \end{pmatrix}$

and $\begin{pmatrix} 4 & 2 \\ 1 & 3 \end{pmatrix}\begin{pmatrix} 2 \\ 1 \end{pmatrix} = \begin{pmatrix} 10 \\ 5 \end{pmatrix} = 5\begin{pmatrix} 2 \\ 1 \end{pmatrix}$

> The eigenvalues and eigenvectors of $\begin{pmatrix} 4 & 2 \\ 1 & 3 \end{pmatrix}$ were found on page 32.

can be combined into the single statement:

$$\begin{pmatrix} 4 & 2 \\ 1 & 3 \end{pmatrix}\begin{pmatrix} -1 & 2 \\ 1 & 1 \end{pmatrix} = \begin{pmatrix} -1 & 2 \\ 1 & 1 \end{pmatrix}\begin{pmatrix} 2 & 0 \\ 0 & 5 \end{pmatrix}$$

which you may write as $\mathbf{MS} = \mathbf{S\Lambda}$

where $\mathbf{M} = \begin{pmatrix} 4 & 2 \\ 1 & 3 \end{pmatrix}$, $\mathbf{S} = \begin{pmatrix} -1 & 2 \\ 1 & 1 \end{pmatrix}$ and $\mathbf{\Lambda} = \begin{pmatrix} 2 & 0 \\ 0 & 5 \end{pmatrix}$.

In just the same way if any 2×2 matrix has eigenvectors $\mathbf{s}_1$, $\mathbf{s}_2$ corresponding to eigenvalues λ_1, λ_2, then $\mathbf{Ms}_1 = \lambda_1\mathbf{s}_1$ and $\mathbf{Ms}_2 = \lambda_2\mathbf{s}_2$ so that $\mathbf{MS} = \mathbf{S\Lambda}$, where $\mathbf{S} = (\mathbf{s}_1 \quad \mathbf{s}_2)$, the 2×2 matrix which has the eigenvectors as columns, and $\mathbf{\Lambda} = \begin{pmatrix} \lambda_1 & 0 \\ 0 & \lambda_2 \end{pmatrix}$, a matrix with the corresponding eigenvalues on the leading diagonal and zeros elsewhere.

If $\mathbf{S}$ is non-singular, $\mathbf{S}^{-1}$ exists, and pre-multiplying $\mathbf{MS} = \mathbf{S\Lambda}$ by $\mathbf{S}^{-1}$ gives $\mathbf{S}^{-1}\mathbf{MS} = \mathbf{\Lambda}$; we then say that $\mathbf{M}$ has been *reduced to diagonal form* or that $\mathbf{M}$ has been *diagonalised*.

Although there are square matrices which cannot be reduced to diagonal form, being able to reduce $\mathbf{M}$ to diagonal form $\mathbf{\Lambda}$ helps if you want to raise $\mathbf{M}$ to a power. Post-multiplying $\mathbf{MS} = \mathbf{S\Lambda}$ by $\mathbf{S}^{-1}$ gives:

$$\mathbf{M} = \mathbf{S\Lambda S}^{-1}$$

so that $\mathbf{M}^4 = (\mathbf{S\Lambda S}^{-1})(\mathbf{S\Lambda S}^{-1})(\mathbf{S\Lambda S}^{-1})(\mathbf{S\Lambda S}^{-1})$

$= \mathbf{S\Lambda}(\mathbf{S}^{-1}\mathbf{S})\mathbf{\Lambda}(\mathbf{S}^{-1}\mathbf{S})\mathbf{\Lambda}(\mathbf{S}^{-1}\mathbf{S})\mathbf{\Lambda S}^{-1}$

$= \mathbf{S\Lambda}^4\mathbf{S}^{-1}$.

> This simplification makes extensive use of the associative property of matrix multiplication together with properties of inverses and identities.

Similarly: $\mathbf{M}^n = \mathbf{S\Lambda}^n\mathbf{S}^{-1}$, or, more formerly by induction.

Since $\mathbf{\Lambda} = \begin{pmatrix} \lambda_1 & 0 \\ 0 & \lambda_2 \end{pmatrix}$, $\mathbf{\Lambda}^n = \begin{pmatrix} \lambda_1^n & 0 \\ 0 & \lambda_2^n \end{pmatrix}$ and you can evaluate $\mathbf{M}^n$ readily, doing only two matrix multiplications whatever the value of n.

EXAMPLE

Find $\mathbf{M}^n$ where $\mathbf{M} = \begin{pmatrix} 4 & 2 \\ 1 & 3 \end{pmatrix}$.

Solution

We have already shown that $\begin{pmatrix} -1 \\ 1 \end{pmatrix}$ and $\begin{pmatrix} 2 \\ 1 \end{pmatrix}$ are eigenvectors, with eigenvalues

2 and 5 respectively, so we take $\mathbf{S} = \begin{pmatrix} -1 & 2 \\ 1 & 1 \end{pmatrix}$ and $\mathbf{\Lambda} = \begin{pmatrix} 2 & 0 \\ 0 & 5 \end{pmatrix}$.

> We could use any non-zero multiples of $\begin{pmatrix} -1 \\ 1 \end{pmatrix}$, $\begin{pmatrix} 2 \\ 1 \end{pmatrix}$ but choose these as the simplest.

Then $\mathbf{S}^{-1} = \begin{pmatrix} -\frac{1}{3} & \frac{2}{3} \\ \frac{1}{3} & \frac{1}{3} \end{pmatrix}$ and $\mathbf{\Lambda}^n = \begin{pmatrix} 2^n & 0 \\ 0 & 5^n \end{pmatrix}$.

Therefore $\mathbf{M}^n = \mathbf{S}\mathbf{\Lambda}^n\mathbf{S}^{-1} = \begin{pmatrix} -1 & 2 \\ 1 & 1 \end{pmatrix} \begin{pmatrix} 2^n & 0 \\ 0 & 5^n \end{pmatrix} \begin{pmatrix} -\frac{1}{3} & \frac{2}{3} \\ \frac{1}{3} & \frac{1}{3} \end{pmatrix}$

$$= \frac{1}{3} \begin{pmatrix} 2 \times 5^n + 2^n & 2 \times 5^n - 2^{n+1} \\ 5^n - 2^n & 5^n + 2^{n+1} \end{pmatrix}.$$

Again the work with 3×3 and other square matrices follows the same pattern, though the calculations are more complicated.

The Cayley–Hamilton theorem

We have already shown that the characteristic equation for the matrix

$\mathbf{M} = \begin{pmatrix} 4 & 2 \\ 1 & 3 \end{pmatrix}$ is $\det(\mathbf{M} - \lambda\mathbf{I}) = 0 \quad \Leftrightarrow \quad \begin{vmatrix} 4 - \lambda & 2 \\ 1 & 3 - \lambda \end{vmatrix} = 0$

$$\Leftrightarrow \quad (4 - \lambda)(3 - \lambda) - 2 = 0$$

$$\Leftrightarrow \quad 10 - 7\lambda + \lambda^2 = 0.$$

Notice that $\mathbf{M}^2 = \begin{pmatrix} 4 & 2 \\ 1 & 3 \end{pmatrix}\begin{pmatrix} 4 & 2 \\ 1 & 3 \end{pmatrix} = \begin{pmatrix} 18 & 14 \\ 7 & 11 \end{pmatrix}$ and that

$$10\mathbf{I} - 7\mathbf{M} + \mathbf{M}^2 = 10\begin{pmatrix} 1 & 0 \\ 0 & 1 \end{pmatrix} - 7\begin{pmatrix} 4 & 2 \\ 1 & 3 \end{pmatrix} + \begin{pmatrix} 18 & 14 \\ 7 & 11 \end{pmatrix} = \begin{pmatrix} 0 & 0 \\ 0 & 0 \end{pmatrix}$$

so that $10\mathbf{I} - 7\mathbf{M} + \mathbf{M}^2 = \mathbf{O}$, the zero matrix.

This illustrates the *Cayley–Hamilton theorem* which states:

> 'Every square matrix $\mathbf{M}$ satisfies its own characteristic equation.'

Note that $\mathbf{I}$ and $\mathbf{O}$ have to be inserted appropriately so that the equation makes sense.

You can readily prove this result for the general 2×2 matrix by direct multiplication, but the proof below, though written for 3×3 matrices, shows a style of argument that can be applied to all other square matrices.

When $\mathbf{M}$ is 3×3, notice that $\det(\mathbf{M} - \lambda\mathbf{I}) = \begin{vmatrix} a_1 - \lambda & b_1 & c_1 \\ a_2 & b_2 - \lambda & c_2 \\ a_3 & b_3 & c_3 - \lambda \end{vmatrix}$

which may be written as $d_0 + d_1\lambda + d_2\lambda^2 - \lambda^3$ where d_0, d_1, d_2 are independent of λ. Then the characteristic equation is

$$d_0 + d_1\lambda + d_2\lambda^2 - \lambda^3 = 0$$

and the Cayley–Hamilton theorem states that

$$d_0\mathbf{I} + d_1\mathbf{M} + d_2\mathbf{M}^2 - \mathbf{M}^3 = \mathbf{O}.$$

Since the elements of the 3×3 matrix $\text{adj}(\mathbf{M} - \lambda\mathbf{I})$ are 2×2 determinants, each element of $\text{adj}(\mathbf{M} - \lambda\mathbf{I})$ is (at most) quadratic in λ. We may therefore write

$$\text{adj}(\mathbf{M} - \lambda\mathbf{I}) = \mathbf{A}_0 + \lambda\mathbf{A}_1 + \lambda^2\mathbf{A}_2$$

where $\mathbf{A}_0$, $\mathbf{A}_1$ and $\mathbf{A}_2$ are 3×3 matrices with elements that are independent of λ.

On page 24 we proved that $\mathbf{M}^{-1} = \dfrac{1}{\det\mathbf{M}}\text{adj}\mathbf{M}$ for $\det\mathbf{M} \neq 0$, from which it follows that:

$$\mathbf{M}(\text{adj}\mathbf{M}) = (\det\mathbf{M})\mathbf{I}$$

> This is also true when $\det\mathbf{M} = 0$ (see Question 9 on page 28).

Substituting $(\mathbf{M} - \lambda\mathbf{I})$ for $\mathbf{M}$ gives:

$$(\mathbf{M} - \lambda\mathbf{I})\,\text{adj}(\mathbf{M} - \lambda\mathbf{I}) = \det(\mathbf{M} - \lambda\mathbf{I})\mathbf{I}$$
$$= (d_0 + d_1\lambda + d_2\lambda^2 - \lambda^3)\mathbf{I};$$

therefore $\quad (\mathbf{M} - \lambda\mathbf{I})(\mathbf{A}_0 + \lambda\mathbf{A}_1 + \lambda^2\mathbf{A}_2) = (d_0 + d_1\lambda + d_2\lambda^2 - \lambda^3)\mathbf{I}$

$$\Rightarrow \mathbf{M}\mathbf{A}_0 + \lambda(\mathbf{M}\mathbf{A}_1 - \mathbf{A}_0) + \lambda^2(\mathbf{M}\mathbf{A}_2 - \mathbf{A}_1) - \lambda^3\mathbf{A}_2 = (d_0 + d_1\lambda + d_2\lambda^2 - \lambda^3)\mathbf{I}$$

$$\Rightarrow \begin{cases} \mathbf{M}\mathbf{A}_0 = d_0\mathbf{I} \\ \mathbf{M}\mathbf{A}_1 - \mathbf{A}_0 = d_1\mathbf{I} \\ \mathbf{M}\mathbf{A}_2 - \mathbf{A}_1 = d_2\mathbf{I} \\ -\mathbf{A}_2 = -\mathbf{I} \end{cases} \text{so that} \begin{cases} d_0\mathbf{I} = \mathbf{M}\mathbf{A}_0 \\ d_1\mathbf{M} = \mathbf{M}^2\mathbf{A}_1 - \mathbf{M}\mathbf{A}_0 \\ d_2\mathbf{M}^2 = \mathbf{M}^3\mathbf{A}_2 - \mathbf{M}^2\mathbf{A}_1 \\ -\mathbf{M}^3 = \quad - \mathbf{M}^3\mathbf{A}_2 \end{cases}$$

and adding these four results gives $d_0\mathbf{I} + d_1\mathbf{M} + d_2\mathbf{M}^2 - \mathbf{M}^3 = \mathbf{O}$ confirming that $\mathbf{M}$ satisfies its own characteristic equation.

EXAMPLE Given $\mathbf{M} = \begin{pmatrix} 4 & 2 \\ 1 & 3 \end{pmatrix}$ use the Cayley–Hamilton theorem to find $\mathbf{M}^8$.

Solution

As before, the characteristic equation is $10 - 7\lambda + \lambda^2 = 0$.
By the Cayley–Hamilton theorem:

$$10\mathbf{I} - 7\mathbf{M} + \mathbf{M}^2 = \mathbf{O}$$

$$\Rightarrow \quad \mathbf{M}^2 = 7\mathbf{M} - 10\mathbf{I}$$

$$\Rightarrow \quad \mathbf{M}^4 = (7\mathbf{M} - 10\mathbf{I})^2$$

$$= 49\mathbf{M}^2 - 140\mathbf{M} + 100\mathbf{I}$$

$$= 49(7\mathbf{M} - 10\mathbf{I}) - 140\mathbf{M} + 100\mathbf{I}$$

$$= 203\mathbf{M} - 390\mathbf{I}$$

$$\Rightarrow \quad \mathbf{M}^8 = (203\mathbf{M} - 390\mathbf{I})^2$$

$$= 41\,209\mathbf{M}^2 - 158\,340\mathbf{M} + 152\,100\mathbf{I}$$

$$= 41\,209(7\mathbf{M} - 10\mathbf{I}) - 158\,340\mathbf{M} + 152\,100\mathbf{I}$$

$$= 130\,123\mathbf{M} - 259\,990\mathbf{I};$$

Thus $\mathbf{M}^8 = 130\,123\begin{pmatrix} 4 & 2 \\ 1 & 3 \end{pmatrix} - 259\,990\begin{pmatrix} 1 & 0 \\ 0 & 1 \end{pmatrix} = \begin{pmatrix} 260\,502 & 260\,246 \\ 130\,123 & 130\,379 \end{pmatrix}$.

(Check this against $\mathbf{M}^n = \dfrac{1}{3}\begin{pmatrix} 2 \times 5^n + 2^n & 2 \times 5^n - 2^{n+1} \\ 5^n - 2^n & 5^n + 2^{n+1} \end{pmatrix}$ with $n = 8$.

See page 38.)

Activity

Use the Cayley–Hamilton theorem to show that:

$$\mathbf{M} = \begin{pmatrix} 5 & 2 \\ 3 & 3 \end{pmatrix} \quad \Rightarrow \quad \mathbf{M}^{n+2} = 8\mathbf{M}^{n+1} - 9\mathbf{M}^n$$

This expression, giving $\mathbf{M}^{n+2}$ in terms of $\mathbf{M}^{n+1}$ and $\mathbf{M}^n$, is an example of a *recurrence relation*.

For Discussion

The Cayley–Hamilton theorem states that a matrix $\mathbf{M}$ satisfies its own characteristic equation. The characteristic equation may be written as $\det(\mathbf{M} - \lambda\mathbf{I}) = 0$. Replacing λ by $\mathbf{M}$ produces a determinant consisting entirely of zeros. Is this sufficient proof of the theorem?

HISTORICAL NOTE

The Cayley–Hamilton theorem was first announced by Arthur Cayley in 'A Memoir on the Theory of Matrices' in 1858, in which he proved the theorem for 2 × 2 matrices and checked it for 3 × 3 matrices. Amazingly he went on to say, 'I have not thought it necessary to undertake the labour of a formal proof of the theorem in the general case of a matrix of any degree.' Essentially the same property was contained in Sir William Hamilton's 'Lectures on Quaternions' in 1853, with a proof covering 4 × 4 matrices. The name 'characteristic equation' is attributed to Augustin Louis Cauchy (1789–1857), and the first general proof of the theorem was supplied in 1878 by Georg Frobenius (1849–1917), complete with modifications to take account of the problems caused by repeated eigenvalues.

Exercise 1F

1. Find matrices S and Λ such that $M = S\Lambda S^{-1}$.

(i) $M = \begin{pmatrix} 5 & 4 \\ 3 & 6 \end{pmatrix}$ (ii) $M = \begin{pmatrix} 7 & -10 \\ 3 & -4 \end{pmatrix}$

(iii) $M = \begin{pmatrix} 0.5 & 0.5 \\ 0.3 & 0.7 \end{pmatrix}$

2. Express $M = \begin{pmatrix} 1.9 & -1.5 \\ 0.6 & 0 \end{pmatrix}$ in the form $S\Lambda S^{-1}$ and hence find M^4. What can you say about M^n when n is very large?

3. Calculate the following.

(i) $\begin{pmatrix} 6 & -6 \\ 2 & -1 \end{pmatrix}^5$ (ii) $\begin{pmatrix} 3 & -1 \\ -1 & 3 \end{pmatrix}^{10}$

(iii) $\begin{pmatrix} 0.7 & 0.3 \\ 0.6 & 0.4 \end{pmatrix}^4$

4. Find examples of 2×2 matrices to illustrate the following.

(i) M has repeated eigenvalues and cannot be diagonalised.

(ii) M has repeated eigenvalues and can be diagonalised.

(iii) M has 0 as an eigenvalue and cannot be diagonalised.

(iv) M has 0 as an eigenvalue and can be diagonalised.

5. Demonstrate that $\begin{pmatrix} 1 & 2 \\ -1 & 4 \end{pmatrix}$ satisfies its own characteristic equation.

6. Prove the Cayley–Hamilton theorem for $M = \begin{pmatrix} a & c \\ b & d \end{pmatrix}$ by calculating M^2 and substituting directly into the characteristic equation.

7. You are given the matrix $M = \begin{pmatrix} 0 & -1 & 1 \\ 6 & -2 & 6 \\ 4 & 1 & 3 \end{pmatrix}$.

(i) Show that -2 and -1 are eigenvalues of M, and find the other eigenvalue.

(ii) Show that $\begin{pmatrix} 1 \\ 1 \\ -1 \end{pmatrix}$ is an eigenvector corresponding to the eigenvalue -2, and find an eigenvector corresponding to the eigenvalue -1.

(iii) Using the Cayley–Hamilton theorem, or otherwise:

(a) show that $M^4 = 11M^2 + 18M + 8I$

(b) find the values of p, q and r such that $M^{-1} = pM^2 + qM + rI$.

(iv) Evaluate $M^{-1} \begin{pmatrix} 1 \\ 1 \\ -1 \end{pmatrix}$. [MEI]

Exercise 1F continued

8. Show that $\begin{pmatrix} 1 \\ 1 \\ 0 \end{pmatrix}$ is an eigenvector of

$$\mathbf{M} = \begin{pmatrix} 1 & 4 & -1 \\ -1 & 6 & -1 \\ 2 & -2 & 4 \end{pmatrix} \text{ and state the}$$

corresponding eigenvalue. By finding the other eigenvalues and their eigenvectors express $\mathbf{M}$ in the form $\mathbf{S\Lambda S}^{-1}$.

9. The Fibonacci sequence $1, 1, 2, 3, \ldots$ is defined by $f_1 = f_2 = 1, f_{n+1} = f_n + f_{n-1}$.

Let $\mathbf{u}_n = \begin{pmatrix} f_{n+1} \\ f_n \end{pmatrix}$.

(i) Show that $\mathbf{u}_{n+1} = \mathbf{M}\mathbf{u}_n$ where

$$\mathbf{M} = \begin{pmatrix} 1 & 1 \\ 1 & 0 \end{pmatrix}.$$

(ii) Show that the eigenvalues of $\mathbf{M}$ are $\lambda_1 = \frac{1}{2}(1 + \sqrt{5}), \lambda_2 = \frac{1}{2}(1 - \sqrt{5})$, with

associated eigenvectors $\begin{pmatrix} \lambda_1 \\ 1 \end{pmatrix}, \begin{pmatrix} \lambda_2 \\ 1 \end{pmatrix}$.

(iii) Deduce that

$$\mathbf{M} = \frac{1}{\lambda_1 - \lambda_2} \begin{pmatrix} \lambda_1 & \lambda_2 \\ 1 & 1 \end{pmatrix} \begin{pmatrix} \lambda_1 & 0 \\ 0 & \lambda_2 \end{pmatrix} \begin{pmatrix} 1 & -\lambda_2 \\ -1 & \lambda_1 \end{pmatrix}$$

and hence show that

$$f_n = \frac{1}{\sqrt{5}} \left(\left(\frac{1 + \sqrt{5}}{2} \right)^n - \left(\frac{1 - \sqrt{5}}{2} \right)^n \right).$$

10. (i) $\mathbf{A}$ and $\mathbf{B}$ are 2×2 matrices with eigenvalues α_1, α_2 and β_1, β_2 respectively. Are the eigenvalues of the product $\mathbf{AB}$ the product of the eigenvalues of $\mathbf{A}$ and $\mathbf{B}$? Justify your answer.

(ii) Find the fallacy in this 'proof'.

$\mathbf{A}$ has eigenvalue λ and $\mathbf{B}$ has eigenvalue μ

$\Rightarrow \mathbf{AB}\mathbf{s} = \mathbf{A}\mu\mathbf{s}$

$\Rightarrow \mathbf{AB}\mathbf{s} = \mu\mathbf{A}\mathbf{s}$

$\Rightarrow \mathbf{AB}\mathbf{s} = \mu\lambda\mathbf{s}$

$\Rightarrow \mathbf{AB}$ has eigenvalue $\lambda\mu$.

11. Matrix $\mathbf{M}$ is $n \times n$. Matrix $\mathbf{S}$ is such that $\mathbf{MS} = \mathbf{SD}$, where $\mathbf{D}$ is a diagonal matrix with $\lambda_1, \lambda_2, \ldots \lambda_n$ on the leading diagonal and zeros

elsewhere. Suppose the first column of $\mathbf{S}$ is $\mathbf{s}_1$. By considering the first columns of the products $\mathbf{SD}$ and $\mathbf{MS}$ show that λ_1 is an eigenvalue of $\mathbf{M}$ and that $\mathbf{s}_1$ is a corresponding eigenvector.

(Similar arguments apply to the other columns of $\mathbf{S}$ and $\lambda_2, \lambda_3, \ldots \lambda_n$ showing that when searching for $\mathbf{S}$ and $\mathbf{D}$ such that $\mathbf{M} = \mathbf{SDS}^{-1}$ where $\mathbf{D}$ is diagonal, the leading diagonal of $\mathbf{D}$ is composed of eigenvalues, and $\mathbf{S}$ is composed of corresponding eigenvectors in the correct order.)

12. As usual $\mathbf{s}_1$, $\mathbf{s}_2$ and $\mathbf{s}_3$ are eigenvectors corresponding to distinct eigenvalues λ_1, λ_2 and λ_3 of the 3×3 matrix $\mathbf{M}$, and $\mathbf{S}$ is the matrix $(\mathbf{s}_1 \; \mathbf{s}_2 \; \mathbf{s}_3)$.

(i) By pre-multiplying $a\mathbf{s}_1 + b\mathbf{s}_2 = \mathbf{0}$ by $\mathbf{M}$ show that $\mathbf{s}_1$ cannot be parallel to $\mathbf{s}_2$.

(ii) Extend the argument to show that $\mathbf{s}_1$, $\mathbf{s}_2$ and $\mathbf{s}_3$ cannot be coplanar.

(iii) Show that $\mathbf{S}$ is non-singular and deduce that $\mathbf{M}$ can be diagonalised.

(All $n \times n$ matrices with n distinct eigenvalues can be diagonalised.)

13. The 3×3 matrix $\mathbf{A}$ is said to be *similar* to the 3×3 matrix $\mathbf{B}$ if a non-singular matrix $\mathbf{P}$ exists such that $\mathbf{A} = \mathbf{P}^{-1}\mathbf{BP}$.

(i) Prove that every 3×3 matrix is similar to itself.

(ii) Prove that:
$\mathbf{A}$ is similar to $\mathbf{B} \Rightarrow \mathbf{B}$ is similar to $\mathbf{A}$.

(iii) Prove that if $\mathbf{A}$ is similar to $\mathbf{B}$ and $\mathbf{B}$ is similar to $\mathbf{C}$ then $\mathbf{A}$ is similar to $\mathbf{C}$.

(iv) Prove that similar matrices have the same characteristic equation and deduce that similar matrices have the same eigenvalues.

14. The *trace* of matrix $\mathbf{A}$ is $\text{tr}(\mathbf{A})$, defined as the sum of the elements on the leading diagonal of $\mathbf{A}$. For 3×3 matrices prove that:

(i) $\text{tr}(\mathbf{A} + \mathbf{B}) = \text{tr}(\mathbf{A}) + \text{tr}(\mathbf{B})$

(ii) $\text{tr}(k\mathbf{A}) = k\text{tr}(\mathbf{A})$, where k is a scalar

(iii) $\text{tr}(\mathbf{AB}) = \text{tr}(\mathbf{BA})$

(iv) if there exists a non-singular matrix $\mathbf{P}$ such that $\mathbf{P}^{-1}\mathbf{AP} = \begin{pmatrix} \alpha & 0 & 0 \\ 0 & \beta & 0 \\ 0 & 0 & \gamma \end{pmatrix}$ then

$\text{tr}(\mathbf{A}) = \alpha + \beta + \gamma$

(v) $\text{tr}(\mathbf{A})$ is the sum of the three eigenvalues of $\mathbf{A}$.

15. Matrices $\mathbf{A}$ and $\mathbf{B}$ can be diagonalised. Assuming neither matrix has repeated eigenvalues, show that:

$\mathbf{A}$ and $\mathbf{B}$ share the same eigenvectors $\Leftrightarrow \mathbf{AB} = \mathbf{BA}$.

(This result, useful in Quantum Mechanics, also holds if eigenvalues are repeated.)

16. A *non-negative* matrix is one which contains no negative elements. In this question, λ_1 is the numerically larger eigenvalue of non-negative 2×2 matrix $\mathbf{M}$.

(i) Prove that λ_1 is positive.

(ii) By considering
$(\mathbf{I} - \mathbf{M})(\mathbf{I} + \mathbf{M} + \mathbf{M}^2 + \mathbf{M}^3 + \ldots + \mathbf{M}^n)$
show that if $\lambda_1 > 1$ then
$(\mathbf{I} - \mathbf{M})^{-1} = \mathbf{I} + \mathbf{M} + \mathbf{M}^2 + \mathbf{M}^3 + \ldots$ and
deduce that $(\mathbf{I} - \mathbf{M})^{-1}$ is non-negative.

(All non-negative matrices exhibit similar properties. They are important when applying mathematical models to economics.)

Investigation

Using a computer, investigate how to find the inverses of matrices.

(i) Program a spreadsheet to find the inverses of:

(a) 3×3 matrices without using the spreadsheet's built-in functions

(b) matrices using the spreadsheet's built-in functions.

(ii) Find out how to use a computer algebra system to invert matrices.

KEY POINTS
- The shortest distance from the point P to the line through A parallel to vector $\mathbf{d}$ is $|\overrightarrow{PA} \times \hat{\mathbf{d}}|$.
- The shortest distance from (x_1, y_1) to the line $ax + by + c = 0$ is
$\left| \dfrac{ax_1 + by_1 + c}{\sqrt{a^2 + b^2}} \right|.$
- The shortest distance from (x_1, y_1, z_1) to the plane $ax + by + cz + d = 0$ is $\left| \dfrac{ax_1 + by_1 + cz_1 + d}{\sqrt{a^2 + b^2 + c^2}} \right|.$
- The scalar triple product $\mathbf{a} \cdot (\mathbf{b} \times \mathbf{c})$ is the (signed) volume of a parallelepiped with edges given by the vectors $\mathbf{a}$, $\mathbf{b}$ and $\mathbf{c}$.

- Cyclic interchange of the vectors does not affect the value of the scalar triple product, but non-cyclic interchange of the vectors multiplies the product by -1.

- The shortest distance between the lines $\mathbf{r} = \mathbf{a} + t\mathbf{d}$ and $\mathbf{r} = \mathbf{b} + t\mathbf{e}$ is $\left| \dfrac{(\mathbf{a} - \mathbf{b}) \cdot (\mathbf{d} \times \mathbf{e})}{|\mathbf{d} \times \mathbf{e}|} \right|$.

  The lines intersect if and only if $(\mathbf{a} - \mathbf{b}) \cdot (\mathbf{d} \times \mathbf{e}) = 0$.

- The volume of the tetrahedron OABC is $\frac{1}{6}|\mathbf{a} \cdot (\mathbf{b} \times \mathbf{c})|$.

- The vectors $\mathbf{p}$, $\mathbf{q}$ and $\mathbf{r}$ (in that order):

  form a right-handed set $\Leftrightarrow \mathbf{p} \cdot (\mathbf{q} \times \mathbf{r}) > 0$

  form a left-handed set $\Leftrightarrow \mathbf{p} \cdot (\mathbf{q} \times \mathbf{r}) < 0$.

- $\mathbf{M} = \begin{pmatrix} a_1 & b_1 & c_1 \\ a_2 & b_2 & c_2 \\ a_3 & b_3 & c_3 \end{pmatrix} \Rightarrow \det\mathbf{M} = \Delta = |\mathbf{M}| = \mathbf{a} \cdot (\mathbf{b} \times \mathbf{c})$

$$= a_1 A_1 + a_2 A_2 + a_3 A_3,$$

$$\text{adj}\mathbf{M} = \begin{pmatrix} A_1 & A_2 & A_3 \\ B_1 & B_2 & B_3 \\ C_1 & C_2 & C_3 \end{pmatrix}, \quad \mathbf{M}\,\text{adj}\mathbf{M} = (\det\mathbf{M})\mathbf{I}$$

$$\text{and } \mathbf{M}^{-1} = \frac{1}{\det\mathbf{M}}\,\text{adj}\mathbf{M}, \text{ provided } \det\mathbf{M} \neq 0,$$

where $A_3 = $ the cofactor of $a_3 = +\begin{vmatrix} b_1 & c_1 \\ b_2 & c_2 \end{vmatrix}$,

$C_2 = $ the cofactor of $c_2 = -\begin{vmatrix} a_1 & b_1 \\ a_3 & b_3 \end{vmatrix}$, etc.

- $\det(\mathbf{MN}) = \det\mathbf{M} \times \det\mathbf{N}$ provided $\mathbf{M}$ and $\mathbf{N}$ are both $n \times n$.

- $(\mathbf{MN})^{-1} = \mathbf{N}^{-1}\mathbf{M}^{-1}$, provided both $\mathbf{M}$ and $\mathbf{N}$ are $n \times n$ and non-singular.

- $(\mathbf{MN})^{\mathrm{T}} = \mathbf{N}^{\mathrm{T}}\mathbf{M}^{\mathrm{T}}$, provided $\mathbf{M}$ is $m \times n$ and $\mathbf{N}$ is $n \times p$.

- An *eigenvector* of square matrix $\mathbf{M}$ is a non-zero vector $\mathbf{s}$ such that $\mathbf{Ms} = \lambda\mathbf{s}$; the scalar λ is the corresponding *eigenvalue*.

- The characteristic equation of $\mathbf{M}$ is $\det(\mathbf{M} - \lambda\mathbf{I}) = 0$.

- If $\mathbf{s}$ is the matrix formed of the eigenvectors of $\mathbf{M}$ and $\boldsymbol{\Lambda}$ is the diagonal matrix formed of the corresponding eigenvalues then $\mathbf{MS} = \mathbf{S}\boldsymbol{\Lambda}$, $\boldsymbol{\Lambda} = \mathbf{S}^{-1}\mathbf{MS}$, $\mathbf{M}^n = \mathbf{S}\boldsymbol{\Lambda}^n\mathbf{S}^{-1}$.

- The Cayley–Hamilton theorem:

  'Every square matrix $\mathbf{M}$ satisfies its own characteristic equation.'

2

Limiting processes

Like most mathematicians, he first obtained his results by totally unrigorous methods, and then polished them up into a decent proof.

Ian Stewart, writing (in 1996) about Archimedes

What is a limit?

Activity

The graph shows the value of $\left(1+\dfrac{1}{n}\right)^{n}$ for $n = 1, 2, 3, \ldots, 10$.

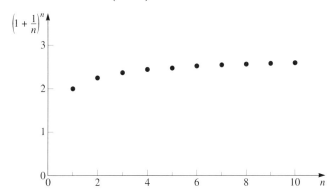

Use your calculator to evaluate this function for several values of n. What appears to be happening if n increases without limit?

(**Be warned:** Calculators treat $1+\dfrac{1}{n}$ as exactly 1 if n is very large.)

In this section we look at the different ways a function may behave as its *argument* (i.e. the independent variable) increases without limit. We start by looking only at sequences, that is functions of which the domain is the set of positive integers. Later in the chapter we shall look at the behaviour of functions of a continuous variable.

Notation

We use $f(1), f(2), f(3), \ldots$ or $u_1, u_2, u_3, \ldots$ to represent the terms of a sequence. The nth term of the sequence will be denoted by $f(n)$ or u_n. With the latter notation, (u_n) is used to denote the whole sequence, particularly when we wish to distinguish between the sequence as a whole, and the nth member of that sequence.

Your calculations in the activity above will have shown you that as n increases without limit, the value of $\left(1 + \dfrac{1}{n}\right)^n$ appears to approach the number e. Another example is the sequence $f(n) \equiv 2 - \dfrac{1}{n}$: as n increases, the value of $f(n)$ approaches 2. You can make $f(n)$ as close to 2 as you like for all sufficiently large n: if (for example) you want $f(n)$ to differ from 2 by less than 0.01 you may use any value for n provided $n \geqslant 101$. However, the terms of the sequence $g(n) \equiv n - \dfrac{1}{n}$ can be made as large as you like for all sufficiently large n: if (for example) you want $g(n)$ to exceed 2000 you may use any value for n provided $n \geqslant 2001$.

Informal definitions

- If $f(n)$ is a sequence in which the terms differ from ℓ by as little as we like for all sufficiently large n we say that $f(n)$ *converges* (or *tends*) to ℓ as n tends to infinity. Then ℓ is known as the *limit* of $f(n)$ as n tends to infinity. We write:

$$f(n) \to \ell \quad \text{as} \quad n \to \infty \quad \text{or} \quad \lim_{n \to \infty} f(n) = \ell.$$

Using this notation, the opening activity suggests that:

$$\left(1 + \frac{1}{n}\right)^n \to e \quad \text{as} \quad n \to \infty$$

and this may alternatively be written as:

$$\lim_{n \to \infty} \left(1 + \frac{1}{n}\right)^n = e.$$

The proof that this sequence converges as suggested is deferred to page 54.

- If $f(n)$ is a sequence in which the terms can be made as large as we like for all sufficiently large n it is customary to say that $f(n)$ tends to infinity as n tends to infinity and to write:

$$f(n) \to \infty \quad \text{as} \quad n \to \infty.$$

Notice carefully that such a sequence does not have a limit. Beware of treating ∞ as though it were a number. The expression $f(n) \to -\infty$ as $n \to \infty$ is defined similarly.

If you plot sequence $f(n)$ against n you may obtain a graph something like those shown in figure 2.1; both graphs illustrate sequences tending to a limit ℓ as n tends to ∞. They show that $f(n)$ is within ε of ℓ for all sufficiently large n where ε is a small positive number. (The symbol ε is the Greek letter *epsilon*.) Here 'all sufficiently large n' means that n may be any integer greater than or equal to some number N. Choosing a smaller value for ε generally requires a larger value for N. No matter how small you choose to make the positive number ε there is a number N (usually dependent on ε) such that $f(n)$ is within ε of ℓ for all $n \geqslant N$.

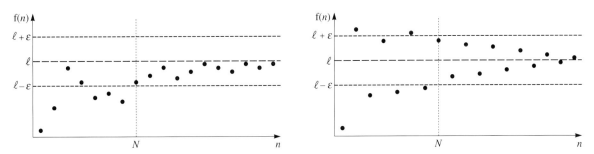

Figure 2.1

The graph in figure 2.2 illustrates the sequence:

$$f(n) \equiv \begin{cases} 1 + \dfrac{1}{n}, & n \text{ is not a multiple of } 10 \\ 1.001, & n \text{ is a multiple of } 10 \end{cases}$$

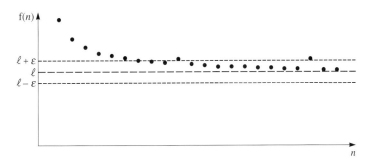

Figure 2.2

Here it is not true that $f(n)$ can be made as close to 1 as we like for all sufficiently large values of n. For example, if ε is chosen to be 0.0001 then $f(n)$ is within ε of 1 for most values of n which exceed 10 000, but not for values of n which are multiples of 10.

Formal definitions

Generalising the ideas above leads to the following formal definitions.

- We write $f(n) \to \ell$ as $n \to \infty$ or $\lim\limits_{n \to \infty} f(n) = \ell$ if and only if, given a positive number ε, however small, there is a number N such that $|f(n) - \ell| < \varepsilon$ for all $n \geqslant N$.

  (Generally the value of N depends on the value of ε.)

- We write $f(n) \to \infty$ as $n \to \infty$ if and only if, given K, however large, there is a number N such that $f(n) > K$ for all $n \geqslant N$.

  (Generally the value of N depends on the value of K.)

Limiting processes

The table below indicates some ways sequences may behave as $n \to \infty$.

	Behaviour	Example
1.	$f(n)$ converges to some limit ℓ as $n \to \infty$; we write $f(n) \to \ell$ as $n \to \infty$ or $\lim_{n \to \infty} f(n) = \ell$.	$f(n) \equiv 2 - \dfrac{1}{n} \to 2$ as $n \to \infty$
2.	$f(n)$ can be made as large as you like for all sufficiently large n; we write $f(n) \to \infty$ as $n \to \infty$.	$f(n) \equiv n^2 \to \infty$ as $n \to \infty$
3.	$f(n)$ can be made negative and numerically as large as we like for all sufficiently large n; we write $f(n) \to -\infty$ as $n \to \infty$.	$f(n) \equiv -2n \to -\infty$ as $n \to \infty$
4.	$f(n)$ oscillates finitely.	$f(n) \equiv 2 + (-1)^n$
5.	$f(n)$ oscillates infinitely.	$f(n) \equiv 2 + n(-1)^n$

All sequences which do not converge to a (finite) limit are described as *divergent*.

Exercise 2A

1. (i) For each of the following use your calculator to evaluate $f(n)$ for several (large) values of n, then make a conjecture about the behaviour of $f(n)$ as $n \to \infty$.

  (a) $f(n) \equiv n^{0.5}$ (b) $f(n) \equiv 0.5^n$

  (c) $f(n) \equiv 2^{1/n}$ (d) $f(n) \equiv 0.5^{1/n}$

  (e) $f(n) \equiv n^{1/n}$ (f) $f(n) \equiv (-n)^n$

  (g) $f(n) \equiv \dfrac{\cos n}{n}$ (h) $f(n) \equiv \dfrac{n^2 - n + 1}{2n^2 + 2n - 1}$

  (ii) For each of the functions evaluated in (i):

  if you think $f(n) \to \infty$ as $n \to \infty$ give a value of N such that $f(n) > 1000$ if $n \geqslant N$;

  if you think $f(n)$ converges to a limit ℓ as $n \to \infty$ give a value of N such that $f(n)$ is within 0.01 of ℓ if $n \geqslant N$.

2. By evaluating the following for large values of n predict their behaviour as $n \to \infty$.

  (i) $\left(1 + \dfrac{3}{n}\right)^n$ (ii) $\left(1 + \dfrac{1}{n}\right)^{\sqrt{n}}$ (iii) $\left(1 - \dfrac{1}{n}\right)^n$

  (iv) $\left(1 + \dfrac{1}{\sqrt{n}}\right)^n$ (v) $n(20^{1/n} - 1)$ (vi) $n^3(0.99)^n$

3. What is $\lim_{n \to \infty} \sqrt[n]{p^n + q^n}$ where p and q are unequal positive numbers?

4. The sequence (a_n) is defined by $a_n = k^n$. Describe the limiting behaviour of the sequence (a_n) as n tends to infinity, distinguishing carefully between the various possible values of the constant k.

5. Take a long strip of paper. Let l and u represent the lower and upper edges of the strip of paper.

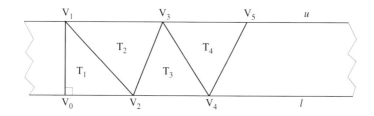

Exercise 2A continued

By folding the strip of paper form

- the line V_0V_1 perpendicular to l and u, and close to one end of the strip

- the line V_1V_2 which bisects the right angle between V_0V_1 and u

- the line V_2V_3 which bisects the obtuse angle between V_1V_2 and l

- the line V_3V_4 which bisects the obtuse angle between V_2V_3 and u.

These lines form triangles T_1, T_2 and T_3 as shown. Continue this process to form further triangles T_4, T_5, and so on.

(i) Show that all the triangles T_1, T_2, T_3, T_4, ... are isosceles.

(ii) Given that angle $V_{n-1}V_nV_{n+1} = (60 + d_n)°$ show that $d_{n+1} = -\frac{1}{2}d_n$ and deduce that, as $n \to \infty$, T_n tends to an equilateral triangle.

6.

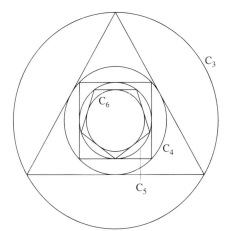

Circle C_3 has radius $10\,\text{cm}$. C_4 is the largest circle that can be drawn inside the equilateral triangle that has its vertices on C_3. C_5 is the largest circle that can be drawn inside the square that has its vertices on C_4. Generalising, C_{n+1} is the largest circle that can be drawn inside the n-sided regular polygon that has its vertices on C_n. The whole diagram forms a *polygonal well*. The radius of C_n appears to converge to a limit as $n \to \infty$. Use a computer or programmable calculator to show that this limit is about $1.15\,\text{cm}$.

7. (i) (a) Starting with $x_0 = 4$ find the first few values generated by the iterative formula $x_{n+1} = \frac{1}{2}\left(x_n + \frac{20}{x_n}\right)$.

(b) Use graphs of $y = \frac{1}{2}\left(x + \frac{20}{x}\right)$ and $y = x$ to show that the sequence converges to a limit. Solve $x = \frac{1}{2}\left(x + \frac{20}{x}\right)$ to find that limit.

(ii) (a) Show that the only possible limits of the iteration $x_{n+1} = x_n^2$ are 0 and 1.

(b) Show graphically that the iteration diverges when $x_0 > 1$, and find out what happens when $0 < x_0 < 1$.

8. For constant a, if the iterative formula

$$x_{n+1} = ax_n(1 - x_n) \qquad \text{①}$$

converges to a limit, then that limit is a root of the quadratic equation $x = ax(1 - x)$. Start with $x_0 = 0.7$ in each of the following.

(i) Show that, for $a = 2.8$, x_n converges to a limit as $n \to \infty$, and find that limit.

(ii) Show that, for $a = 3.2$, x_n eventually alternates between two values, which you should find.

(iii) What happens when
(a) $a = 3.5$ (b) $a = 3.83$ (c) $a = 4$?

(Equation ① is known as the *logistic equation*; it can be derived from Verhulst's 1845 model for restricted population growth, with x_n representing the size of a population at time n (sealed down so that 1 represents the largest population the environment can support) and a representing the fertility of the population.)

Combining limits

In looking at the behaviour of sequences as $n \to \infty$ in Exercise 2A you will have evaluated the terms of the sequences for various large values of n. In this section you will use algebraic methods: knowing how simple sequences behave can often help you identify the behaviour of more complex sequences.

Here are two ways of finding the limit of $f(n) \equiv \dfrac{3n^2 - n + 2}{n^2 - 1}$ as $n \to \infty$.

1. We can express $f(n)$ in partial fractions:

$$f(n) \equiv \frac{3n^2 - n + 2}{n^2 - 1}$$

$$\equiv 3 - \frac{n - 5}{(n - 1)(n + 1)}$$

$$\equiv 3 - \frac{3}{n - 1} + \frac{2}{n - 1} \to 3 \quad \text{as} \quad n \to \infty$$

since $\dfrac{3}{n + 1} \to 0$ and $\dfrac{2}{n - 1} \to 0$ as $n \to \infty$ assuming that we may add (and subtract) limits.

2. Alternatively: $\quad f(n) \equiv \dfrac{3n^2 - n + 2}{n^2 - 1}$

> Dividing numerator and denominator by n^2.

$$\equiv \frac{3 - \dfrac{1}{n} + \dfrac{2}{n^2}}{1 - \dfrac{1}{n^2}} \to \frac{3 - 0 + 0}{1 - 0} \quad \text{as} \quad n \to \infty$$

showing that $f(n) \to 3$ as $n \to \infty$, assuming that the limit of a sum is the sum of the limits, and that the limit of a quotient is the quotient of the limits.

You can confirm that both methods have led us to the correct result, as (for large n) the difference between $f(n)$ and 3 is

$$\left| \frac{3n^2 - n + 2}{n^2 - 1} - 3 \right| = \left| \frac{-n + 5}{n^2 - 1} \right|$$

$$= \left| \frac{n - 5}{n^2 - 1} \right|$$

$$< \left| \frac{n + 1}{n^2 - 1} \right|$$

$$= \left| \frac{1}{n - 1} \right|$$

which we can make as small as we like by making n sufficiently large.

Properties of limits

You will find the following properties of limits useful. (We prove them in the Appendix on page 247.)

In what follows, $f_1(n) \to \ell_1$ as $n \to \infty$ and $f_2(n) \to \ell_2$ as $n \to \infty$.

1. **Multiplying a function by a constant multiplies the limit by that constant**

   $af_1(n) \to a\ell_1$ as $n \to \infty$ where a is a constant.

2. **The limit of a sum (or difference) is the sum (or difference) of the limits**

   $f_1(n) + f_2(n) \to \ell_1 + \ell_2$ and $f_1(n) - f_2(n) \to \ell_1 - \ell_2$ as $n \to \infty$.

3. **The limit of a product is the product of the limits**

   $f_1(n) \times f_2(n) \to \ell_1\ell_2$ as $n \to \infty$.

4. **The limit of a quotient is the quotient of the limits**

   $\dfrac{f_1(n)}{f_2(n)} \to \dfrac{\ell_1}{\ell_2}$ as $n \to \infty$ provided $\ell_2 \neq 0$.

EXAMPLE Find $\lim\limits_{n \to \infty} \dfrac{\sqrt{n}+2}{n-5}$.

Solution

$$\frac{\sqrt{n}+2}{n-5} = \frac{\dfrac{1}{\sqrt{n}}+\dfrac{2}{n}}{1-\dfrac{5}{n}} \to \frac{0+0}{1-0} \quad \text{as} \quad n \to \infty$$

so that $\lim\limits_{n \to \infty} \dfrac{\sqrt{n}+2}{n-5} = 0$, applying properties **2** and **4**.

This example illustrates again how helpful it can be to divide the numerator and denominator by the highest power of n present in the denominator.

You will also find the so-called *sandwich theorem* very useful:

if $f(n) \leqslant g(n) \leqslant h(n)$ for all sufficiently large n

and both $f(n)$ and $h(n) \to \ell$ as $n \to \infty$

then $g(n) \to \ell$ as $n \to \infty$.

Limiting processes

Put differently: if g(n) is between f(n) and h(n), and f(n) and h(n) converge to the same limit, then g(n) also converges to that limit. The theorem is intuitively obvious as, if we have two sequences with terms that eventually get very close to ℓ (see figure 2.3), then a third sequence with terms that are eventually squeezed between those of the first two must also get very close to ℓ. It is quite common for $f(n) \equiv 0$ or some other constant.

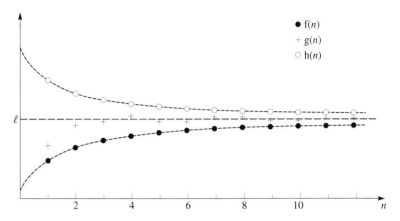

Figure 2.3

Standard results

The following table lists the behaviour of several standard and important sequences. Knowing (and quoting) them can save much effort.

	Limits	Validity
1.	$\lim_{n \to \infty} \dfrac{1}{n^p} = 0$	$p > 0$
2.	$\lim_{n \to \infty} \sqrt[n]{x} = 1$	$x > 0$
3.	$\lim_{n \to \infty} x^n = 0$	$|x| < 1$
4.	$\lim_{n \to \infty} n^p q^n = 0$	$p > 0$ and $|q| < 1$
5.	$\lim_{n \to \infty} \left(1 + \dfrac{r}{n}\right)^n = e^r$	all r
6.	$\lim_{n \to \infty} \sqrt[n]{n} = 1$	
7.	$\lim_{n \to \infty} \dfrac{x^n}{n!} = 0$	all x

The next three worked examples provide proofs of **1**, **3** and **5**; you will prove the others in Questions 6, 7, 8 and 10 of Exercise 2B.

EXAMPLE

(i) Find N so that $\left|\dfrac{1}{n^p}\right| < 0.01$ whenever $n > N$ (a) when $p = 2$ (b) when $p = \frac{1}{3}$.

(ii) Prove that $\displaystyle\lim_{n \to \infty} \dfrac{1}{n^p} = 0$, provided $p > 0$.

Solution

(i) Since n is a positive number $\dfrac{1}{n^p}$ is positive. Therefore:

$$\left|\frac{1}{n^p}\right| < 0.01 \iff \frac{1}{n^p} < \frac{1}{100} \iff n^p > 100 \iff n > \sqrt[p]{100}$$

(a) Putting $p = 2$:

$$\left|\frac{1}{n^p}\right| < 0.01$$

whenever $n > \sqrt{100}$
so we may take $N = 10$.

$\dfrac{1}{9^2} \approx 0.0123 > 0.01$

$\dfrac{1}{11^2} \approx 0.0083 < 0.01$

(b) Putting $p = \frac{1}{3}$:

$$\left|\frac{1}{n^p}\right| < 0.01 \text{ whenever } n^{\frac{1}{3}} > 100$$

$$\iff n > 100^3$$

so we may take $N = 1\,000\,000$.

(ii) The calculations in (i) illustrate
the general method.

To prove that $\displaystyle\lim_{n \to \infty} \dfrac{1}{n^p} = 0 \ (p > 0)$

we need to show that $\left|\dfrac{1}{n^p} - \ell\right| < \varepsilon$

for sufficiently large n, where $\ell = 0$.

If ε is any (small) positive number,
for positive n:

$$\left|\frac{1}{n^p}\right| < \varepsilon \iff \frac{1}{n^p} < \varepsilon \iff n^p > \varepsilon^{-1} \iff n > \sqrt[p]{\varepsilon^{-1}}$$

so we can be sure that $\dfrac{1}{n^p}$ is within ε of 0 whenever $n > \sqrt[p]{\varepsilon^{-1}}$.

We conclude that $\displaystyle\lim_{n \to \infty} \dfrac{1}{n^p} = 0$ since by choosing n sufficiently large we can
make $\dfrac{1}{n^p}$ as close to 0 as we like.

EXAMPLE

Prove that $\displaystyle\lim_{n \to \infty} x^n = 0$ provided $|x| < 1$.

Solution

Clearly $\displaystyle\lim_{n \to \infty} x^n = 0$ if $x = 0$.

If $x \neq 0$ we can write $|x| = \dfrac{1}{1 + y}$ where $y > 0$ since $|x| < 1$.

Then $|x|^n = \dfrac{1}{(1+y)^n} = \dfrac{1}{1 + ny + \dfrac{n(n-1)}{2!}y^2 + \cdots + y^n} < \dfrac{1}{ny}$.

All the terms in this binomial
expansion are positive. We
discard all except one of them.

But $|x|^n < \dfrac{1}{ny} \Rightarrow -\dfrac{1}{ny} < x^n < \dfrac{1}{ny}$

so that x^n is sandwiched between the sequences $-\dfrac{1}{ny}$ and $\dfrac{1}{ny}$.

Both $-\dfrac{1}{ny}$ and $\dfrac{1}{ny} \to 0$ as $n \to \infty$ so that $x^n \to 0$ as $n \to \infty$ by the sandwich theorem.

In the activity at the start of this chapter you evaluated $\left(1 + \dfrac{1}{n}\right)^n$ for various

(large) values of n and noticed that it looked as though $\left(1 + \dfrac{1}{n}\right)^n \to e$ as $n \to \infty$.

This is a special case of the important result proved in the next example.

EXAMPLE

Prove that $\displaystyle\lim_{n \to \infty} \left(1 + \dfrac{r}{n}\right)^n = e^r$, where r is any number.

Solution

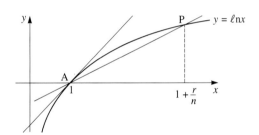

$A(1, 0)$ and $P\left(1 + \dfrac{r}{n}, \ln\left(1 + \dfrac{r}{n}\right)\right)$ are points on the curve $y = \ln x$.

(The diagram is for positive r, but the following argument also applies when r is negative, provided that n is large enough to make $1 + \dfrac{r}{n} > 0$.)

The gradient of the chord AP is

$$\frac{\ln\left(1 + \dfrac{r}{n}\right)}{\dfrac{r}{n}} = \frac{n}{r} \ln\left(1 + \frac{r}{n}\right) = \frac{1}{r}\ln\left(\left(1 + \frac{r}{n}\right)^n\right).$$

For fixed r, as $n \to \infty$, $1 + \dfrac{r}{n} \to 1$ and the chord AP approaches the tangent at A, whose gradient is 1.

Therefore $\dfrac{1}{r}\ln\left(\left(1 + \dfrac{r}{n}\right)^n\right) \to 1$ as $n \to \infty$

$\Rightarrow \quad \ln\left(\left(1 + \dfrac{r}{n}\right)^n\right) \to r$ as $n \to \infty$

$\Rightarrow \quad \left(1 + \dfrac{r}{n}\right)^n \to e^r$ as $n \to \infty$.

Exercise 2B

1. Find the limit (if there is one) of each of the following as $n \to \infty$.

   (i) $\dfrac{2n - 5}{3n + 7}$ (ii) $\dfrac{1 - n}{\sqrt{n}}$ (iii) $\dfrac{1}{n}\cos\dfrac{1}{n}$

   (iv) $\dfrac{\cos(n^2\pi)}{n}$ (v) $\left(1 + \dfrac{1}{n}\right)^{3n}$

2. (i) Show that
   $$(\sqrt{n^2 + 100} - n)(\sqrt{n^2 + 100} + n) \equiv 100$$
   and hence deduce that
   $$(\sqrt{n^2 + 100} - n) \to 0 \text{ as } n \to \infty.$$

   (ii) Adapt the method to find
   $$\lim_{n \to \infty} (\sqrt{n^2 + n} - n).$$

3. Find the limit (if there is one) of each of the following as $n \to \infty$.

   (i) $\dfrac{n!}{n^n}$ (ii) $\dfrac{n^4}{3^n}$ (iii) $\dfrac{n^2}{\sqrt{n^3 - 10}}$

   (iv) $\dfrac{a^n}{n^a}$ (v) $\dfrac{\sqrt{4n^2 + n - 20} - 5n}{n + 3}$

4. (i) Given that $a_n \to \infty$ as $n \to \infty$, find $\displaystyle\lim_{n \to \infty} \dfrac{1}{a_n}$ and justify your answer.

   (ii) Given that $b_n \to 0$ as $n \to \infty$, what can you say about $\dfrac{1}{b_n}$ as $n \to \infty$? Justify your answer.

5. The sequences (a_n) and (b_n) both increase without limit as $n \to \infty$. New sequences (c_n) and (d_n) are defined by $c_n = a_n - b_n$ and $d_n = \dfrac{a_n}{b_n}$.

   Find examples of (a_n) and (b_n) such that, as $n \to \infty$, (i) $c_n \to -\infty$ (ii) $c_n \to 0$ (iii) $c_n \to 3$ (iv) $c_n \to \infty$ (v) c_n oscillates infinitely (vi) $d_n \to 0$ (vii) $d_n \to 5$ (viii) $d_n \to \infty$.

6. (i) Let $y_n = \sqrt[n]{x} - 1$ where $x > 1$.

   (a) Use a binomial expansion to show that $x > 1 + ny_n$.

   (b) Explain why $0 < y_n < \dfrac{x - 1}{n}$ and deduce that $y_n \to 0$ as $n \to \infty$.

   (c) Explain why $\displaystyle\lim_{n \to \infty} \sqrt[n]{x} = 1$ when $x > 1$.

   (ii) By considering $\sqrt[n]{\dfrac{1}{x}}$ explain why
   $$\lim_{n \to \infty} \sqrt[n]{x} = 1 \text{ when } 0 < x < 1.$$

   (iii) Complete the proof that $\displaystyle\lim_{n \to \infty} \sqrt[n]{x} = 1$ provided $x > 0$.

7. In this question $a_n = n^p q^n$.

   (i) Take $p = 3$, $q = 0.5$ and let k be the mean of q and 1.

   (a) Show that $\dfrac{a_{n+1}}{a_n} < k \Leftrightarrow \left(1 + \dfrac{1}{n}\right)^3 < \dfrac{k}{q}$ and deduce that $\dfrac{a_{n+1}}{a_n} < k$ for all $n > 7$.

   (b) Show that $a_{r+7} < k^r a_7$ and deduce that, for $p = 3$ and $q = 0.5$,
   (1) $a_{r+7} \to 0$ as $r \to \infty$
   (2) $a_n \to 0$ as $n \to \infty$.

   (ii) Generalise the argument in (i) to show that $a_n \to 0$ as $n \to \infty$ when $p > 0$ and $0 < q < 1$.

   (iii) Complete the proof that $\displaystyle\lim_{n \to \infty} n^p q^n = 0$ provided $p > 0$ and $|q| < 1$.

8. (i) Write out the first three terms of the binomial expansion of
   $$\left(1 + \dfrac{2}{\sqrt{n}}\right)^n$$
   and deduce that
   $$n < \left(1 + \dfrac{2}{\sqrt{n}}\right)^n$$
   where n is a positive integer.

   (ii) Write out the binomial expansion of
   $$\left(1 + \dfrac{1}{n}\right)^n$$
   and show that $\left(1 + \dfrac{1}{n}\right)^n < n$ provided $n > 3$.

   (iii) Use these results to help you prove that
   $$\lim_{n \to \infty} \sqrt[n]{n} = 1.$$

9. (i) By considering the area of a suitable region under the graph of $y = \dfrac{1}{x}$, show that $n(t^{\frac{1}{n}} - 1) \to \ln t$ as $n \to \infty$.

   (ii) The following is an algorithm for obtaining an approximation for $\ln t$ when using a basic calculator with only the four 'functions' and a square root key.

   • Enter the value of t, and press the square root key n times.

   • Subtract 1, and then multiply by 2 n times.

   Try it out, and explain why it works.

Exercise 2B continued

10. In this question $a_n = \dfrac{x^n}{n!}$ where x is a constant.

(i) Assume that x is positive.

(a) By considering $\dfrac{a_{n+1}}{a_n} < \dfrac{1}{2}$ show that

$$a_{N+r} < \left(\dfrac{1}{2}\right)^r a_N \text{ where } N \text{ is any}$$

integer greater than $2x$ and r is a positive integer.

(b) Explain why $a_{N+r} \to 0$ as $r \to \infty$

and deduce that $\dfrac{x^n}{n!} \to 0$ as $r \to \infty$

when x is positive.

(ii) Complete the proof that $\displaystyle\lim_{n \to \infty} \dfrac{x^n}{n!} = 0$

whatever the value of the constant x.

11. In this question $f(n) \equiv na^n$ where a is a constant.

(i) For $a = 0.95$, show that:

(a) $f(n) > f(n+1)$ if and only if $n > 19$

(b) $f(20 + r) = f(20) \times \left(1 + \dfrac{r}{20}\right)(0.95)^r$

provided $r \geqslant 1$

(c) $\left(1 + \dfrac{r}{20}\right) < \left(1 + \dfrac{1}{20}\right)^r$ provided r is

an integer greater than 1

(d) $f(n) \to 0$ as $n \to \infty$.

(ii) Adapt the method of (i) to prove that $f(n) \to 0$ as $n \to \infty$ when $0 \leqslant a < 1$.

(iii) Complete the proof that $f(n) \to 0$ as $n \to \infty$ when $|a| < 1$.

12. In this question ℓ represents the limit to which an iterative process $x_{r+1} = f(x_r)$ is converging and $e_r = x_r - \ell$ represents the 'error' involved in taking x_r as ℓ.

(i) (a) Starting from $x_0 = 2$ use the iteration

$$x_{r+1} = \sqrt{\dfrac{a}{x_r}} \text{ to find } \sqrt[3]{6} \text{ to two decimal}$$

places, tabulating x_r, e_r and $\dfrac{e_{r+1}}{e_r}$.

(b) Show that $e_{r+1} = -\dfrac{\ell^2 e_r}{x_r(x_{r+1} + \ell)}$ and

deduce that provided x_r is close to ℓ the ratio of successive absolute errors is approximately $\frac{1}{2}$.

(ii) (a) Starting from $x_0 = 1$, use the iteration

$$x_{r+1} = \dfrac{1}{2}\left(\dfrac{a}{x_r} + x_r\right) \text{ to find } \sqrt{20} \text{ to}$$

three decimal places, tabulating x_r and e_r.

(b) Show that $e_{r+1} = \dfrac{e_r^2}{2x_r}$.

(iii) Explain why the iteration in (ii) converges more rapidly than the iteration in (i).

(The iteration in (ii) is an example of *second-order convergence*, where $e_{r+1} = ke_r^2$, with k independent of e_r. The iteration in (i) is an example of *first-order convergence*, where $e_{r+1} = ke_r$ with (again) k independent of e_r. The Newton–Raphson process, of which the iteration in (ii) is an example, always gives second-order convergence.)

Convergence of series

An athlete runs at $10\,\mathrm{m\,s}^{-1}$:

in the first second he covers 10 metres

in the next $\frac{1}{2}$ second he covers 5 metres

in the next $\frac{1}{4}$ second he covers 2.5 metres.

The sum of the first n terms of the geometric progression $10, 5, 2.5, \ldots$ is

$\dfrac{10(1 - (\frac{1}{2})^n)}{1 - \frac{1}{2}}$ which simplifies to $20(1 - (\frac{1}{2})^n)$, and this never quite reaches 20.

How is this compatible with the fact that the athlete can run 20 metres in finite time? The answer lies in the sequence of times we were using: since the sum of the first n terms of the geometric progression $1, \frac{1}{2}, \frac{1}{4}, \ldots$ never quite reaches 2, we restricted ourselves to looking at the distance the athlete had covered in times which were always less than 2 seconds. No wonder we failed to see him travel 20 metres or more!

No matter how many terms you take, the sum of the geometric progression $1, \frac{1}{2}, \frac{1}{4}, \ldots$ never quite reaches 2. However, you are about to see that the sequence $1, \frac{1}{2}, \frac{1}{3}, \frac{1}{4}, \frac{1}{5}, \ldots$, though it may look similar, behaves in a totally different way.

The series $S_n = 1 + \dfrac{1}{2} + \dfrac{1}{3} + \dfrac{1}{4} + \cdots + \dfrac{1}{n}$ is known as the *harmonic series*. We now show that by making n large enough we can make S_n as large as we like. Let n be a power of 2: i.e. $n = 2^k$ where k is an integer. Then:

$$S_n = 1 + \frac{1}{2} + \left(\frac{1}{3} + \frac{1}{4} \right) + \left(\frac{1}{5} + \frac{1}{6} + \frac{1}{7} + \frac{1}{8} \right)$$

> After the first two terms, the remaining terms form groups of 2, 4, 8, ... fractions.

$$+ \cdots + \left(\frac{1}{2^{k-1}+1} + \frac{1}{2^{k-1}+2} + \cdots + \frac{1}{2^k} \right)$$

> Replace terms by the smallest term in the group.

$$> 1 + \frac{1}{2} + \left(\frac{1}{4} + \frac{1}{4} \right) + \left(\frac{1}{8} + \frac{1}{8} + \frac{1}{8} + \frac{1}{8} \right)$$

$$+ \cdots + \left(\frac{1}{2^k} + \frac{1}{2^k} + \cdots + \frac{1}{2^k} \right)$$

> The groups have $2, 4, 8, \ldots, 2^{k-1}$ identical terms.

$$= 1 + \frac{1}{2} + 2 \times \frac{1}{4} + 4 \times \frac{1}{8} + \cdots + 2^{k-1} \times \frac{1}{2^k}$$

$$= 1 + \frac{1}{2} \times k$$

$$= 1 + \frac{k}{2} \to \infty \quad \text{as } k \to \infty$$

showing that $S_n \to \infty$ as $n \to \infty$.

Solving $1 + \dfrac{k}{2} = 20$ gives $k = 38$ showing that no more than 2^{38} ($\approx 2.75 \times 10^{11}$) terms are needed for S_n to exceed 20. A considerably improved estimate of the number of terms needed (see page 94) shows that it lies between e^{19} ($\approx 1.78 \times 10^8$) and e^{20} ($\approx 4.85 \times 10^8$). Thus, although S_n tends to infinity, it does so extremely slowly. This section examines the question of whether the sum of an infinite series exists or not.

Notation and language

It is usual to use $u_1, u_2, u_3, \ldots, u_n, \ldots$ for the terms of an infinite sequence. The sums:

$$S_1 = u_1, \quad S_2 = u_1 + u_2, \quad S_3 = u_1 + u_2 + u_3, \quad \ldots, \quad S_n = \sum_{r=1}^{n} u_r, \quad \ldots$$

form a new sequence, called the sequence of *partial sums*. If S_n tends to a limit, usually denoted by S, as $n \to \infty$ then we say that the *series* $u_1 + u_2 + u_3 + \ldots$ *converges* to S, and write:

$$\sum_{r=1}^{\infty} u_r = S.$$

In all other cases we say that the series *diverges*.

Tests for convergence

There are several ways of testing whether a series $u_1 + u_2 + u_3 + \ldots$ converges or not, some tests providing conclusive answers with some series, but not with others. The next few pages will introduce you to five of these tests.

1. Consider the limit of u_n as $n \to \infty$

An infinite geometric progression with common ratio r converges if $|r| < 1$ but not otherwise. You may have noticed that in a convergent geometric progression (a geometric progression with a sum to infinity) the absolute value of the terms tends to 0 as n increases. Generalising, if a series $S_n = \sum_{r=1}^{n} u_r$ is convergent then $u_n \to 0$ as $n \to \infty$. For if $S_n \to S$ as $n \to \infty$ then $S_{n-1} \to S$ as $n \to \infty$; but $u_n = S_n - S_{n-1} \to S - S$ as $n \to \infty$ so that $\lim_{n \to \infty} u_n = 0$.

This essential feature of all convergent series gives rise to the following test:

if u_n does not tend to zero as n tends to infinity the series $\sum_{r=1}^{n} u_r$ diverges.

But (as shown by the harmonic series) the fact that $u_n \to 0$ as $n \to \infty$ does not guarantee convergence. Thus $u_n \to 0$ as $n \to \infty$ is a necessary but not a sufficient condition for $\sum_{r=1}^{n} u_r$ to converge.

Activity

How you can tell that: (i) $\displaystyle\sum_{r=1}^{n} \frac{r-2}{2r}$

(ii) $\displaystyle\sum_{r=1}^{n} k$, where k is a non-zero constant, are divergent?

2. **Consider the limit of $S_n = \sum_{r=1}^{n} u_r$ as $n \to \infty$**

Sometimes you can find an explicit expression for the partial sum, S_n, from which you can decide whether or not the series converges.

- With the usual notation for geometric progressions, $S_n = \dfrac{a(1 - r^n)}{1 - r}$ which tends to the limit $\dfrac{a}{1 - r}$ provided $|r| < 1$, as mentioned above.

- Another example of a convergent series is given by $u_n = \dfrac{1}{n(n + 1)}$ and in this case:

$$S_n = \sum_{r=1}^{n} \frac{1}{r(r + 1)} = \sum_{r=1}^{n} \left(\frac{1}{r} - \frac{1}{r + 1} \right)$$

$$= \left(1 - \frac{1}{2} \right) + \left(\frac{1}{2} - \frac{1}{3} \right) + \cdots + \left(\frac{1}{n} - \frac{1}{n + 1} \right) = 1 - \frac{1}{n + 1}$$

so $S_n \to 1$ as $n \to \infty$ giving $\sum_{r=1}^{n} \dfrac{1}{r(r + 1)} = 1$.

- If $u_n = n$ then $S_n = \sum_{r=1}^{n} r = \frac{1}{2}n(n + 1)$

  which can be made as large as we please by choosing n sufficiently large, so the series diverges to infinity.

> Considering the limit of u_n as $n \to \infty$ would also lead you to conclude that both these series diverge.

- If $u_n = (-1)^n$ then $S_n = \begin{cases} -1 & \text{if } n \text{ is odd} \\ 0 & \text{if } n \text{ is even} \end{cases}$

  and as S_n does not tend to a limit the series diverges, oscillating finitely.

Activity

Describe the behaviour of $S_n = \sum_{r=1}^{n} x^{r-1}$ as $n \to \infty$ in the following cases.

(i) $x < -1$　(ii) $x = -1$　(iii) $-1 < x < 1$　(iv) $x = 1$　(v) $x > 1$

3. **Is S_n increasing and bounded above or decreasing and bounded below?**

If all the terms of a sequence are positive then the partial sums increase as we include more terms. In this case the sequence of partial sums $S_1, S_2, S_3, \ldots, S_n, \ldots$ behaves in one of two ways.

> Notice carefully that we are talking about the behaviour of S_n, not u_n which is positive.

- One possibility is that each partial sum $S_1, S_2, S_3, \ldots, S_n, \ldots$ is less than some number K. We then say that the sequence is *bounded above* by K. The situation is illustrated in figure 2.4, and intuition suggests (correctly) that S_n tends to some limit, S, as $n \to \infty$, where $S \leqslant K$. The proof is too difficult to include here; you should assume this property from now on.

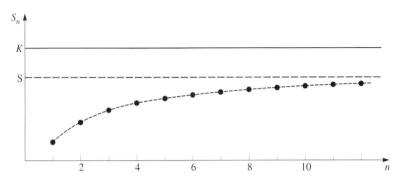

Figure 2.4

- The alternative possibility is that the sequence of partial sums is not bounded above, in which case it diverges to infinity. This is because for any K (however large) there exists a number N such that $S_N > K$. Since S_n increases as n increases, $n > N \Rightarrow S_n > K$ so that $S_n \to \infty$ as $n \to \infty$.

Activity

A sequence consists entirely of negative terms. Describe the behaviour of the partial sums:

(i)   if the partial sums are bounded below

(ii) if the partial sums are not bounded below.

4. Comparison test

The example below illustrates an important method of deciding whether or not the sum of a series is bounded: we compare the terms of a series (or its partial sums) with the corresponding terms (or partial sums) of another series, the behaviour of which is known.

EXAMPLE

Decide whether the following series converge or diverge.

(i) $x + \dfrac{x^2}{2} + \dfrac{x^3}{3} + \cdots + \dfrac{x^r}{r} + \cdots$, where $0 < x < 1$.

(ii) $1 + 2^k + 3^k + 4^k + \cdots$, (a) when $k \geqslant -1$ (b) when $k < -1$.

Solution

(i) Each term in the series $x + \dfrac{x^2}{2} + \dfrac{x^3}{3} + \cdots + \dfrac{x^{n-1}}{n-1}$ is positive and less than the corresponding term in the geometric series $x + x^2 + x^3 + \cdots + x^{n-1}$. The geometric series has a sum to infinity since the common ratio is x and $0 < x < 1$.

Therefore $x + \dfrac{x^2}{2} + \dfrac{x^3}{3} + \cdots + \dfrac{x^r}{r} + \cdots$ also converges.

NB: This method does not tell us the limit to which the series converges, merely that it converges. In fact this series converges to $-\ln(1-x)$.

(ii) (a) When $k \geqslant -1$, each term of $1^k + 2^k + 3^k + 4^k + \cdots$ is greater than or equal to the corresponding term in the harmonic series $1 + \dfrac{1}{2} + \dfrac{1}{3} + \cdots + \dfrac{1}{r} + \cdots$.

You already know that the harmonic series diverges (see page 57).

Therefore $1^k + 2^k + 3^k + 4^k + \cdots$ also diverges to infinity provided $k \geqslant -1$.

(b) When $k < -1$ we let n be one less than a power of 2: i.e. $n = 2^p - 1$ where p is an integer. Then:

$$S_n = 1^k + (2^k + 3^k) + (4^k + 5^k + 6^k + 7^k) + (8^k + \cdots + 15^k) + \cdots$$
$$+ ((2^{p-1})^k + \cdots + (2^p - 1)^k)$$

> In each group replace smaller terms by the largest term in the group.

$$< 1 + (2^k + 2^k) + (4^k + 4^k + 4^k + 4^k) + (8^k + \cdots + 8^k) + \cdots$$
$$+ ((2^{p-1})^k + \cdots + (2^{p-1})^k)$$

> 2^{p-1} terms

> 8 terms

$$= 1 + 2 \times 2^k + 4 \times 4^k + 8 \times 8^k + \cdots + (2^{p-1}) \times (2^{p-1})^k$$
$$= 1 + 2^{k+1} + 4^{k+1} + 8^{k+1} + \cdots + (2^{p-1})^{k+1}$$

This is a geometric progression with a sum to infinity, since its common ratio is 2^{k+1} which is positive but less than 1 when $k < -1$.

Thus S_n is less than this sum to infinity as p and hence $n \to \infty$.

We conclude that $1^k + 2^k + 3^k + 4^k + \cdots$ converges when $k < -1$.

Comparison with a convergent geometric series is frequently used to prove the convergence of another series. Comparison with the harmonic series is frequently used to prove divergence.

5. Alternating series

An *alternating series* consists of terms which are alternately positive and negative, as in $1 - \dfrac{1}{2} + \dfrac{1}{3} - \dfrac{1}{4} + \cdots + \dfrac{(-1)^r}{r} + \cdots$. In Exercise 2C, Question 5 you will prove that an alternating series $u_1 + u_2 + u_3 + \cdots$ converges provided:

$$u_n \to 0 \text{ as } n \to \infty \quad \text{and} \quad |u_n| > |u_{n+1}|$$

for all n.

This implies that the alternating series $1 - \dfrac{1}{2} + \dfrac{1}{3} - \dfrac{1}{4} + \cdots + \dfrac{(-1)^r}{r} + \cdots$ converges. You may have recognised it as the Maclaurin expansion for $\ln(1 + x)$ with $x = 1$.

Activity

A series of positive terms converges. Prove that the corresponding alternating series also converges.

For Discussion

(i) Justify the following statements.

    (a) Deleting a finite number of terms at the start of a series does not affect whether the series is convergent or not.

    (b) Inserting and/or deleting a finite number of finite terms anywhere in a series does not affect whether the series is convergent or not.

These results mean that when discussing the convergence of a series of finite terms the behaviour of u_n or S_n for 'small' n is irrelevant; what matters is the behaviour of u_n or S_n for all $n > N$ where N is sufficiently large.

(ii) (a) Show that:

        (1) the series $(1 - 1) + (1 - 1) + (1 - 1) + \cdots$ converges to 0

        (2) the series $1 - (1 - 1) - (1 - 1) - \cdots$ converges to 1

        (3) the series $1 - 1 + 1 - 1 + 1 - 1 + \cdots$ diverges.

    (b) A series converges to sum S. Show that if the terms of this series are grouped in brackets without changing the order of the terms, the series so formed is also convergent, with sum S.

Exercise 2C

1. Find an expression for the partial sum S_n and hence decide whether the infinite series converges or not. If it converges state its sum to infinity.

    (i) $\displaystyle\sum_{r=1}^{\infty} \frac{1}{r^2 - \frac{1}{4}}$ (ii) $\displaystyle\sum_{r=1}^{\infty} \frac{2}{r(r+1)(r+2)}$

    (iii) $\displaystyle\sum_{r=1}^{\infty} \ln\left(\frac{r+1}{r}\right)$ (iv) $\displaystyle\sum_{r=1}^{\infty} (-1)^{r-1} \frac{2r+1}{r(r+1)}$

2. By comparing the first series with the second series decide whether the first series converges or diverges. If the method is inconclusive, say so.

    (i) $\displaystyle\sum_{r=1}^{\infty} \frac{1}{2r - 1}; \quad \sum_{r=1}^{\infty} \frac{1}{r}$

    (ii) $\displaystyle\sum_{r=1}^{\infty} \frac{1}{r\sqrt{r+1}}; \quad \sum_{r=1}^{\infty} r^{-3/2}$

    (iii) $\displaystyle\sum_{r=2}^{\infty} \frac{1}{\ln r}; \quad \sum_{r=2}^{\infty} \frac{1}{r}$

    (iv) $\displaystyle\sum_{r=1}^{\infty} \frac{1}{r^2 + 5}; \quad \sum_{r=1}^{\infty} \frac{1}{r^2}$

3. By replacing each term by the smallest term show that the series $\sum_{r=1}^{n} \dfrac{1}{\sqrt{r}}$ is not bounded above as n increases. What do you conclude?

4. Prove that $\sum_{r=1}^{\infty} e^{-rx}$ diverges for $x \leqslant 0$ but converges for $x > 0$.

5. The alternating series $u_1 + u_2 + u_3 + \cdots$ is such that:

   (A) $u_n \to 0$ as $n \to \infty$

   and (B) $|u_n| > |u_{n+1}|$ for all n.

   Assume that u_1 is positive. Using the usual notation for partial sums prove that:

   (i) $S_1, S_3, S_5, \ldots$ form a decreasing sequence which is bounded below by 0

   (ii) $S_2, S_4, S_6, \ldots$ form an increasing sequence which is bounded above by u_1

   (iii) both sequences tend to the same limit and deduce that $u_1 + u_2 + u_3 + \cdots$ converges.

   Hint: $S_{2n+1} = S_{2n} + u_{2n+1}$.

6. What difference (if any) does it make to the conclusions in Question 5:

   (i) if condition (A) is not satisfied

   (ii) if u_1 is negative

   (iii) if $u_1, u_2, u_3, \ldots, u_{30}$ are all positive, but thereafter the signs of the terms alternate, and $|u_n| > |u_{n+1}|$ for all $n > 100$ but not necessarily for $n \leqslant 100$?

7. Use the ideas of Question 5 to show that:

   (i) $1 + x + \dfrac{x^2}{2!} + \dfrac{x^3}{3!} + \cdots$ converges for $x < 0$

   (ii) $1 + 2x + 3x^2 + 4x^3 + \cdots$ converges for $-1 < x < 0$.

8. Prove that $x - \dfrac{x^3}{3!} + \dfrac{x^5}{5!} - \cdots$ and $1 - \dfrac{x^2}{2!} + \dfrac{x^4}{4!} - \cdots$ converge for all x.

   (In Exercise 2D we shall indicate how to prove that the limits to which these series converge are $\sin x$ and $\cos x$.)

9. Explain what is wrong with the following argument.

$$S = 2 + 4 + 8 + 16 + \cdots$$
$$\Rightarrow \tfrac{1}{2}S = 1 + 2 + 4 + 8 + \cdots$$
$$\Rightarrow \tfrac{1}{2}S = 1 + S$$
$$\Rightarrow S = -2$$

10. (i) Show that:

$$\ln 2 - \ln 1\tfrac{1}{2} + \ln 1\tfrac{1}{3} - \ln 1\tfrac{1}{4}$$
$$= \ln 2 - \ln 1 - \ln 3 + \ln 2 + \ln 4$$
$$- \ln 3 - \ln 5 + \ln 4.$$

   (ii) Explain why the series
   $\ln 2 - \ln 1\tfrac{1}{2} + \ln 1\tfrac{1}{3} - \ln 1\tfrac{1}{4} + \cdots$ converges but the series $\ln 2 - \ln 1 - \ln 3 + \ln 2 + \ln 4 - \ln 3 - \ln 5 + \ln 4 + \cdots$ diverges.

11. (i) Prove that if u_n is positive for all n and
   $\dfrac{u_{n+1}}{u_n} < k < 1$ for all n, then the error in taking $\sum_{r=1}^{n} u_r$ as the value of $\sum_{r=1}^{\infty} u_r$ is less than $\dfrac{ku_n}{1-k}$.

   (ii) Prove that $\sum_{r=1}^{\infty} \dfrac{1}{2^r r}$ converges and that the error in taking the sum of the first five terms as the sum of the infinite series is less than 0.01.

12. In this question (a_n) and (b_n) are sequences of positive terms and $\lim_{n \to \infty} \dfrac{a_n}{b_n} = \ell \neq 0$.
   Show that:

$$\sum_{r=1}^{\infty} a_r \text{ converges} \iff \sum_{r=1}^{\infty} b_r \text{ converges}.$$

13. (i) Given that $u_r \geqslant 0$ for all r, prove that:

$$\sum_{r=1}^{\infty} u_r \text{ converges} \implies \sum_{r=1}^{\infty} u_r^2 \text{ converges}.$$

   (ii) If u_r is not restricted in this way, does the result still hold? Justify your answer.

14. Show that if $\sum_{r=1}^{\infty} u_r$ converges, then
   $$\sum_{r=1}^{\infty} \dfrac{r+10}{r} u_r \text{ also converges}.$$

Investigation

This investigation introduces *d'Alembert's ratio test*, a more advanced test for convergence of a series of positive terms.

(i)   The sequence $u_1, u_2, u_3, \ldots$ consists of positive terms; there is a fixed positive number k and a positive integer N such that $\dfrac{u_{n+1}}{u_n} < k < 1$ for all $n \geqslant N$.

   (a) Given any positive integer m, explain why $u_{N+m} < k^m u_N$ and
$$\sum_{r=N}^{N+m} u_r < \frac{u_N}{1-k}.$$

   (b) Show that, for $n > N$, $\displaystyle\sum_{r=1}^{n} u_r < \sum_{r=1}^{N-1} u_r + \frac{u_N}{1-k}$, and deduce that $\displaystyle\sum_{r=1}^{\infty} u_r$ converges.

This proves the following form of d'Alembert's ratio test:

   if $u_1, u_2, u_3, \ldots$ are positive and there is a fixed positive number k

   and a positive integer N such that $\dfrac{u_{n+1}}{u_n} < k < 1$ for all $n \geqslant N$,

   then $\displaystyle\sum_{r=1}^{\infty} u_r$ converges.

The test is named after the French mathematician Jean le Rond d'Alembert, 1717–83, though it was first used by the Cambridge mathematician Edward Waring in 1776.

(ii) Show that if $u_n = \dfrac{x^n}{n!}$ then $\dfrac{u_{n+1}}{u_n} < \dfrac{1}{2}$ provided $n > 2x - 1$, and hence deduce

   that the series $1 + x + \dfrac{x^2}{2!} + \dfrac{x^3}{3!} + \cdots + \dfrac{x^r}{r!} + \cdots$ converges for all positive x.

(iii) Show that if $u_n = \dfrac{1}{\sqrt{n}}$ then $\dfrac{u_{n+1}}{u_n} < 1$ for all n but that $\displaystyle\sum_{r=1}^{\infty} u_r$ diverges.

   (This demonstrates the important point that for $\displaystyle\sum_{r=1}^{\infty} u_r$ to be convergent it is

   not sufficient merely to show that $\dfrac{u_{n+1}}{u_n} < 1$ for all sufficiently large n.)

The following alternative form of d'Alembert's ratio test is often easier to use:

   if $u_1, u_2, u_3, \ldots$ are positive and $\dfrac{u_{n+1}}{u_n} \to \ell$ as $n \to \infty$

   then $\ell < 1 \Rightarrow \displaystyle\sum_{r=1}^{\infty} u_r$ converges.

(iv) (a) Explain why the alternative form works and prove that
$$\ell > 1 \Rightarrow \sum_{r=1}^{\infty} u_r \text{ diverges.}$$

   (b) Find examples to show that the test is inconclusive when $\ell = 1$, so that
$$\sum_{r=1}^{\infty} u_r \text{ may converge or diverge.}$$

Convergence and validity of the Maclaurin series

In *Pure Mathematics 5* you met the Maclaurin series for various functions and had, at that stage, to take on trust statements about the values of x for which the various expansions are valid. You can now justify many of those statements: we start by showing one way of proving that the Maclaurin series for e^x converges to e^x for all x.

Proof that the Maclaurin series for e^x is valid for all x

The integral $\int_0^x e^{-t} \, dt$ can be evaluated directly, obtaining $\left[-e^{-t} \right]_0^x = 1 - e^{-x}$.

Alternatively we can integrate by parts.

$$\int_0^x e^{-t} \, dt = \int_0^x 1 \times e^{-t} \, dt = \left[t \, e^{-t} \right]_0^x + \int_0^x t \, e^{-t} \, dt = x \, e^{-x} + \int_0^x t \, e^{-t} \, dt$$

Equating these two results and rearranging gives:

$$1 = e^{-x} + x \, e^{-x} + \int_0^x t \, e^{-t} \, dt$$

then integrating by parts again gives:

$$1 = e^{-x} + x \, e^{-x} + \frac{x^2}{2} e^{-x} + \int_0^x \frac{t^2}{2} e^{-t} \, dt$$

and we can prove by induction that:

$$1 = e^{-x} \left(1 + x + \frac{x^2}{2!} + \frac{x^3}{3!} + \cdots + \frac{x^{n-1}}{(n-1)!} \right) + \int_0^x \frac{t^n}{(n-1)!} \frac{1}{} e^{-t} \, dt.$$

Multiplying by e^x gives:

$$e^x = 1 + x + \frac{x^2}{2!} + \frac{x^3}{3!} + \cdots + \frac{x^{n-1}}{(n-1)!} + R_n(x)$$

where $R_n(x) = e^x \int_0^x \frac{t^{n-1}}{(n-1)!} e^{-t} \, dt$.

The argument above is valid for all values of x. We have expressed e^x as the sum of $R_n(x)$ and the first n terms of the Maclaurin series for e^x (i.e. as far as the term in x^{n-1}). Since $R_n(x)$ is the difference between e^x and the sum of the truncated Maclaurin series it is known as the *remainder* or the *error term* or the *truncation error* after n terms. We show below that $R_n(x) \to 0$ as $n \to \infty$, completing the proof that the Maclaurin series for e^x converges to a limit, and that that limit is e^x.

The technique of showing that an error term tends to zero is commonly used when proving the convergence of Maclaurin and other power series.

Since t is between 0 and x:

$$x > 0 \Rightarrow 0 < t$$

$$\Rightarrow 0 < e^{-t} < 1$$

$$\Rightarrow 0 < \frac{t^{n-1}}{(n-1)!} e^{-t} < \frac{t^{n-1}}{(n-1)!}$$

$$\Rightarrow 0 < R_n(x) = e^x \int_0^x \frac{t^{n-1}}{(n-1)!} e^{-t} \, dt$$

$$< e^x \int_0^x \frac{t^{n-1}}{(n-1)!} \, dt = e^x \frac{x^n}{n!} \to 0 \quad \text{as } n \to \infty$$

which proves that $R_n(x) \to 0$ as $n \to \infty$ when $x > 0$.

When $x < 0$ it is helpful to substitute s for $-t$. Then $ds = -dt$ and $t = x \Rightarrow s = -x$ so that:

$$R_n(x) = e^x \int_0^x \frac{t^{n-1}}{(n-1)!} e^{-t} \, dt = -(-1)^{n-1} e^x \int_0^{-x} \frac{s^{n-1}}{(n-1)!} e^s \, ds$$

$$0 < s < -x$$
$$\Rightarrow 0 < e^s < e^{-x}$$
$$\Rightarrow 0 < \frac{s^{n-1}}{(n-1)!} e^s < \frac{s^{n-1}}{(n-1)!} e^{-x}$$

$$\Rightarrow 0 < |R_n(x)| = e^x \int_0^{-x} \frac{s^{n-1}}{(n-1)!} e^s \, ds$$

$$< e^x \int_0^{-x} \frac{s^{n-1}}{(n-1)!} e^{-x} \, ds = \frac{(-x)^n}{n!} \to 0 \quad \text{as } n \to \infty$$

which proves that $R_n(x) \to 0$ as $n \to \infty$ when $x < 0$.

These two results (together with the obvious convergence of the series when $x = 0$) show that the Maclaurin series for e^x is valid for all x.

Other Maclaurin series

Exercise 2D guides you to prove the convergence and validity of the Maclaurin series for $\arctan x$, $\ln(1+x)$, $\sin x$ and $\cos x$. The following theorem describes the general situation, but we merely quote it, the proof being beyond the scope of this book.

If $f(t)$ and all its derivatives exist for $0 \leqslant t \leqslant x$ then

$$f(x) = f(0) + x\,f'(0) + \frac{x^2}{2!} f''(0) + \cdots + \frac{x^{n-1}}{(n-1)!} f^{(n-1)}(0) + R_n(x)$$

where $R_n(x) = \frac{x^n}{n!} f^{(n)}(\theta x)$, where θ is a number between 0 and 1.

If $R_n(x) \to 0$ as $n \to \infty$ for a range of values of x then the (infinite) Maclaurin series is valid for that range of values of x; not only does the series converge to a limit, but that limit is $f(x)$. Though very rare, there are examples in which the Maclaurin series for $f(x)$ converges, but not to $f(x)$: see Exercise 2F, Question 10.

Exercise 2D

Each question in this exercise is structured to guide you through a proof. These proofs rarely feature in A level courses, and the student in a hurry may safely omit the exercise.

1. **Proof that the Maclaurin series for $\arctan x$ is valid for $|x| \leqslant 1$ only**

   (i) By summing a geometric progression show that:

   $$\frac{1}{1+x} = 1 - x + x^2 - x^3 + \cdots$$

   $$+ (-x)^{n-1} + \frac{(-x)^n}{1+x}$$

   (ii) Substitute t^2 for x in (i) and integrate between the limits 0 and x to show that:

   $$\arctan x = x - \frac{x^3}{3} + \frac{x^5}{5} - \frac{x^7}{7} + \cdots$$

   $$+ (-1)^{n-1} \frac{x^{2n-1}}{2n-1} + R_n(x)$$

   where $R_n(x) = \displaystyle\int_0^x \frac{(-1)^n t^{2n}}{1+t^2} \, dt.$

   (iii) Explain why $R_n(x) = E_{2n-1}(x) = E_{2n}(x)$ where $E_k(x)$ is the truncation error after the term in x^k.

   (iv) Complete the proof that the Maclaurin series for $\arctan x$ is convergent for $|x| \leqslant 1$ but not for $|x| > 1$ by explaining why:

   (a) $0 < \dfrac{t^{2n}}{1+t^2} < t^{2n}$ for $0 < t < x$

   (b) $|R_n(x)| < \dfrac{|x^{2n+1}|}{2n+1}$

   (c) $R_n(x) \to 0$ as $n \to \infty$ if $|x| \leqslant 1$ but not if $|x| > 1$.

   Hint: When $|x| > 1$ let $|x| = 1 + h$.

2. **Proof that the Maclaurin series for $\ln(1+x)$ is valid for $-1 < x \leqslant 1$ only.**

   (i) Prove that:

   $$\ln(1+x) = x - \frac{x^2}{2} + \frac{x^3}{3} - \frac{x^4}{4} + \cdots$$

   $$+ (-1)^n \frac{x^{n-1}}{n-1} + R_n(x)$$

   where $R_n(x) = \displaystyle\int_0^x \frac{(-t)^n}{1+t} \, dt.$

   (ii) Prove that $|R_n(x)| \to 0$ as $n \to \infty$ for $0 \leqslant x \leqslant 1$.

   (iii) For $-1 < x < 0$, prove that

   $$|R_n(x)| < \int_0^{-x} \frac{s^n}{1+x} \, ds$$

   and deduce that

   $$|R_n(x)| \to 0 \quad \text{as} \quad n \to \infty.$$

   (iv) Complete the proof that the Maclaurin series for $\ln(1+x)$ is valid for $-1 < x \leqslant 1$ only, by explaining why it is not convergent for:

   (a) $x > 1$ (b) $x \leqslant -1$.

3. **Proof that the Maclaurin series for $\sin x$ and $\cos x$ are valid for all x**

   This proof does not use error terms, but squeezes $\sin x$ between two sequences.

   (i) For $x > 0$: show that:

   (a) $\sin x < x$

   (b) $\cos x > 1 - \dfrac{x^2}{2!}$

   Hint: Substitute t for x in (a) and integrate from 0 to x.

   (c) $\sin x > x - \dfrac{x^3}{3!}$

   (d) $x - \dfrac{x^3}{3!} < \sin x < x$

   (e) $\sin x$ is between

   $$x - \frac{x^3}{3!} + \frac{x^5}{5!} - \cdots - \frac{x^{4k-1}}{(4k-1)!} \text{ and}$$

   $$x - \frac{x^3}{3!} + \frac{x^5}{x!} - \cdots - \frac{x^{4k-1}}{(4k-1)!} + \frac{x^{4k+1}}{(4k+1)!}$$

   and deduce that the difference between the two sequences tends to zero as the number of terms tends to infinity. (This shows that the two sequences converge to the same limit and that that limit is $\sin x$.)

   (ii) Adapt the argument to show that the Maclaurin series for $\sin x$ is valid when $x < 0$.

   (iii) Adapt the argument to show that the Maclaurin series for $\cos x$ is valid for all x.

Exercise 2D continued

4. Proof that e is irrational

Suppose that e is the rational number $\frac{a}{b}$ where a and b are positive integers.

(i)   Show that $b!e(= b! \times e)$ is a positive integer.

(ii) Show that
$$b!e = b!\left(1 + 1 + \frac{1}{2!} + \frac{1}{3!} + \cdots + \frac{1}{b!}\right) + E_b,$$
where:
$$E_b = \frac{1}{b+1} + \frac{1}{(b+1)(b+2)} + $$
$$\frac{1}{(b+1)(b+2)(b+3)} + \cdots .$$

(iii) By comparing E_b with a suitable geometric series show that $0 < E_b < \frac{1}{b}$ and $b!e$ is not an integer.

(iv) Use the conclusions in (i) and (iii) to deduce that e is irrational.

The limit of a function of a continuous variable

So far we have restricted our study of limits to functions which are infinite sequences: in each case the domain has been the set of positive integers. In this section we extend the concept so that we can analyse the behaviour of functions of a continuous variable. Now our domains will consist of real numbers, not just integers.

As an example consider $f(x) \equiv \dfrac{3x+2}{x-1}$. By dividing $3x+2$ by $x-1$ we can express $f(x)$ in the form $f(x) \equiv 3 + \dfrac{5}{x-1}$ from which it is clear that if x is large $f(x) \approx 3$. The fact that we can make $f(x)$ differ from 3 by as little as we like by choosing any value for x which is sufficiently large means that we may write $\dfrac{3x+2}{x-1} \to 3$ as $x \to \infty$, or alternatively $\displaystyle\lim_{x \to \infty} \dfrac{3x+2}{x-1} = 3$.

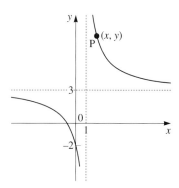

Figure 2.5

We can now also discuss how a function behaves as x approaches a specified (finite) value. Figure 2.5 shows the graph of $y = \dfrac{3x+2}{x-1}$. Let P be a point on the curve with coordinates (x, y) where $x > 1$. As the value of x is decreases towards 1 (that is, as x approaches 1 from the right) the value of y increases without limit. We can make y as large as we like by choosing any value for x which is sufficiently close to 1 (but larger than 1). We write $\dfrac{3x + 2}{x - 1} \to +\infty$ as $x \to 1^+$.

The graph also illustrates that

$$\frac{3x + 2}{x - 1} \to -\infty \quad \text{as} \quad x \to 1^-$$

Although $\lim\limits_{x \to 2^-} f(x) = \lim\limits_{x \to 2^+} f(x) = f(2) = 8$ for the function illustrated in figure 2.5, it is quite possible for a function to converge on different (finite) limits as x tends to some value a from the left and from the right, as illustrated in figure 2.6, which shows the graph of $y = f(x)$, where $f(x) = \dfrac{|x|}{x}$, $x \neq 0$. Here $\lim\limits_{x \to 0^+} f(x) = 1$, and $\lim\limits_{x \to 0^-} f(x) = -1$.

Figure 2.6

Formal definitions

The formal definitions follow those used for the limits of sequences.

- We write $f(x) \to \ell$ as $x \to \infty$ or $\lim\limits_{x \to \infty} f(x) = \ell$ if and only if, given a positive number ε, however small, there is a number X such that $|f(x) - \ell| < \varepsilon$ for all $x \geqslant X$.

- We write $f(x) \to \ell$ as $x \to a^+$ or $\lim\limits_{x \to a^+} f(x) = \ell$ if and only if, given a positive number ε, however small, there exists δ such that $|f(x) - \ell| < \varepsilon$ whenever $0 < x - a < \delta$.

  (A similar definition is used for $\lim\limits_{x \to a^-} f(x)$.)

Because $\lim\limits_{x \to a^-} f(x)$ may not be the same as $\lim\limits_{x \to a^+} f(x)$ we need to be careful when we talk about $\lim\limits_{x \to a} f(x)$, which is defined as follows.

- If $\lim\limits_{x \to a^-} f(x) = \lim\limits_{x \to a^+} f(x) = \ell$ then the common limit ℓ is denoted by $\lim\limits_{x \to a} f(x)$; $\lim\limits_{x \to a} f(x)$ is undefined (i.e. does not exist) if $\lim\limits_{x \to a^-} f(x) \neq \lim\limits_{x \to a^+} f(x)$.

Activity

The following is sometimes given as an alternative definition of $\lim_{x \to a} f(x)$:

$\lim_{x \to a} f(x) = \ell$ if and only if, given a positive number ε, however small, there exists δ such that $|f(x) - \ell| < \varepsilon$ whenever $|x - a| < \delta$.

Show that the two definitions are equivalent.

EXAMPLE Given $f(x) = \begin{cases} x, & x < a \\ 6 - x, & x \geqslant a \end{cases}$ find the value of a such that $\lim_{x \to a} f(x)$ exists.

Solution

We need a such that $\lim_{x \to a^-} f(x) = \lim_{x \to a^+} f(x)$, i.e. such that as we approach $x = a$ from the left along the line $y = x$ we get the same value as when we approach $x = a$ from the right along the line $y = 6 - x$.

From the graph it is clear that this requires $a = 3$.

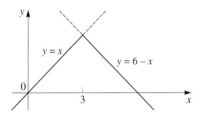

EXAMPLE Sketch the graph of $y = f(x) = \dfrac{x^2 - 3x + 2}{x - 1}$ and find $\lim_{x \to 1} f(x)$.

Solution

$$f(x) = \frac{x^2 - 3x + 2}{x - 1} = \frac{(x - 1)(x - 2)}{x - 1} = x - 2, \quad x \neq 1$$

So the graph of $y = f(x)$ is identical to the graph of $y = x - 2$ except that $y = f(x)$ has a 'hole' (known as a *singularity*) at $x = 1$.

Since $\lim_{x \to 1^-} f(x) = \lim_{x \to 1^+} f(x) = -1$, $\lim_{x \to 1} f(x) = -1$.

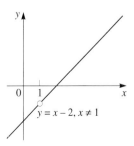

For Discussion

Sketch graphs to show that:

(i) $\lim\limits_{x \to a} f(x)$ may be defined even when $f(a)$ is not defined

(ii) $\lim\limits_{x \to a} f(x)$ and $f(a)$ may be defined but $\lim\limits_{x \to a} f(x) \neq f(a)$.

Properties of limits

The usual properties of limits, proved in the Appendix on page 247, but summarised below, continue to apply. In the following: $f_1(x) \to \ell_1$ as $x \to a$ and $f_2(x) \to \ell_2$ as $x \to a$, and a may be replaced by a^+, a^-, ∞, or $-\infty$.

1. **Multiplying a function by a constant multiplies the limit by that constant**

   $kf_1(x) \to k\ell_1$ as $x \to a$ where k is a constant.

2. **The limit of a sum (or difference) is the sum (or difference) of the limits**

   $f_1(x) + f_2(x) \to \ell_1 + \ell_2$ and $f_1(x) - f_2(x) \to \ell_1 - \ell_2$ as $x \to a$.

3. **The limit of a product is the product of the limits**

   $f_1(x) \times f_2(x) \to \ell_1\ell_2$ as $x \to a$.

4. **The limit of a quotient is the quotient of the limits**

   $\dfrac{f_1(x)}{f_2(x)} \to \dfrac{\ell_1}{\ell_2}$ as $x \to a$ provided $\ell_2 \neq 0$.

The sandwich theorem also continues to apply.

   If $f(x) \leqslant g(x) \leqslant h(x)$ for all x sufficiently close to a

   and both $f(x)$ and $h(x) \to \ell$ as $x \to a$

   then $g(x) \to \ell$ as $x \to a$.

Activity

$N(x)$ and $D(x)$ are polynomials. Use property 4 and your knowledge of the behaviour of polynomial functions to show that the limit of the rational function $\dfrac{N(x)}{D(x)}$ as $x \to a$ may be found by evaluating $\dfrac{N(a)}{D(a)}$ provided the denominator $D(a) \neq 0$.

Standard results

Knowledge of a number of standard results can save you a lot of effort. The example below proves that $\lim\limits_{x \to 0} \left(\dfrac{\sin x}{x} \right) = 1$, with which you are probably already familiar, and you will then move on to other important limits.

EXAMPLE

Show that $\lim\limits_{x \to 0} \left(\dfrac{\sin x}{x} \right) = 1$.

Solution

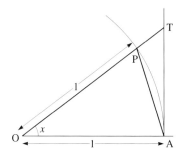

We need to show that:

$$\lim\limits_{x \to 0^+} \left(\frac{\sin x}{x} \right) = \lim\limits_{x \to 0^-} \left(\frac{\sin x}{x} \right) = 1.$$

The diagram shows the sector OAP of a circle of radius 1. Angle AOP is x radians, where $0 < x < \dfrac{\pi}{2}$. The tangent at A meets OP extended at T. Then:

area of triangle OAP < area of sector OAP < area of triangle OAT

$\Rightarrow \quad \frac{1}{2}\sin x < \frac{1}{2}x < \frac{1}{2}\tan x$

$\Rightarrow \quad 1 < \dfrac{x}{\sin x} < \dfrac{1}{\cos x} \quad$ on multiplying by $\dfrac{2}{\sin x}$ which is positive

$\Rightarrow \quad \cos x < \dfrac{\sin x}{x} < 1 \quad$ on taking reciprocals of positives and reversing inequalities.

Since $\lim\limits_{x \to 0^+} (\cos x) = 1$, by the sandwich theorem $\lim\limits_{x \to 0^+} \left(\dfrac{\sin x}{x} \right) = 1$.

To find the limit as x tends to 0^- let $x = -t$, where $0 < t < \dfrac{\pi}{2}$.

Then $\dfrac{\sin x}{x} = \dfrac{\sin(-t)}{-t} = \dfrac{\sin t}{t}$ and $\lim\limits_{x \to 0^-} \left(\dfrac{\sin x}{x} \right) = \lim\limits_{t \to 0^+} \left(\dfrac{\sin t}{t} \right) = 1$,

completing the proof that $\lim\limits_{x \to 0} \left(\dfrac{\sin x}{x} \right) = 1$.

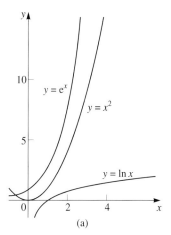

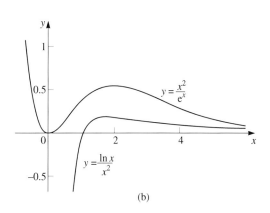

(a)   (b)

Figure 2.7

The graphs of $y = e^x$, $y = x^2$ and $y = \ln x$ (see figure 2.7(a)) suggest that, as $x \to \infty$, e^x tends to ∞ faster than x^2 which itself tends to ∞ faster than $\ln x$.

This hypothesis is strengthened by the graphs of $y = \dfrac{\ln x}{x^2}$ and $y = \dfrac{x^2}{e^x}$ (see figure 2.7(b)), which suggest that both $\dfrac{\ln x}{x^2}$ and $\dfrac{x^2}{e^x} \to 0$ as $x \to \infty$.

Activity

(i) Using a computer or otherwise, draw graphs of $y = \dfrac{\ln x}{x^p}$ for various small values of the positive constant p and show that these graphs suggest that $\dfrac{\ln x}{x^p} \to 0$ as $x \to \infty$ no matter how small p may be.

(ii) Using a computer or otherwise, draw graphs of $y = \dfrac{x^p}{e^x}$ for various large values of the positive constant p and show that these graphs suggest that $\dfrac{x^p}{e^x} \to 0$ as $x \to \infty$ no matter how large p may be.

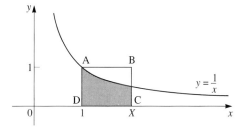

Figure 2.8

To put these observations on a firm footing notice that $\ln X = \displaystyle\int_1^X \dfrac{1}{x}\,dx$, the area under the graph of $y = \dfrac{1}{x}$ from $x = 1$ to $x = X$ (see figure 2.8). Provided $X > 1$, this area is positive, but less than $X - 1$, the area of rectangle ABCD. That is:

$$X > 1 \quad \Rightarrow \quad 0 < \ln X < 1 - X$$

$$\Rightarrow \quad 0 < \frac{\ln X}{X^2} < \frac{X - 1}{X^2},$$

sandwiching $\dfrac{\ln X}{X^2}$ between 0 and $\dfrac{X - 1}{X^2}$.

But $\dfrac{X - 1}{X^2} \to 0$ as $X \to \infty$, so that $\displaystyle\lim_{X \to \infty} \dfrac{\ln X}{X^2} = 0$ by the sandwich theorem.

(We say that $\ln X$ tends to infinity more slowly than X^2.)

Substituting x^p for X^2, where p is any positive constant, gives:

$$\frac{\ln X}{X^2} = \frac{\ln(x^{p/2})}{x^p} = \frac{p}{2} \times \frac{\ln x}{x^p}$$

so that $\dfrac{\ln x}{x^p} = \dfrac{2}{p} \times \dfrac{\ln X}{X^2} \to 0$ as x (and X) $\to \infty$.

So $\displaystyle\lim_{x \to \infty} \dfrac{\ln x}{x^p} = 0$, where p is any positive constant, however small.

(We say that $\ln x$ tends to infinity more slowly than any positive power of x.)

Activity

By writing $\dfrac{(\ln x)^p}{x^q} = \left(\dfrac{\ln x}{x^{q/p}} \right)^p$ prove that $\dfrac{(\ln x)^p}{x^q} \to 0$ as $x \to \infty$, where p and q are positive constants, and hence deduce that, for $p > 0$, $\displaystyle\lim_{x \to \infty} \dfrac{x^p}{e^x} = 0$.

(We say that e^x tends to infinity faster than any positive power of x.)

Activity

By replacing x by $\dfrac{1}{t}$ in $\dfrac{\ln x}{x^p}$ show that, for $p > 0$, $\displaystyle\lim_{x \to 0^+} (x^p \ln x) = 0$.

(The tendency of $|\ln x|$ to become large as x tends to 0 is overcome by any positive power of x, however small.)

You can save a lot of time by remembering (and quoting) the four results just proved.

- $\displaystyle\lim_{x \to 0} \left(\frac{\sin x}{x} \right) = 1$

- $\displaystyle\lim_{x \to \infty} \frac{\ln x}{x^p} = 0$ for $p > 0$

- $\displaystyle\lim_{x \to \infty} \frac{x^p}{e^x} = 0$ for all p

- $\displaystyle\lim_{x \to 0^+} (x^p \ln x) = 0$ for $p > 0$

Continuous functions

Whether a function is continuous or not is inextricably linked to limits. If $f(x)$ is defined for values on both sides of $x = c$ then $f(x)$ is *continuous* at $x = c$ if all three of the following conditions are satisfied:

(i) $f(c)$ is defined,   (ii) $\lim\limits_{x \to c} f(x)$ exists,   (iii) $f(c) = \lim\limits_{x \to c} f(x)$.

(When dealing with end-points conditions (ii) and (iii) are changed to one-sided limits.) If one or more of these conditions is not satisfied we say that $f(x)$ has a *discontinuity* at $x = c$. A *continuous function* is a function which is continuous at all values of x in its domain. This definition and the properties of limits mean that sums, differences, products and (provided you restrict the domain so that you are not dividing by zero) quotients of continuous functions are also continuous functions. Polynomial functions, rational functions (with suitably restricted domains), $\sin x$, $\cos x$,

$$\tan x \left(x \neq k\pi \pm \frac{\pi}{2} \right), e^x, \text{ and } \ln x (x > 0) \text{ are all examples of continuous functions.}$$

Exercise 2E

1. For each of the following functions, if the limit exists for the stated value of a, find (a) $\lim\limits_{x \to a^-} f(x)$, (b) $\lim\limits_{x \to a^+} f(x)$, (c) $\lim\limits_{x \to a} f(x)$. If there is no limit, say so, giving your reasons.

   (i) $f(x) = \dfrac{x-1}{x-2}$, $a = 2$

   (ii) $f(x) = \left| \dfrac{x}{2-x} \right|$, $a = 3$

   (iii) $f(x) = \sin \dfrac{1}{x}$, $a = 0$

   (iv) $f(x) = x \cos \dfrac{1}{x}$, $a = 0$

   (v) $f(x) = \sqrt[3]{x} \sin \dfrac{1}{x}$, $a = 0$

   (vi) $f(x) = \dfrac{x^3 - 1}{x - 1}$, $a = 1$

   (vii) $f(x) = \dfrac{|x| + x}{x}$, $a = 0$

2. In this question $f(x) \equiv [x]$, the largest integer which does not exceed x. Sketch the graph of $y = f(x)$ and find (if possible) $\lim\limits_{x \to a^-} f(x)$,

   $\lim\limits_{x \to a^+} f(x)$ and $\lim\limits_{x \to a} f(x)$ when:

   (i) $a = 5$ (ii) $a = \pi$ (iii) $a = -1.9$.

3. Sketch the graph of $y = x \sin \dfrac{1}{x}$ and find $\lim\limits_{x \to 0} \left(x \sin \dfrac{1}{x} \right)$.

4. In each of the following you are given $f(x)$ and the number a; find $\ell = \lim\limits_{x \to a} f(x)$, and use your calculator to find the largest value of δ (correct to 3 s.f.) that ensures that $f(x)$ is within 0.01 of ℓ whenever x is within δ of a.

   (i) $f(x) = e^x$, $a = 1$

   (ii) $f(x) = \begin{cases} 6 - x, & x < 2 \\ x^2, & x \geqslant 2 \end{cases}$, $a = 2$

   (iii) $f(x) = \dfrac{x}{e^x}$, $a = 0$

   (iv) $f(x) = \dfrac{\sin 2x}{x}$, $a = 0$

   (v) $f(x) = x \ln|x|$, $a = 0$

5. By considering $\lim\limits_{x \to 0} \left(x \times \dfrac{1}{x} \right)$ show that $\lim\limits_{x \to 0} \dfrac{1}{x}$ does not exist.

6. Is it true that $\lim\limits_{x \to a} f(x) = 0$ implies that $\lim\limits_{x \to a} \sqrt{f(x)} = 0$? Justify your answer.

Exercise 2E continued

7. Find counter-examples to show that the following statements are false.

   (i)   If $f(x)$ gets closer to ℓ as x approaches a then $\lim\limits_{x \to a} f(x) = \ell$.

   (ii) If, given a positive number δ, there exists a positive number ε such that $|f(x) - \ell| < \varepsilon$ whenever $|x - a| < \delta$ then $\lim\limits_{x \to a} f(x) = \ell$.

8. Sketch the graph of $y = x^{1/x}$ for $x > 0$ and find $\lim\limits_{x \to 0^+} \left(x^{1/x}\right)$ and $\lim\limits_{x \to \infty} \left(x^{1/x}\right)$.

9. Given that $f(x)$ is the product of any polynomial and e^{-x}, prove that $\lim\limits_{x \to \infty} f(x) = 0$.

10. This question concerns the function
$$y = \left(1 + \frac{1}{x}\right)^x \text{ for positive values of } x.$$

   (i)   Describe the behaviour of y for large values of x.

   (ii) Use the fact that $\lim\limits_{x \to 0}(x \ln x) = 0$ to show that $y \to e$ as $x \to \infty$.

   (iii) Find the limit of $\dfrac{dy}{dx}$ as $x \to \infty$.

   (iv) Sketch the graph of y for positive values of x.   [SMP]

The foundations of calculus

Differentiation from first principles

Figure 2.9 shows the graph of $y = f(x)$ and the points P $(a, f(a))$ and Q $(a + h, f(a + h))$. The gradient of the chord PQ is $\dfrac{f(a + h) - f(a)}{h}$ and as Q slides along the curve towards P the gradient of the chord PQ approaches the gradient of the tangent at P. We now apply the concepts and notation of limits and define the function f as *differentiable* at a if and only if $\lim\limits_{h \to 0} \dfrac{f(a + h) - f(a)}{h}$ exists.

If the limit exists it is known as the derivative of f at a, and is denoted by $f'(a)$.

In other words: $f'(a) = \lim\limits_{h \to 0} \dfrac{f(a + h) - f(a)}{h}$

or equivalently: $f'(a) = \lim\limits_{x \to a} \dfrac{f(x) - f(a)}{x - a}$

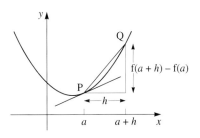

Figure 2.9

EXAMPLE

Explain why the function $f(x) = |x|$ is not differentiable at $x = 0$.

Solution

If h is positive:

$$\frac{f(h) - f(0)}{h} = \frac{h}{h} = 1 \to 1 \quad \text{as} \quad h \to 0^+.$$

If h is negative:

$$\frac{f(h) - f(0)}{h} = \frac{-h}{h} = -1 \to -1 \quad \text{as} \quad h \to 0^-.$$

Because the limits as h tends to zero from left and right are not identical, the value of $\displaystyle\lim_{h \to 0} \frac{f(h) - f(0)}{h}$ is not defined and therefore $f(x) = |x|$ is not differentiable at $x = 0$.

$y = |x|$

EXAMPLE

Show that $\dfrac{d}{dx} \sin x = \cos x$.

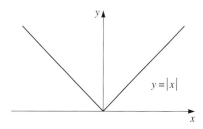

$\dfrac{d}{dx} f(x)$ is another way of writing $f'(x)$.

Solution

Let $f(x) = \sin x$. Then

$$\frac{f(x + h) - f(x)}{h} = \frac{\sin(x + h) - \sin x}{h}$$

$$= \frac{2 \sin \dfrac{h}{2} \cos\left(x + \dfrac{h}{2}\right)}{h}$$

$$= \frac{\sin \dfrac{h}{2}}{\dfrac{h}{2}} \cos\left(x + \dfrac{h}{2}\right).$$

Since $\dfrac{h}{2} \to 0$ as $h \to 0$ $\qquad \displaystyle\lim_{h \to 0} \frac{\sin \dfrac{h}{2}}{\dfrac{h}{2}} = 1$

$\displaystyle\lim_{x \to 0} \frac{\sin x}{x} = 1.$

and $\qquad \displaystyle\lim_{h \to 0} \cos\left(x + \frac{h}{2}\right) = \cos x$

The limit of a product is the product of the limits.

so that $\dfrac{f(x + h) - f(x)}{h} \to \cos x$ as $h \to 0$ proving that $\dfrac{d}{dx} \sin x = \cos x$.

L'Hôpital's rule

We have just used the fact that $\lim\limits_{x \to 0} \dfrac{\sin x}{x} = 1$. Notice that you obtain the same answer by differentiating the numerator and denominator (getting $\cos x$ and 1) and evaluating the quotient of the derivatives at $x = 0$: $\dfrac{\cos 0}{1} = 1$.

This is no accident, for, provided $g'(a) \neq 0$:

$$f(a) = g(a) = 0 \Rightarrow \frac{f'(a)}{g'(a)} = \frac{\lim\limits_{x \to a} \dfrac{f(x) - f(a)}{x - a}}{\lim\limits_{x \to a} \dfrac{g(x) - g(a)}{x - a}} = \frac{\lim\limits_{x \to a} \dfrac{f(x)}{x - a}}{\lim\limits_{x \to a} \dfrac{g(x)}{x - a}}$$

$$= \lim_{x \to a} \frac{f(x)/(x - a)}{g(x)/(x - a)}$$

$$= \lim_{x \to a} \frac{f(x)}{g(x)}$$

This important result is known as L'Hôpital's rule. It is useful when you want to find the limit of a quotient as the variable tends to a value which would make both the numerator and the denominator zero. The Marquis de L'Hôpital is reputed to have purchased the result from its discoverer, Jean Bernoulli (1667–1748), one of a Swiss family noted over several centuries for its mathematicians.

EXAMPLE

Find (i) $\lim\limits_{x \to 1} \dfrac{\sqrt{3 + x} - 2}{\sqrt{x} - 1}$ (ii) $\lim\limits_{x \to 0} \dfrac{1 - \cos x}{x^2}$.

Solution

(i) Using $f(x)$ to represent the numerator, $\sqrt{3 + x} - 2$
and $g(x)$ to represent the denominator, $\sqrt{x} - 1$,

> It is essential to check that numerator and denominator are both zero for the relevant value of x, in this case $x = 1$.

$f(1) = 0$ and $g(1) = 0$ so we can apply L'Hôpital's rule:

$$f'(x) = \frac{1}{2\sqrt{3 + x}} \quad \text{so} \quad f'(1) = \frac{1}{4}$$

$$g'(x) = \frac{1}{2\sqrt{x}} \quad \text{so} \quad g'(1) = \frac{1}{2}$$

Therefore $\lim\limits_{x \to 1} \dfrac{\sqrt{3 + x} - 2}{\sqrt{x} - 1} = \dfrac{f'(1)}{g'(1)} = \dfrac{\frac{1}{4}}{\frac{1}{2}} = \dfrac{1}{2}$.

(ii) The numerator is $f(x) = 1 - \cos x$, and the denominator is $g(x) = x^2$.
Then $f(0) = 0$ and $g(0) = 0$ so we can apply L'Hôpital's rule:

$$f'(x) = \sin x \quad \text{so} \quad f'(0) = 0$$

$$g'(x) = 2x \quad \text{so} \quad g'(0) = 0$$

Since $g'(0) = 0$ we cannot evaluate $\dfrac{f'(0)}{g'(0)}$ directly. However, because

$f'(0) = g'(0) = 0$, we can apply L'Hôpital's rule again:

$$\lim_{x \to 0} \frac{f'(x)}{g'(x)} = \frac{f''(0)}{g''(0)} = \frac{\cos 0}{2} = \frac{1}{2}$$

so that $\displaystyle\lim_{x \to 0} \frac{1 - \cos x}{x^2} = \frac{1}{2}$.

HISTORICAL NOTE

The processes of differentiation and integration were developed (independently) in the seventeenth century by Isaac Newton (1642–1727) in England and Gottfried Wilhelm Leibniz (1646–1716) in Germany. Most of Newton's major discoveries were made in 1665–6, during the closure (due to plague) of the university at Cambridge, but he published no account of any of this work until 1687. Leibniz developed his calculus around 1674, but did not publish his findings until 1684. These new ideas came out of their studies of topics such as sums of infinite series, motion and the area enclosed by a curve. As their methods supplied easy and accurate solutions to many practical problems they were welcomed by many, but the logical foundations under-pinning the subject came in for major criticism, provoking much thought about the nature of a function and the concept of a limit. The Marquis de L'Hôpital (1661–1704) published the first textbook on calculus in Paris in 1696. Swiss mathematician Leonard Euler (1707–83) recognised the difficulties inherent in the geometrical basis of calculus and emphasised the importance of the concept of a function, work which was later developed by Joseph Louis Lagrange (1736–1813), though his functions were strictly algebraic – graphs of these functions could not have discontinuities or sharp corners. It was Jean le Rond d'Alembert (1717–83) who first recognised the need for a satisfactory theory of limits, though that development had to wait for Augustin Louis Cauchy (1789–1857) who tied together both functions and limits in an approach to calculus which is essentially the one in use today. This study of limiting processes is known as 'Analysis', from the Greek word meaning 'to unravel'.

1. Find a function which is continuous for all x and which is differentiable for all x except $x = 1$ and $x = 2$.

2. This question concerns the function $f(x) = x|x|$.

    (i) Prove that $f(x)$ is an odd function.

    (ii) Prove that $f(x)$ is differentiable at $x = 0$.

    (iii) Sketch the graph of $y = f(x)$.

    (iv) Find $f'(x)$.

    (v) Is $f'(x)$ differentiable at $x = 0$? Justify your answer.

3. Find (i) $\lim\limits_{x \to 0} \dfrac{\sin 4x}{2x}$ (ii) $\lim\limits_{x \to 0} \dfrac{e^x - 1}{\sin x}$

    (iii) $\lim\limits_{x \to 0} \dfrac{x^3}{x - \sin x}$ (iv) $\lim\limits_{x \to k} \dfrac{\sqrt[3]{x} - \sqrt[3]{k}}{x - k}$

    (v) $\lim\limits_{x \to 0} \dfrac{\sin x^\circ}{x}$ (vi) $\lim\limits_{x \to \pi/2} \dfrac{\sin 2x}{4x^2 - \pi^2}$

4. Use L'Hôpital's rule to find $\lim\limits_{x \to 0} \dfrac{\ln(1 + x)}{x}$ and hence find $\lim\limits_{x \to 0}((1 + x)^{1/x})$. Use this result to deduce that $\lim\limits_{n \to \infty}\left(\left(1 + \dfrac{1}{n}\right)^n\right) = e$.

5. (i) What is wrong with the following?

    $$\lim_{x \to 0} \frac{x^2 - 2x}{x^2 - \sin x} = \lim_{x \to 0} \frac{2x - 2}{2x - \cos x}$$

    $$= \lim_{x \to 0} \frac{2}{2 + \sin x} = 1$$

    (ii) Evaluate $\lim\limits_{x \to 0} \dfrac{x^2 - 2x}{x^2 - \sin x}$ correctly.

6. Using L'Hôpital's rule, or otherwise, find the limiting value of $\dfrac{\sin(a + x) - \sin(a - x)}{\sqrt{a + x} - \sqrt{a - x}}$ as $x \to 0$, where a is a positive constant.

    [MEI]

7. (i) Show that $f'(a)$ exists $\Rightarrow f(x)$ is continuous at $x = a$.

    (ii) Is the converse true or false? Justify your answer.

8. In this question u_0 and v_0 are functions of x and we use u_r to represent $\dfrac{\mathrm{d}^r u_0}{\mathrm{d}x^r}$, defining v_r similarly.

    (i) Explain why $\dfrac{\mathrm{d}}{\mathrm{d}x}(u_0 v_0) = u_0 v_1 + u_1 v_0$ and deduce that

    $$\frac{\mathrm{d}}{\mathrm{d}x}(u_r v_{n-r}) = u_r v_{n+1-r} + u_{r+1} v_{n-r}.$$

    (ii) Show that $^nC_r + {}^nC_{r-1} = {}^{n+1}C_r$.

    (iii) Use induction to prove that:

    $$\frac{\mathrm{d}^n}{\mathrm{d}x^n}(u_0 v_0) = u_0 v_n + n u_1 v_{n-1} + {}^nC_2\, u_2 v_{n-2}$$

    $$+ \cdots + {}^nC_r\, u_r v_{n-r} + \cdots + u_n v_0.$$

    (This is called Leibniz's theorem. Note the resemblance to the binomial theorem.)

9. Suppose $f(x) \to \infty$ and $g(x) \to \infty$ as $x \to a$. By considering $\dfrac{G'(a)}{F'(a)}$, where $F(x)$ and $G(x)$ are the reciprocals of $f(x)$ and $g(x)$ respectively, show that if $\dfrac{f'(x)}{g'(x)} \to \ell$ as $x \to a$ then $\dfrac{f(x)}{g(x)} \to \ell$ as $x \to a$.

10. In this question $\psi(x) = \begin{cases} e^{-1/x^2}, & x \neq 0 \\ 0, & x = 0. \end{cases}$

    (i) Prove that $\psi(x)$ is a continuous function, and sketch the graph of $y = \psi(x)$.

    (ii) Use induction to prove that $\psi^{(n)}(x) = P_n\left(\dfrac{1}{x}\right)e^{-1/x^2}$ for $x \neq 0$, where P_n is a polynomial of degree n. Deduce that $\psi^{(n)}(x) \to 0$ as $x \to 0$.

    (iii) Prove that $\psi^{(n)}(0) = 0$.

    (iv) Show that the Maclaurin expansion for $\psi(x)$ converges for all x, but only converges to $\psi(x)$ when $x = 0$.

What is a definite integral?

To find the area of the region between the curve $y = f(x)$ and the x-axis, from $x = a$ to $x = b$, divide the interval $a \leqslant x \leqslant b$ into n subintervals:

$$x_0 \leqslant x \leqslant x_1, \text{ where } x_0 = a$$

$$x_1 \leqslant x \leqslant x_2$$

$$\vdots$$

$$x_{n-1} \leqslant x \leqslant x_n, \text{ where } x_n = b$$

as shown in figure 2.10. For each subinterval, $x_k \leqslant x \leqslant x_{k+1}$, calculate $f(X_k)(x_{k+1} - x_k)$, where X_k is any value between x_k and x_{k+1}, and let

$$S_n = \sum_{k=0}^{n-1} f(X_k)(x_{k+1} - x_k).$$

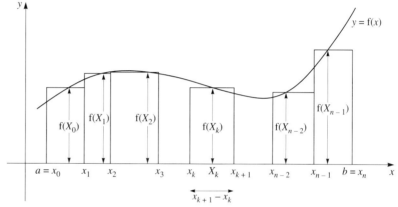

Figure 2.10

If S_n tends to a limit as $\begin{cases} n \to \infty & \text{and} \\ \text{the length of all subintervals} & \to 0 \end{cases}$

and that limit is independent of how the interval $a \leqslant x \leqslant b$ is subdivided and how the X_k are chosen then that limit is known as the *definite integral* of $f(x)$ from a to b and is denoted by $\displaystyle\int_a^b f(x)\,dx$. The values a and b are known as *limits* – do not confuse this use of the word 'limits' with the subject of this chapter; a is the *lower limit* and b is the *upper limit*. The *integrand* is $f(x)$, the function being integrated.

Since the definition above has assumed that the lower limit is less than or equal to the upper limit, we additionally define $\displaystyle\int_b^a f(x)\,dx$ as $-\displaystyle\int_a^b f(x)\,dx$ where $a \leqslant x \leqslant b$.

It can be shown that $\displaystyle\int_a^b f(x)\,dx$ exists if $f(x)$ is continuous for all x in $a \leqslant x \leqslant b$.

Continuity is a sufficient but not a necessary condition for the convergence of S_n. It is, however, essential that the integrand be *bounded* in $a \leqslant x \leqslant b$; this means that we require that there be numbers, m and M, such that $m \leqslant f(x) \leqslant M$ for all x in $a \leqslant x \leqslant b$, as illustrated in figure 2.11.

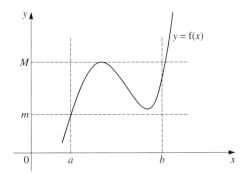

Figure 2.11

The definition we have given is a slightly simplified version of what is known as the *Riemann integral*, named in honour of Georg Bernhard Riemann (1826–66), a German mathematician who was deeply involved in moves to make mathematics more abstract. There are other ways of defining the definite integral, but all acceptable definitions give the same result for the area under the graph of a continuous function.

Activity

The definite integral $\int_a^b f(x)\,\mathrm{d}x$ has been defined as $\displaystyle\lim_{n \to \infty}\left(\sum_{k=0}^{n-1} f(X_k)(x_{k+1} - x_k)\right)$, if it exists and is independent of how the X_k are chosen, where $a = x_0 \leqslant x_1 \leqslant x_2 \leqslant \ldots \leqslant x_{n-1} \leqslant x_n = b$, $x_k \leqslant X_k \leqslant x_{k+1}$, provided all subintervals, $x_{k+1} - x_k$, tend to zero. In this activity you will use this definition to evaluate definite integrals, for convenience taking subintervals of equal width, with X_k at the left-hand end of its subinterval.

(i) The diagram shows the graph of $y = x^2$, for $0 \leqslant x \leqslant p$, together with a set of rectangles standing on the x-axis. These rectangles are of equal width, and the top left vertex of each rectangle is on the curve $y = x^2$.

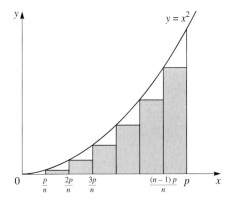

(a) Write down the common width of the rectangles and explain why the height of the $(k - 1)$th rectangle is $\left(\dfrac{kp}{n}\right)^2$. Deduce that

$$\int_0^p x^2\,\mathrm{d}x = \lim_{n \to \infty}\left(\sum_{k=0}^{n-1}\left(\frac{p}{n}\left(\frac{kp}{n}\right)^2\right)\right),\ \text{if it exists.}$$

(b) Explain why $\displaystyle\sum_{k=0}^{n-1}\left(\frac{p}{n}\left(\frac{kp}{n}\right)^2\right) = \frac{p^3}{n^3}\sum_{k=0}^{n-1}k^2$ and deduce that

$$\sum_{k=0}^{n-1}\left(\frac{p}{n}\left(\frac{kp}{n}\right)^2\right) = \frac{p^3}{6n^2}(n-1)(2n-1).$$

$$\left[\textbf{Reminder}: \sum_{r=1}^{n}r^2 = \tfrac{1}{6}n(n+1)(2n+1)\right]$$

(c) Complete the proof that $\displaystyle\int_0^p x^2\,dx = \frac{p^3}{3}$.

(ii) Prove in a similar way that (a) $\displaystyle\int_0^p x\,dx = \frac{p^2}{2}$ and (b) $\displaystyle\int_0^p x^3\,dx = \frac{p^4}{4}$.

$$\left[\textbf{Reminder}: \sum_{r=1}^{n}r^3 = (\tfrac{1}{2}n(n+1))^2\right]$$

It is sometimes possible to evaluate a definite integral as the limit of a sum, as in the activity above, but it is usually much easier to use the *fundamental theorem of calculus*, which we shall prove shortly.

Notice the following properties of the definite integral.

1. $\displaystyle\int_a^b f(x)\,dx + \int_b^c f(x)\,dx = \int_a^c f(x)\,dx.$

2. $\displaystyle\int_a^b kf(x)\,dx = k\int_a^b f(x)\,dx.$

3. $\displaystyle\int_a^b f(x)\,dx + \int_a^b g(x)\,dx = \int_a^b (f(x)+g(x))\,dx.$

4. $\displaystyle\int_a^b 1\,dx = b-a.$

5. If $f(x) > g(x)$ for $a \leqslant x \leqslant b$ then $\displaystyle\int_a^b f(x)\,dx > \int_a^b g(x)\,dx.$

Activity

(i)   Draw sketch graphs to illustrate each of properties 1 to 5 above.

(ii) Use properties of sums and limits to justify properties 1 to 4 above.

(iii) Show that $f(x) > 0$ for $a \leqslant x \leqslant b \;\Rightarrow\; \displaystyle\int_a^b f(x)\,dx > 0$ and justify **5** above by

considering $\displaystyle\int_a^b (f(x) - g(x))\,dx.$

(iv) By considering $\displaystyle\int_a^b (-f(x))\,dx$ show that $f(x) < 0$ for $a \leqslant x \leqslant b \;\Rightarrow\; \displaystyle\int_a^b f(x)\,dx < 0.$

The following example shows how property **5** can be used to estimate the value of a definite integral which you cannot find analytically.

EXAMPLE

Explain why $\dfrac{1}{\sqrt{1-x^3}} < \dfrac{1}{\sqrt{1-x^2}}$ when $0 < x < 1$ and hence prove that

$$0.5 < \int_0^{\frac{1}{2}} \frac{1}{\sqrt{1-x^3}}\, dx < 0.53.$$

Solution

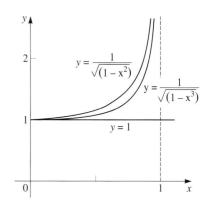

$$0 < x < 1 \;\Rightarrow\; 0 < x^3 < x^2$$
$$\Rightarrow\; 1 > 1 - x^3 > 1 - x^2$$
$$\Rightarrow\; 1 < \frac{1}{\sqrt{1-x^3}} < \frac{1}{\sqrt{1-x^2}}.$$

As shown in the graph this means that, between $x = 0$ and $x = \frac{1}{2}$:

the area below $y = 1 <$ the area below $y = \dfrac{1}{\sqrt{1-x^3}} <$ the area below $y = \dfrac{1}{\sqrt{1-x^2}}$.

Therefore $\displaystyle\int_0^{\frac{1}{2}} 1\, dx < \int_0^{\frac{1}{2}} \frac{1}{\sqrt{1-x^3}}\, dx < \int_0^{\frac{1}{2}} \frac{1}{\sqrt{1-x^2}}\, dx.$

Now $\displaystyle\int_0^{\frac{1}{2}} 1\, dx = 0.5$ and $\displaystyle\int_0^{\frac{1}{2}} \frac{1}{\sqrt{1-x^2}}\, dx = \Big[\arcsin x\Big]_0^{\frac{1}{2}} = \frac{\pi}{6} \approx 0.524 < 0.53.$

Therefore $0.5 < \displaystyle\int_0^{\frac{1}{2}} \frac{1}{\sqrt{1-x^3}}\, dx < 0.53.$

Activity

(i)   Use the substitution $x = -u$ to show that $\displaystyle\int_{-a}^{0} f(x)\, dx = \int_0^a f(-u)\, du.$

(ii) Deduce that $\displaystyle\int_{-a}^{a} f(x)\, dx = \int_0^a (f(-x) + f(x))\, dx.$

(iii) Explain why:

   (a) $f(x)$ is an odd function $\Rightarrow \displaystyle\int_{-a}^{a} f(x)\, dx = 0$

   (b) $f(x)$ is an even function $\Rightarrow \displaystyle\int_{-a}^{a} f(x)\, dx = 2\int_0^a f(x)\, dx$

   and illustrate these results in terms of areas on sketch graphs.

(iv) Show that $I = \displaystyle\int_{-3}^{3} \frac{(x+3)^3}{x^2+3}\, dx = \int_{-3}^{3} \frac{x^3+27x}{x^2+3}\, dx + \int_{-3}^{3} \frac{9x^2+27}{x^2+3}\, dx.$

   Explain why $\displaystyle\int_{-3}^{3} \frac{x^3+27x}{x^2+3}\, dx = 0$ and deduce that $I = 54.$

   (When the limits of an integration are symmetrically arranged on either side of zero, expressing the integrand as the sum of an odd function and an even function can save much effort!)

The fundamental theorem of calculus

You have met two different types of integration:

- indefinite integration is a form of 'anti-differentiation' by means of which you undo the process of differentiation
- definite integration, as on the last few pages, finds the area under a curve.

Now that you have a definition of the integral in terms of area it is important to relate the two processes. This vital step, known as the 'fundamental theorem of calculus', is proved below.

First notice that the definite integral $\int_a^b f(x)\,dx$ is a number which does not depend on x. This implies that $\int_a^b f(t)\,dt$ and $\int_a^b f(\theta)\,d\theta$ are different ways of writing the same number. The variables x, t and θ are used here as *dummy variables*; it does not matter which dummy variable you use. We avoid using the same letter for both the dummy variable and the limit.

Now consider $\int_a^x f(t)\,dt$, where $f(t)$ is defined and continuous for all t in $a \leqslant t \leqslant b$. We regard a as a constant, but we are using the variable x as the upper limit of the integration, so the value we obtain depends on the variable x. Let $F(x)$ represent $\int_a^x f(t)\,dt$. By definition, the derivative of $F(x)$ with respect to x is

$$\frac{d}{dx}\left(\int_a^x f(t)\,dt\right) = F'(x) = \lim_{h \to 0} \frac{F(x+h) - F(x)}{h}$$

$$= \lim_{h \to 0}\left(\frac{1}{h}\left(\int_a^{x+h} f(t)\,dt - \int_a^x f(t)\,dt\right)\right)$$

$$= \lim_{h \to 0}\left(\frac{1}{h}\int_x^{x+h} f(t)\,dt\right).$$

Let m and M, respectively, be the least and greatest value of $f(t)$ in $x \leqslant t \leqslant x+h$.

Then $\dfrac{1}{h}\displaystyle\int_x^{x+h} m\,dt \leqslant \dfrac{1}{h}\int_x^{x+h} f(t)\,dt \leqslant \dfrac{1}{h}\int_x^{x+h} M\,dt$

$\Rightarrow m \leqslant \dfrac{1}{h}\displaystyle\int_x^{x+h} f(t)\,dt \leqslant M$, since m and M do not depend on t.

Provided $f(x)$ is continuous at x both m and M tend to $f(x)$ as $h \to 0^+$, so by the sandwich theorem $\displaystyle\lim_{h \to 0^+}\left(\frac{1}{h}\int_x^{x+h} f(t)\,dt\right) = f(x).$

Similarly $\displaystyle\lim_{h \to 0^-}\left(\frac{1}{h}\int_x^{x+h} f(t)\,dt\right) = f(x).$

Thus $\displaystyle\lim_{h \to 0}\left(\frac{1}{h}\int_x^{x+h} f(t)\,dt\right) = f(x)$, proving that $\dfrac{d}{dx}\left(\displaystyle\int_a^x f(t)\,dt\right) = f(x).$

This highly significant result, known as the *fundamental theorem* of calculus, establishes the link between differential calculus (the study of rates of change) and integral calculus (the study of areas under curves). Although it was first noticed by Newton's tutor, Isaac Barrow (1630–77), in about 1664, it was Newton who first realised the implications of the discovery.

Activity

(i) The diagram shows the graph of $y = f(x)$. Let $F(x)$ denote the area, shown shaded, between $y = f(x)$ and the x-axis, from the fixed line $x = a$ to the vertical line through $P(x, f(x))$. The diagram also shows the graph of $y = F(x)$, and the point $Q(x, F(x))$ on that graph.

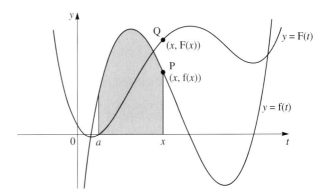

Observe that, if drawn accurately, the y-coordinate of P is the gradient of $y = F(x)$ at Q as x varies.

(ii) Choose your own function $f(x)$ and draw its graph and the graph of $F(x)$, the area under $y = f(x)$. Check that the gradient of $y = F(x)$ is the y-coordinate of the corresponding point on $y = f(x)$. (If you have access to software which has the facility to draw such graphs automatically, you will find it useful here.)

Definite integration

We can now justify the usual routine for evaluating a definite integral.
If $a \leqslant x_1 \leqslant x_2 \leqslant b$ then:

$$\int_{x_1}^{x_2} f(x)\,dx = \int_{a}^{x_2} f(x)\,dx - \int_{a}^{x_1} f(x)\,dx$$

$$= F(x_2) - F(x_1)$$

> We use $\left[F(x)\right]_a^b$ to stand for $F(b) - F(a)$.

Notice that $F(x)$ is any function of which the derivative is $f(x)$; any such function is known as a *primitive* of $f(x)$. If $F_1(x)$ and $F_2(x)$ are different primitives of $f(x)$:

$$F_1'(x) = F_2'(x) = f(x) \Rightarrow F_1'(x) - F_2'(x) = 0$$

$$\Rightarrow \frac{d}{dx}(F_1(x) - F_2(x)) = 0$$

$$\Rightarrow F_1(x) - F_2(x) = c, \quad \text{a constant.}$$

So primitives of $f(x)$ can differ only by a constant.

The process of finding a primitive is known as *indefinite integration*; we usually write $\int f(x)\,dx = F(x) + c$, where c is an arbitrary constant.

1. Sketch the graph of $y = |x^2 - 2|$ and evaluate $\int_0^2 |x^2 - 2| \, dx$.

2. Sketch the graph of $y = 3x + |x|$ and evaluate $\int_{-1}^2 (3x + |x|) \, dx$.

3. Find all the values of c which make $\int_0^c (x - 1)(x - 4) \, dx = 0$ and use a sketch graph of $y = (x - 1)(x - 4)$ to explain the geometrical significance of these values.

4. Evaluate the following integrals.

(i) $\int_0^\pi |1 - 2\sin x| \, dx$

(ii) $\int_{-\pi/2}^{\pi/2} \sin^9 x \, dx$

(iii) $\int_{-a}^a \dfrac{x^7}{a^4 + x^2} \, dx$

(iv) $\int_{1/e}^e e^{|\ln x|} \, dx$

(v) $\int_{-1/2}^{1/2} \dfrac{5x^3 - 12x + 4}{\sqrt{1 - x^2}} \, dx$

(vi) $\int_{-\pi/2}^{\pi/2} (\cos x + 2\sin x)^3 \, dx$

5. Show that $x\left(1 - \dfrac{x^2}{11}\right)^2 < x - \dfrac{x^3}{6}$ for $0 < x < 1$.

Hence, using the fact that $x - \dfrac{x^3}{6} < \sin x < x$ for $x > 0$, show that $\sqrt{x}\left(1 - \dfrac{x^2}{11}\right) < \sqrt{\sin x} < \sqrt{x}$

for $0 < x < 1$ and deduce that:

$\dfrac{148}{231} < \int_0^1 \sqrt{\sin x} \, dx < \dfrac{2}{3}$.

6. Show that $x \neq 1 \Rightarrow 1 < \sqrt{1 + x^4} < 1 + x^2$ and deduce that $1 < \int_0^1 \sqrt{1 + x^4} \, dx < \dfrac{4}{3}$.

7. Prove that $\left| \int_a^b f(x) \, dx \right| \leqslant \int_a^b |f(x)| \, dx$.

8. In this question $f(x) = \int_0^x e^{-t^2} \, dt$.

(i) Differentiate with respect to x:
(a) $f(x)$, (b) $f(\sqrt{x})$.

(ii) Using integration by parts, show that $\int_0^X f(x) \, dx = X f(X) + \tfrac{1}{2} e^{-X^2} - \tfrac{1}{2}$.

9. This question is based on the following result.

$$\int_0^a f(x) \, dx = \int_0^a f(a - x) \, dx \qquad \text{①}$$

(i) Prove the result labelled ①.

Hint: Use the substitution $u = a - x$.

(ii) By sketching $y = f(x)$ and $y = f(a - x)$ interpret ① geometrically.

(iii) You are given that $I = \displaystyle\int_0^{\pi/2} \sin^2 \theta \, d\theta$.

(a) Show that $I = \displaystyle\int_0^{\pi/2} \cos^2 \theta \, d\theta$.

(b) Explain why $I = \dfrac{1}{2}\displaystyle\int_0^{\pi/2} 1 \, d\theta$ and deduce that $I = \dfrac{\pi}{4}$.

(iv) Use similar methods to evaluate:

(a) $\displaystyle\int_0^\pi x \sin^2 x \, dx$ and

(b) $\displaystyle\int_0^{\pi/4} \ln(1 + \tan x) \, dx$.

10. Proof that π is irrational

Let $I_n = \displaystyle\int_{-1}^1 (1 - x^2)^n \cos\left(\dfrac{\pi x}{2}\right) dx$.

(i) By integrating by parts twice, show that

$$\left(\dfrac{\pi}{2}\right)^2 I_n = 2n(2n - 1)I_{n-1} - 4n(n - 1)I_{n-2}, \quad n \geqslant 2$$

(ii) Show that $\dfrac{\pi}{2} \times I_0 = 2$, $\left(\dfrac{\pi}{2}\right)^3 I_1 = 4$, and

prove by induction that

$$\left(\dfrac{\pi}{2}\right)^{2n+1} I_n = n! \, P_n\left(\dfrac{\pi}{2}\right), \text{ where } P_n \text{ is a}$$

polynomial of degree $\leqslant n$ with integer coefficients which may depend on n.

Exercise 2G continued

(iii) Suppose π is rational so that $\dfrac{\pi}{2} = \dfrac{a}{b}$ where a and b are integers.

(a) Prove that $\dfrac{a^{2n+1}}{n!} I_n = b^{2n+1} P_n\left(\dfrac{a}{b}\right)$

and deduce that $\dfrac{a^{2n+1}}{n!} I_n$ is an integer for all n.

(b) Show that
$$0 < (1 - x^2)^n \cos\left(\dfrac{\pi x}{2}\right) < 1 \text{ for}$$
$-1 < x < 1$, and prove that
$0 < I_n < 2$.

(c) Explain why $0 < \dfrac{a^{2n+1}}{n!} I_n < 1$ if n is sufficiently large and hence deduce that π is irrational.

Improper integrals

Some innocent-looking integrals can present difficulties. The definition of the definite integral $\displaystyle\int_a^b f(x)\,dx$ requires that the integrand, $f(x)$, be bounded (that is, finite) for all values in $a \leqslant x \leqslant b$, and that both limits be finite. Under certain conditions we can relax one or both conditions, obtaining what are known as *improper integrals*.

One limit unbounded

First consider the area under the curve $y = \dfrac{1}{x^2}$ from $x = 1$ to $x = X$ (see figure 2.12). This area is

$$\int_1^X \frac{1}{x^2}\,dx = \left[-\frac{1}{x}\right]_1^X = 1 - \frac{1}{X}.$$

As $X \to \infty$, $1 - \dfrac{1}{X} \to 1$. So although the region bounded by the curve, the line $x = 1$, and the x-axis extends infinitely far to the right, it has a finite area, 1. We say that $\displaystyle\int_1^\infty \frac{1}{x^2}\,dx$ *converges* and we write $\displaystyle\int_1^\infty \frac{1}{x^2}\,dx = 1$.

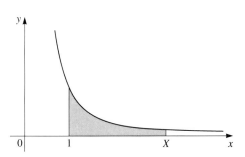

Figure 2.12

At first sight you may think the area under the curve $y = \dfrac{1}{x}$ from $x = 1$ to $x = X$ will behave in the same way (see figure 2.13). But

$$\int_1^X \frac{1}{x}\,dx = \Big[\ln x\Big]_1^X = \ln X$$

which does not tend to a limit as $X \to \infty$; we say that $\displaystyle\int_1^\infty \frac{1}{x}\,dx$ *diverges*.

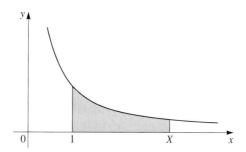

Figure 2.13

In general, if $\displaystyle\int_a^X f(x)\,dx \to \ell$ as $X \to \infty$, where ℓ is a fixed (finite) number, we say that $\displaystyle\int_a^\infty f(x)\,dx$ converges and write $\displaystyle\int_a^\infty f(x)\,dx = \ell$. In all other cases the integral is undefined and we say that $\displaystyle\int_a^\infty f(x)\,dx$ diverges.

EXAMPLE

Evaluate the following, if possible.

(i) $\displaystyle\int_0^\infty \sin x\,dx$ (ii) $\displaystyle\int_0^\infty e^{-x}\sin x\,dx$

Solution

(i) $\displaystyle\int_0^X \sin x\,dx = \Big[-\cos x\Big]_0^X = 1 - \cos X$ which does not tend to a limit as $X \to \infty$. Therefore $\displaystyle\int_0^X \sin x\,dx$ diverges.

(ii) To evaluate $\displaystyle\int_0^X e^{-x}\sin x\,dx$, try integrating by parts twice.

$$\int_0^X e^{-x}\sin x\,dx = \Big[-e^{-x}\cos x\Big]_0^X - \int_0^X e^{-x}\cos x\,dx$$

$$= -e^{-X}\cos X + 1 - \Big[e^{-x}\sin x\Big]_0^X - \int_0^X e^{-x}\sin x\,dx$$

and we can manipulate this to find that:

$$\int_0^X e^{-x} \sin x \, dx = \tfrac{1}{2}(1 - e^{-X} \cos X - e^{-X} \sin X).$$

As $X \to \infty$, $e^{-X} \to 0$ while $|\cos X|$ and $|\sin X|$ never exceed 1 so that $e^{-X} \cos X \to 0$ and $e^{-X} \sin X \to 0$.

Therefore $\int_0^\infty e^{-x} \sin x \, dx = \tfrac{1}{2}$.

Similarly $\int_{-\infty}^b f(x) \, dx$ is defined to be $\lim\limits_{W \to \infty} \int_{-W}^b f(x) \, dx$, if the limit exists.

The improper integral $\int_{-\infty}^\infty f(x) \, dx$ is defined to be $\int_{-\infty}^a f(x) \, dx + \int_a^\infty f(x) \, dx$, where a is any number, provided both integrals converge separately. Such integrals are important in work on probability density functions.

Integrand unbounded

A second type of improper integral can arise when we consider the area between a curve and a vertical asymptote. Figure 2.14 shows the graph of $y = \dfrac{1}{\sqrt{x}}$. The region between the curve, the two axes and $x = 4$ has infinite height, but, as you are about to see, finite area. To evaluate this area you might write down $\int_0^4 x^{-\frac{1}{2}} \, dx$ but you need to use a limiting process to evaluate this integral as the integrand $x^{-\frac{1}{2}}$ is undefined when $x = 0$.

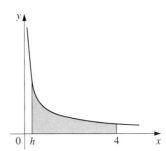

Figure 2.14

The area under the curve $y = \dfrac{1}{\sqrt{x}}$ from $x = h$ to $x = 4$, where $0 < h < 4$, shown shaded in figure 2.14, is $\int_h^4 x^{-\frac{1}{2}} \, dx = \left[2x^{\frac{1}{2}}\right]_h^4 = 4 - 2\sqrt{h}$. As $h \to 0^+$, $4 - 2\sqrt{h} \to 4$ so the area of the region bounded by the curve, the axes and the line $x = 4$ is 4. Although the region has infinite height, it has finite area, and we say that $\int_0^4 x^{-\frac{1}{2}} \, dx$ converges so that $\int_0^4 x^{-\frac{1}{2}} \, dx = 4$.

In general, if $f(x)$ is defined for $a < h \leqslant x \leqslant b$, but undefined for $x = a$, the integrand is undefined at the lower limit. If $\int_h^b f(x)\,dx \to \ell$ as $h \to a^+$, we say that $\int_a^b f(x)\,dx$ converges and write $\int_a^b f(x)\,dx = \ell$.

Similarly, if $f(x)$ is defined for $a \leqslant x \leqslant h < b$, but undefined for $x = b$, the integrand is undefined at the upper limit. If $\int_a^h f(x)\,dx \to \ell$ as $h \to b^-$, we say that $\int_a^b f(x)\,dx$ converges and write $\int_a^b f(x)\,dx = \ell$.

In all other cases we say that $\int_a^b f(x)\,dx$ diverges.

EXAMPLE

Find, if possible, $\displaystyle\int_0^1 \frac{x^2}{\sqrt{1 - x^2}}\,dx$.

Solution

The integrand $\dfrac{x^2}{\sqrt{1 - x^2}}$ is undefined when $x = 1$, so we need to use a limiting process: i.e. we first find $\displaystyle\int_0^h \frac{x^2}{\sqrt{1 - x^2}}\,dx$ and then find the limit of this integral as $h \to 1^-$, if it exists.

To do this let $x = \sin u$ and $h = \sin \alpha$. Then:

$$\int_0^h \frac{x^2}{\sqrt{1 - x^2}}\,dx = \int_0^\alpha \frac{\sin^2 u}{\cos u}\cos u\,du$$

$$= \int_0^\alpha \frac{1}{2}(1 - \cos 2u)\,du$$

$$= \left[\frac{u}{2} - \frac{\sin 2u}{4}\right]_0^\alpha$$

$$= \frac{\alpha}{2} - \frac{\sin 2\alpha}{4}.$$

As $h \to 1^-$, $\alpha \to \dfrac{\pi}{2}$ and $\dfrac{\alpha}{2} - \dfrac{\sin 2\alpha}{4} \to \dfrac{\pi}{4}$.

Therefore $\displaystyle\int_0^1 \frac{x^2}{\sqrt{1 - x^2}}\,dx = \frac{\pi}{4}$.

Both types of improper behaviour

Sometimes integrals involve both types of improper behaviour. By subdividing the domain we can express the original integral as the sum of two or more integrals, each of which is improper in only one way. If all these improper integrals converge individually their sum is the value of the original integral. For the function shown in figure 2.15:

$$\int_{-\infty}^{\infty} f(x)\, dx = \int_{-\infty}^{a} f(x)\, dx + \int_{a}^{b} f(x)\, dx + \int_{b}^{c} f(x)\, dx + \int_{c}^{\infty} f(x)\, dx$$

provided $a < b < c$ and all four integrals converge. The particular values of a and c do not matter.

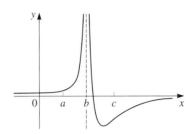

Figure 2.15

Exercise 2H

1. Find the values of m for which:

(i) $\displaystyle\int_{1}^{\infty} x^m\, dx$ converges

(ii) $\displaystyle\int_{0}^{1} x^m\, dx$ converges

(iii) $\displaystyle\int_{0}^{\infty} x^m\, dx$ converges.

2. Investigate these integrals, evaluating those that converge.

(i) $\displaystyle\int_{0}^{\infty} \frac{1}{(x+2)^3}\, dx$
(ii) $\displaystyle\int_{1}^{\infty} \frac{1}{3+x^2}\, dx$

(iii) $\displaystyle\int_{-\infty}^{0} \frac{1}{\sqrt{4-x}}\, dx$
(iv) $\displaystyle\int_{0}^{1} \frac{x}{\sqrt{1-x^2}}\, dx$

(v) $\displaystyle\int_{3}^{\infty} \frac{1}{x(x+3)}\, dx$
(vi) $\displaystyle\int_{0}^{3} \frac{1}{x(x+3)}\, dx$

(vii) $\displaystyle\int_{0}^{\infty} x e^{-x^2}\, dx$
(viii) $\displaystyle\int_{0}^{\pi/2} \frac{\cos x}{\sqrt{\sin x}}\, dx$

(ix) $\displaystyle\int_{-\infty}^{\infty} \frac{1}{x^2 + 4x + 8}\, dx$

3. How can you tell that the answer to this calculation is wrong? Where is the fault?

$$\int_{-1}^{1} \frac{1}{x^2}\, dx = \left[-\frac{1}{x}\right]_{-1}^{1} = -2$$

4. Show that, for $a > 0$,

$$\int_{0}^{\infty} e^{-ax} \cos bx\, dx = \frac{a}{a^2 + b^2}.$$

5. Use the substitution $x = a\cos^2\theta + b\sin^2\theta$ to show that $\displaystyle\int_{a}^{b} \frac{1}{\sqrt{(x-a)(b-x)}}\, dx = \pi$ where $0 < a < b$.

6. (i) By comparing $\displaystyle\int_{3}^{X} \frac{\ln x}{x}\, dx$ and $\displaystyle\int_{3}^{X} \frac{1}{x}\, dx$, for $X > 3$, deduce that $\displaystyle\int_{3}^{\infty} \frac{\ln x}{x}\, dx$ diverges.

(ii) By making suitable comparisons decide which of the following improper integrals converge. (You do not need to evaluate the integrals.)

(a) $\displaystyle\int_{3}^{16} \frac{\sqrt{x}}{x-3}\, dx$
(b) $\displaystyle\int_{2}^{\infty} \frac{x}{x^7 + 1}\, dx$

(c) $\displaystyle\int_{0}^{\pi} \frac{\sin x}{x^2}\, dx$
(d) $\displaystyle\int_{1}^{\infty} e^{-x^2}\, dx$

7. A *probability density function* is a function f such that $f(x) \geqslant 0$ for all x and $\displaystyle\int_{-\infty}^{\infty} f(x)\, dx = 1$. The *variance* of a distribution symmetrical about $x = 0$ is $\displaystyle\int_{-\infty}^{\infty} x^2 f(x)\, dx$.

(i) In the *Laplace distribution* $f(x) \equiv A e^{-\lambda|x|}$, where A and λ are positive constants. Find A in terms of λ and show that the variance is $2\lambda^{-2}$.

Exercise 2H continued

(ii) In the *Cauchy distribution* $f(x) \equiv \dfrac{B}{1+x^2}$,

where B is a positive constant. Find the value of B. What is the variance?

8. (i) Sketch the graph of $y = \ln(\sin t)$ for $0 \leqslant t \leqslant \pi$.

(ii) If $A = \displaystyle\int_0^{\pi/2} \ln(2\cos x)\,dx$ and

$B = \displaystyle\int_0^{\pi/2} \ln(2\sin x)\,dx$, prove that

$A + B = B$. Hence evaluate $\displaystyle\int_0^{\pi/2} \ln\cos x\,dx$.

(Assume that all these integrals converge.)

[MEI, adapted]

9. In this question $f(x) \geqslant 0$ for all x, and

$\displaystyle\int_{-\infty}^{\infty} f(x)\,dx$ is defined. The number σ is the

positive square root of $\displaystyle\int_{-\infty}^{\infty} (x - \mu)^2\,f(x)\,dx$,

where μ is a constant. Explain why:

(i) $\sigma^2 \geqslant \displaystyle\int_{-\infty}^{\mu - k\sigma} (x - \mu)^2\,f(x)\,dx$

$+ \displaystyle\int_{\mu + k\sigma}^{\infty} (x - \mu)^2\,f(x)\,dx$

where k is a positive constant

(ii) $\displaystyle\int_{\mu + k\sigma}^{\infty} (x - \mu)^2\,f(x)\,dx \geqslant k^2\sigma^2 \int_{\mu + k\sigma}^{\infty} f(x)\,dx$

and prove that $\displaystyle\int_{-\infty}^{\mu - k\sigma} f(x)\,dx + \int_{\mu + k\sigma}^{\infty} f(x)\,dx \leqslant \dfrac{1}{k^2}$.

(This is known as *Chebyshev's inequality*. Interpreted statistically, it states that the probability that a random variable differs from its mean by more than k standard deviations does not exceed $\dfrac{1}{k^2}$, no matter what the distribution.)

Series and integrals

The activity below will show you how comparing the nth partial sum of a series and the value of an appropriate definite integral can help you decide whether a series converges or not.

Activity

The diagram shows part of the graph of $y = \dfrac{1}{x}$ together with a sequence of n rectangles, each 1 unit wide.

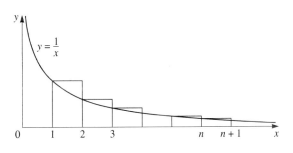

Limiting processes

(i) What feature of the diagram represents:

(a) $\displaystyle\sum_{r=1}^{n}\frac{1}{r}$ (b) $\displaystyle\int_{1}^{n+1}\frac{1}{x}\,dx$?

(ii) Explain why $\displaystyle\int_{1}^{n+1}\frac{1}{x}\,dx < \sum_{r=1}^{n}\frac{1}{r}$ and deduce that $\displaystyle\sum_{r=1}^{n}\frac{1}{r}$ can be made as large as you like by making n sufficiently large.

(iii) Draw another diagram (again with n rectangles, each of width 1 unit) to explain why:

$$\sum_{r=1}^{n}\frac{1}{r} < 1 + \int_{1}^{n}\frac{1}{x}\,dx.$$

(iv) Deduce that:

(a) $\displaystyle\ln(n+1) < \sum_{r=1}^{n}\frac{1}{r} < 1 + \ln n$ (b) for large n, $\displaystyle\sum_{r=1}^{n}\frac{1}{r} \approx \ln n$.

(v) Use (iv)(a) to show that $\displaystyle 0 < \sum_{r=1}^{n}\frac{1}{r} - \ln n < 1$.

(vi) Let u_n represent $\displaystyle\sum_{r=1}^{n}\frac{1}{r} - \ln n$. Show that $u_{n+1} - u_n = \dfrac{1}{n+1} - \displaystyle\int_{n}^{n+1}\frac{1}{x}\,dx$ and deduce that:

(a) u_n steadily decreases (i.e. $m > n \Rightarrow u_m < u_n$)

(b) u_n tends to a limit (denoted by γ) as $n \to \infty$, where $0 \leqslant \gamma < 1$.

(The limit $\gamma \approx 0.557\,256\,6$ is known as *Euler's constant*; it is not yet known whether γ is rational or irrational.)

(vii) Show that $\displaystyle\sum_{r=1}^{n}\frac{1}{r} > 20 \Rightarrow n > e^{19}$ and that $\displaystyle\sum_{r=1}^{n}\frac{1}{r} < 20 \Rightarrow n < e^{20}$.

In parts (ii) and (iii) of the activity above you showed that the partial sum $\displaystyle\sum_{r=1}^{n}\frac{1}{r}$ is between $\displaystyle\int_{1}^{n+1}\frac{1}{x}\,dx$ and $1 + \displaystyle\int_{1}^{n}\frac{1}{x}\,dx$. Trapping the partial sum between two values, usually definite integrals, can enable you to set bounds for the number of terms needed for the partial sum to reach a specified value. Notice that the technique depends on the terms of the series being positive and decreasing (as n increases).

We now generalise the method above. Suppose f(x) is a decreasing function such that f(x) is positive for all $x > 0$, as shown in figure 2.16(a), which also shows a series of n rectangles of width 1 unit. The heights (and areas) of these rectangles are f(1), f(2), f(3), ..., f(n). (It is more convenient to use function notation than sequence notation here.) Positioned as shown these rectangles are known as *upper rectangles*, and it is clear that the sum of the areas of the rectangles is $\displaystyle\sum_{r=1}^{n}f(r)$ and that this exceeds the area under the curve $y =$ f(x) from $x = 1$ to $x = n + 1$, so that:

$$\int_{1}^{n+1}\text{f}(x)\,dx < \sum_{r=1}^{n}\text{f}(r). \qquad \text{①}$$

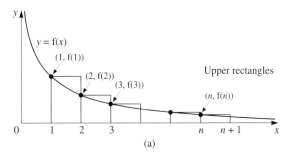

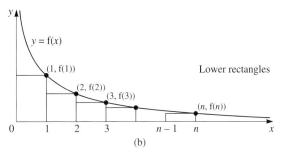

Figure 2.16

Figure 2.16(b) shows a similar arrangement of rectangles, known as *lower rectangles*, from which it is clear that:

$$\sum_{r=1}^{n} f(r) < f(1) + \int_{1}^{n} f(x)\,dx \qquad \textcircled{2}$$

Now $\displaystyle\sum_{r=1}^{n} f(r)$ increases as n increases and $\textcircled{2}$ shows that if $\displaystyle\int_{1}^{\infty} f(x)\,dx$ converges, then $\displaystyle\sum_{r=1}^{n} f(r)$ is bounded above.

$$\text{Thus } \int_{1}^{\infty} f(x)\,dx \text{ converges} \Rightarrow \sum_{r=1}^{\infty} f(r) \text{ converges.}$$

Similarly, from $\textcircled{1}$:

$$\sum_{r=1}^{\infty} f(r) \text{ converges} \Rightarrow \int_{1}^{\infty} f(x)\,dx \text{ converges}$$

since $\displaystyle\int_{1}^{n} f(x)\,dx$ increases as n increases. Therefore:

$$\sum_{r=1}^{\infty} f(r) \text{ converges if and only if } \int_{1}^{\infty} f(x)\,dx \text{ converges.}$$

This provides yet another test for the convergence of an infinite series, known as the *integral test*, or the *Maclaurin–Cauchy test*. If we can evaluate the improper integral, putting $\textcircled{1}$ and $\textcircled{2}$ together gives:

$$\int_{1}^{\infty} f(x)\,dx < \sum_{r=1}^{\infty} f(r) < f(1) + \int_{1}^{\infty} f(x)\,dx$$

providing an interval of width $f(1)$ within which $\displaystyle\sum_{r=1}^{\infty} f(r)$ must lie.

EXAMPLE Use the integral test to decide the convergence of $\sum_{r=1}^{\infty} \frac{1}{r^p}$, for $p > 1$. If it is convergent locate the sum within an interval of width 1.

Solution

Let $f(x) = \frac{1}{x^p}$. Then, for $p > 1$, $f(x)$ is positive for $x > 0$ and decreasing, and so satisfies the conditions needed for us to be able to apply the integral test.

$$\int_1^L \frac{1}{x^p}\,dx = -\frac{1}{p-1}\left[\frac{1}{x^{p-1}}\right]_1^L = \frac{1}{p-1}\left(1 - \frac{1}{L^{p-1}}\right) \to \frac{1}{p-1} \quad \text{as} \quad L \to \infty$$

so $\int_1^\infty \frac{1}{x^p}\,dx$ converges. Therefore, for $p > 1$, $\sum_{r=1}^{\infty} \frac{1}{r^p}$ converges to a limit between

$$\int_1^\infty \frac{1}{x^p}\,dx = \frac{1}{p-1} \quad \text{and} \quad f(1) + \int_1^\infty \frac{1}{x^p}\,dx = 1 + \frac{1}{p-1} = \frac{p}{p-1}.$$

Thus $\frac{1}{p-1} \leqslant \sum_{r=1}^{\infty} \frac{1}{r^p} \leqslant \frac{p}{p-1}$.

Activity

Show that if $f(x)$ is a positive but decreasing function for $x \geqslant k$, where k is a positive integer, then $\int_{k+1}^{k+m+1} f(x)\,dx \leqslant \sum_{r=k+1}^{k+m} f(r) \leqslant \int_k^{k+m} f(x)\,dx$.

If you can find a primitive of $f(x)$ this inequality can be useful:

(i) if you want to know how many terms of a sequence should be summed to be within a stated distance of the sum to infinity, as in Question 7(i) of Exercise 2I

(ii) if you want to set bounds for the number of terms that need to be used for the sum of an ever-increasing divergent series to exceed a stated amount, as in part (vii) of the activity on page 94.

Exercise 2I

1. (i) By sketching appropriate graphs and rectangles show that:

$$\sum_{r=2}^{n} \frac{1}{r \ln r} > \int_2^{n+1} \frac{1}{x \ln x}\,dx$$

(ii) Show that $\int_2^{n+1} \frac{1}{x \ln x}\,dx = \Big[\ln(\ln x)\Big]_2^{n+1}$

and deduce that $\sum_{r=2}^{\infty} \frac{1}{r \ln r}$ diverges.

(iii) Adapt the method to show that $\sum_{r=2}^{\infty} \frac{1}{r(\ln r)^2}$ converges.

2. Sketch the graph of $y = \frac{1}{1+x^2}$ for positive x and hence show that:

$$\int_1^{n+1} \frac{1}{1+x^2}\,dx < \sum_{r=1}^{n} \frac{1}{1+r^2} < \int_0^{n} \frac{1}{1+x^2}\,dx.$$

What does this tell you about $\sum_{r=1}^{\infty} \frac{1}{1+r^2}$?

3. If $f(x)$ is a steadily decreasing function of x in the closed interval $a \leqslant x \leqslant b$ and $b - a = nh$, show that:

$$h \sum_{r=0}^{n-1} f(a+rh) > \int_a^b f(x)\,dx > h \sum_{r=1}^{n} f(a+rh).$$

By taking $f(x) = \dfrac{1}{1+x^2}$ show that if

$$S_n = \sum_{r=1}^{n} \frac{4n}{n^2+r^2} \text{ then } \frac{2}{n} > \pi - S_n > 0 \text{ and hence}$$

that $\lim\limits_{n\to\infty} S_n$ is π. [MEI]

4. Show that:

(i) $\quad 2\sqrt{n+1} - 2 < \displaystyle\sum_{r=1}^{n} \frac{1}{\sqrt{r}} < 2\sqrt{n} - 1$

(ii) $\quad 2\sqrt{1+\dfrac{1}{n}} < \displaystyle\sum_{r=1}^{n} \frac{1}{\sqrt{rn}} + \frac{2}{\sqrt{n}} < 2 - \frac{1}{\sqrt{n}}$

and deduce the value of $\lim\limits_{n\to\infty} \displaystyle\sum_{r=1}^{n} \frac{1}{\sqrt{rn}}$.

5. (i) The functions $f(t)$ and $g(t)$ are defined and continuous in the region $0 \leqslant t \leqslant x$, and $0 \leqslant f(t) \leqslant g(t)$ in this region. With the aid of a diagram explain why:

$$0 \leqslant \int_0^x f(t)\,dt \leqslant \int_0^x g(t)\,dt.$$

Hence, or otherwise, show that, for $x > 0$:

(a) $\quad 0 \leqslant \displaystyle\int_0^x e^{-t}\,dt \leqslant \int_0^x dt$

(b) $\quad 1 - x \leqslant e^{-x} \leqslant 1$

(c) $\quad e - 1 \leqslant \displaystyle\int_0^1 \exp(e^{-t})\,dt \leqslant e$ where

$\exp(x)$ is an alternative notation for e^x.

(ii) By expanding $\exp(e^{-t})$ as a power series in e^{-t} and integrating term by term, show

that $I = \displaystyle\int_0^1 \exp(e^{-t})\,dt = 1 + \sum_{k=1}^{\infty} a_k$ where

$$a_k = \frac{1 - e^{-k}}{k(k!)}.$$

Given that, for $k \geqslant n$, $a_k < \dfrac{1}{n(n!)}\left(\dfrac{1}{n}\right)^{k-n}$,

show that $\displaystyle\sum_{k=n}^{\infty} a_k < \frac{1}{(n-1)(n!)}$.

Hence show that the error in estimating

I as $1 + \displaystyle\sum_{k=1}^{6} a_k$ is less than 10^{-4}.

[O&C]

6. (i) Sketch the curve with equation $y = \sqrt{1+x}$ for $0 \leqslant x \leqslant 3$.

Show on your sketch a set of rectangles

with total area $\dfrac{1}{n} \displaystyle\sum_{r=1}^{3n} \sqrt{1 + \frac{r}{n}}$.

(ii) Find the limiting value of $n^{-\frac{3}{2}} \displaystyle\sum_{r=1}^{3n} \sqrt{n+r}$ as $n \to \infty$.

(iii) Hence estimate the value of

$\sqrt{901} + \sqrt{902} + \sqrt{903} + \cdots + \sqrt{3600}$.

(iv) State with reasons whether the exact

value of $\displaystyle\sum_{r=901}^{3600} \sqrt{r}$ is greater or less than your

estimate in (iii). [MEI]

7. (i) (a) Show that:

$$\int_k^{n+1} \frac{1}{x^2}\,dx < \sum_{r=k}^{n} \frac{1}{r^2} < \int_{k-1}^{n} \frac{1}{x^2}\,dx, \text{ for}$$

$k > 1$, and deduce that

$$\frac{1}{k} < \sum_{r=k}^{\infty} \frac{1}{r^2} < \frac{1}{k-1}.$$

(b) How many terms of $\displaystyle\sum_{r=1}^{\infty} \frac{1}{r^2}$ should be

summed to be sure of obtaining an answer within 0.01 of the sum to infinity?

(ii) (a) Starting from the Maclaurin series for $\sin x$ deduce that $z = (k\pi)^2$ satisfies

$$1 - \frac{z}{3!} + \frac{z^2}{5!} - \frac{z^3}{7!} + \cdots + \frac{(-1)^r z^r}{(2r+1)!} + \cdots = 0$$

when k is any positive integer.

(b) The sum of the reciprocals of the n roots of the equation

$$a_0 + a_1 z + a_2 z^2 + \cdots + a_n z^n = 0 \text{ is } -\frac{a_1}{a_0}.$$

Assuming that this result holds when

n is infinite, show that: $\displaystyle\sum_{r=1}^{\infty} \frac{1}{(r\pi)^2} = \frac{1}{3!}$

and deduce that $\displaystyle\sum_{r=1}^{\infty} \frac{1}{r^2} = \frac{\pi^2}{6}$.

(This result was first proved by Euler in about 1736, using the method outlined above. The problem had been under discussion since 1673 or earlier.)

Investigation

On the Argand diagram the point representing the complex number c is coloured black if the iteration:

$$z_{n+1} = z_n^2 + c, \quad z_0 = 0 \qquad \qquad ①$$

is bounded for all n, i.e. if $|z^n|$ is always less than some positive constant. Points representing c for which ① is not bounded are left white. Black points form the *Mandelbrot set*, first discovered by Benoit Mandelbrot, in 1980.

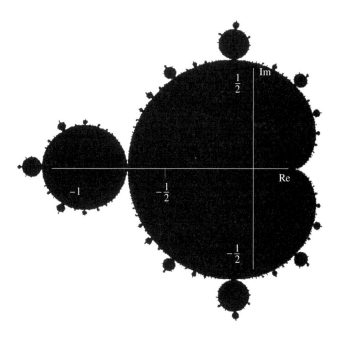

(i) Assuming that if the iteration $z_{n+1} = f(z_n)$ converges to ℓ then $|f'(\ell)| < 1$, show that any limit to which ① converges is in the region $|z| < \frac{1}{2}$.

(ii) Show that the only conceivable limits of ① are $\alpha = \frac{1}{2} - (\frac{1}{4} - c)^{\frac{1}{2}}$ and $\beta = \frac{1}{2} + (\frac{1}{4} - c)^{\frac{1}{2}}$, and deduce that:

   (a) α may be in the region $|z| < \frac{1}{2}$ (b) β is not in the region $|z| < \frac{1}{2}$.

(iii) By finding $\mathrm{Re}(c)$ and $\mathrm{Im}(c)$ when α is on the boundary of $|z| = \frac{1}{2}$ show that the interior of the cardioid $\begin{cases} x = \frac{1}{4} + \frac{1}{2}\cos\theta(1 - \cos\theta) \\ y = \frac{1}{2}\sin\theta(1 - \cos\theta) \end{cases}$ is black.

(iv) Show that if $|c + 1| < \frac{1}{4}$ then ① eventually alternates between the roots (other than α and β) of $z = (z^2 + c)^2 + c$. Hence deduce that the interior of the circle $|c + 1| = \frac{1}{4}$ is black.

(v) Show that if $|z_n| \geq 4$ and $|c| \leq |z_n|$ then $|z_{n+1}| \geq 3|z_n|$ and deduce that, whatever the value of c, if $|z_n| \geq 4$ for some n then the point representing c is white.

(vi) Investigate further.

K E Y   P O I N T S

- We write $f(n) \to \ell$ as $n \to \infty$ or $\lim_{n \to \infty} f(n) = \ell$ if and only if, given a positive number ε, however small, there is a number N such that $|f(n) - \ell| < \varepsilon$ for all $n \geqslant N$.

- We write $f(n) \to \infty$ as $n \to \infty$ if and only if, given K, however large, there is a number N such that $f(n) > K$ for all $n \geqslant N$.

- We write $f(x) \to \ell$ as $x \to a^+$ or $\lim_{n \to a^+} f(x) = \ell$ if and only if, given a positive number ε, however small, there exists δ such that $|f(x) - \ell| < \varepsilon$ whenever $0 < x - a < \delta$; $\lim_{x \to a^-} f(x)$ is defined similarly.

- If $\lim_{n \to a^-} f(x) = \lim_{x \to a^+} f(x)$ then the common limit is denoted by $\lim_{x \to a} f(x)$.

- If $f_1(n) \to \ell_1$ as $n \to \infty$ and $f_2(n) \to \ell_2$ as $n \to \infty$ then:

$$f_1(n) + f_2(n) \to \ell_1 + \ell_2 \quad \text{as} \quad n \to \infty$$

$$f_1(n) \times f_2(n) \to \ell_1 \ell_2 \quad \text{as} \quad n \to \infty$$

$$\frac{f_1(n)}{f_2(n)} \to \frac{\ell_1}{\ell_2} \quad \text{as} \quad n \to \infty \quad \text{provided} \quad \ell_2 \neq 0$$

with similar properties for the other forms of limit.

- The sandwich theorem:

  if $f(n) \leqslant g(n) \leqslant h(n)$ for all $n \geqslant N$ and both $f(n)$ and

  $h(n) \to \ell$ as $n \to \infty$ then $g(n) \to \ell$ as $n \to \infty$.

(Similar results hold for the other forms of limit.)

- If $S_n = \sum_{r=1}^{n} u_r$ tends to a limit, S, as $n \to \infty$ then S_n converges to S, otherwise S_n diverges.

- $\sum_{r=1}^{n} u_r$ diverges if u_n does not tend to zero as n tends to infinity.

- When (u_r) is a sequence of positive terms:

  the partial sums are bounded above $\Rightarrow \sum_{r=1}^{\infty} u_r$ converges.

- The function $f(x)$ is continuous at $x = a$ if $f(a) = \lim_{x \to a} f(x)$.

- A continuous function is continuous at all values of x in its domain.

- Sums, differences, products and quotients of continuous functions are continuous; quotients may have restricted domain.

- $f'(a) = \lim_{h \to 0} \frac{f(a + h) - f(a)}{h} = \lim_{x \to a} \frac{f(x) - f(a)}{x - a}$.

- L'Hôpital's rule:

$$f(a) = g(a) = 0 \Rightarrow \lim_{x \to a} \frac{f(x)}{g(x)} = \frac{f'(a)}{g'(a)}$$

provided $g'(a) \neq 0$.

- The fundamental theorem of calculus:

$$\frac{d}{dx} \left(\int_a^x f(t) \, dt \right) = f(x).$$

- Improper integrals:

$\int_a^\infty f(x) \, dx$ is the limit of $\int_a^X f(x) \, dx$ as $X \to \infty$ if the limit exists.

If $f(x) \to \infty$ as $x \to b^-$ then

$\int_a^b f(x) \, dx$ is the limit of $\int_a^h f(x) \, dx$ as $h \to b^-$ if the limit exists.

- $\sum_{r=1}^{\infty} u_r$ converges if and only if $\int_1^\infty f(x) \, dx$ converges where $f(r) = u_r$ and (u_r) is a decreasing sequence of positive terms.

Important results:

$$\lim_{n \to \infty} \left(1 + \frac{r}{n} \right)^n = e^r; \qquad\qquad p > 0 \Rightarrow \lim_{n \to \infty} \frac{1}{n^p} = 0;$$

$$\lim_{n \to \infty} \sqrt[n]{n} = 1; \qquad\qquad x > 0 \Rightarrow \lim_{n \to \infty} \sqrt[n]{x} = 1;$$

$$\lim_{n \to \infty} \frac{x^n}{n!} = 0; \qquad\qquad |x| < 1 \Rightarrow \lim_{n \to \infty} x^n = 0;$$

$$\lim_{x \to \infty} \frac{x^p}{e^x} = 0 \text{ for all } p; \qquad\qquad p > 0 \Rightarrow \lim_{x \to \infty} \frac{\ln x}{x^p} = 0;$$

$$\lim_{x \to 0} \left(\frac{\sin x}{x} \right) = 1; \qquad\qquad p > 0 \Rightarrow \lim_{x \to 0^+} (x^p \ln x) = 0;$$

$$p > 0 \text{ and } |q| < 1 \Rightarrow \lim_{n \to \infty} n^p q^n = 0.$$

The harmonic series $1 + \frac{1}{2} + \frac{1}{3} + \frac{1}{4} + \cdots$ diverges.

3 Multivariable calculus

O neglectful Nature, wherefore art thou thus partial...?

Leonardo da Vinci 1452–1519

Functions of more than one variable

If you know the radius r of a circle you can easily calculate the area A by using the formula $A = \pi r^2$. Here A depends on the single variable r and this relationship can be written as $A = f(r)$, where $f(r) = \pi r^2$. To find the volume V of a cylinder you need to know both the radius r and the height h, since $V = \pi r^2 h$. In this case V depends on the two variables r and h, and this can be shown by writing $V = g(r, h)$, where $g(r, h) = \pi r^2 h$. The order of the variables matters, since for example $g(2, 5) = 20\pi$ whereas $g(5, 2) = 50\pi$, and these are not the same.

Similarly the formula $\triangle = \frac{1}{2}ab \sin C$ shows how the area $\triangle$ of triangle ABC depends on the sides a, b and the angle C:

$$\triangle = h(a, b, C), \quad \text{where} \quad h(a, b, C) = \frac{1}{2}ab \sin C.$$

In these three examples, f is a function of one variable, g is a function of two variables and h is a function of three variables. An important reason for studying functions of more than one variable is that in many applications the value of the quantity you want to know depends on the values of several other quantities. For example, in weather forecasting, atmospheric pressure depends on position (three space coordinates) and time, a total of four variables, and in agriculture the cost of wheat depends on such things as the yield, demand, costs of seed, fertiliser, labour and subsidies. All these situations may be modelled using functions of more than one variable, and then questions arise about how to deal with local approximations (small changes), rates of change and greatest or least values with such functions. Although a realistic model may use dozens of variables, this chapter will concentrate on functions of two or three variables, since the main conceptual step is from 'one' to 'more than one'.

NOTE

An example of the use of many variables is the Treasury model of the British economy, which started in 1970 with about 200 variables. The number of variables grew to 1275 by 1989, when it was reduced to 530. In 1995 the model underwent further substantial slimming, and it now has 357 variables which are connected by about 30 functions.

Representing a function of two variables

Just as a function f of one variable is often represented by the curve with equation $y = f(x)$, so the most natural way to represent a function g of two variables x, y is as a *surface*. Take mutually perpendicular x-, y- and z-axes with the z-axis vertical. Each pair of values (x, y) can be used to work out $z = g(x, y)$. Then as the point $(x, y, 0)$ varies over the x–y plane (or part of it, if the domain of g is restricted) so the set of points (x, y, z) forms the surface representing the function, as in figure 3.1.

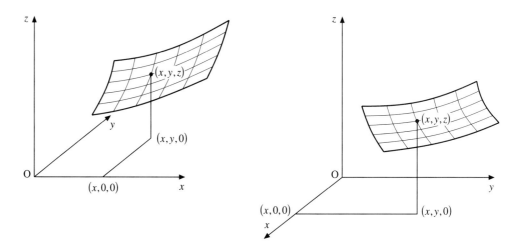

Figure 3.1 Note the two conventional viewing angles

The difficulty is that this representation is three-dimensional, and it is not usually feasible actually to construct the surface. Various compromises are possible.

1 Use a two-dimensional perspective drawing of the surface. Many modern computer packages will do this very realistically, and it may even be possible to animate the display so that you can 'walk around' the surface.

2 Just write the appropriate values of z at points in the x–y plane. This is the way the depth of the ocean is given in naval charts of coastal waters.

3 Use colour or shading to indicate the value of z. This is common in atlases, and also when showing temperature on weather maps. With the use of conventional colour codes (such as 'blue = cold, red = hot') this method is good for giving a quick impression when not much detail is needed. It is also perhaps a more natural way to represent a temperature distribution than the abstract idea of a 'temperature surface'.

4 Join all the points (x, y) for which $f(x, y)$ takes a particular value, to give a *contour* or *level curve*, as on an Ordnance Survey map. This is often done in addition to **2** or **3** to add precision and ease of reading to the graph. The same terms are also used for the set of points (x, y, c) in three dimensions for which $f(x, y) = c$: thus the actual water's edge of a reservoir and its representation on a map are both called *contours*. Some contours have special names, for example an *isobar* is a pressure contour, and an *isotherm* is a temperature contour.

EXAMPLE

Draw contour lines and a perspective sketch of the surface $z = 2\cos x - y^2$.

Solution

The equation of the surface can be rearranged as $y^2 = 2\cos x - z$ and so the contour $z = k$ has equation $y^2 = 2\cos x - k$. This is of the form $y^2 = f(x)$, so we use the method of *Pure Mathematics 4*, page 39: first sketch the graph $y = 2\cos x - k$ (shown with a broken line for $k = 1$ in the diagram below) and use it to draw the contour, which is symmetrical about the x-axis.

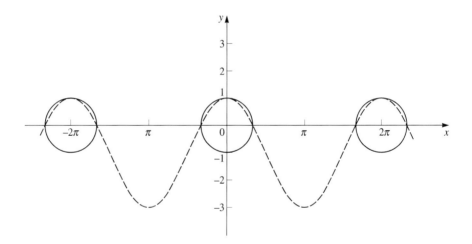

Taking the values $k = 2, 1, 0, -1, -2, -3$ gives the set of contours shown in this diagram.

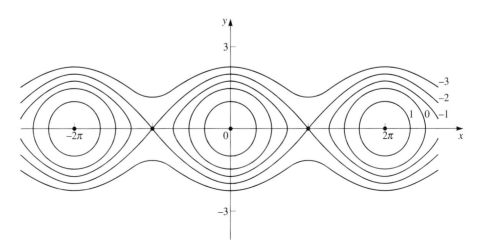

Note that:

(i) there are no points of the surface for which $z > 2$, since this would give $y^2 < 0$, which is impossible

(ii) when $k = 2$ the contour is the set of isolated points $(2m\pi, 0)$

(iii) when $k = -2$ the contour is $y^2 = 2\cos x + 2 = 4\cos^2(\frac{1}{2}x)$

$$\Leftrightarrow \quad y = \pm 2\cos(\tfrac{1}{2}x),$$

a pair of cosine curves crossing on the x-axis

(iv) when $k < -2$ the contours do not meet the x-axis.

From these contours we see that the surface has a chain of peaks at the points $(2m\pi, 0, 2)$, as drawn in perspective here.

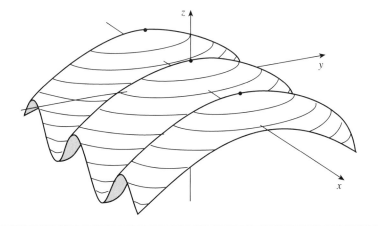

Exercise 3A

1. The following table shows the varying length of day with latitude in the Northern hemisphere. The length of day (i.e. the period between sunrise and sunset, not including twilight) is given in hours, correct to one decimal place, on the 15th day of each month.

Month	Equator	10°	20°	30°	40°	50°	60°	70°	80°	N. pole
Jan	12.1	11.6	11.0	10.4	9.6	8.5	6.6	0.0	0.0	0.0
Feb	12.1	11.8	11.4	11.2	10.7	10.1	9.2	7.3	0.0	0.0
Mar	12.1	12.1	12.0	12.0	11.9	11.8	11.7	11.5	10.9	0.0
Apr	12.1	12.4	12.6	12.9	13.2	13.7	14.5	16.1	24.0	24.0
May	12.1	12.6	13.1	13.6	14.4	15.4	17.1	22.2	24.0	24.0
Jun	12.1	12.7	13.3	14.1	15.0	16.4	18.8	24.0	24.0	24.0
Jul	12.1	12.7	13.3	13.9	14.8	15.6	17.5	24.0	24.0	24.0
Aug	12.1	12.5	12.8	13.3	13.8	14.6	15.8	18.4	24.0	24.0
Sep	12.1	12.2	12.3	12.4	12.5	12.7	13.0	13.6	15.3	24.0
Oct	12.1	11.9	11.7	11.5	11.2	10.8	10.2	9.1	5.2	0.0
Nov	12.1	11.7	11.2	10.7	10.0	9.1	7.6	3.1	0.0	0.0
Dec	12.1	11.5	10.9	10.2	9.3	8.1	5.9	0.0	0.0	0.0

(i) Draw axes representing the date D (ignore differences in the lengths of months) and the latitude λ. Sketch on these the contours for the duration of the day, T hours, from $T = 0$ to $T = 24$ at intervals of 4 hours.

(ii) Draw (or use a computer to draw) a perspective sketch of the surface representing T as a function of D and λ. Comment on any unusual features of this surface, and say why they occur.

2. The diagram shows a contour map of a hillside with two paths, A and B.

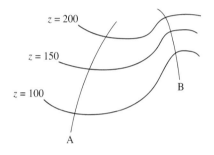

$z = 200$

$z = 150$

$z = 100$

B

A

(i) On which path will you have the steeper climb?

(ii) On which path will you probably have the better views?

(iii) On which path are you more likely to be near a stream?

3. Sketch a contour diagram for the surface $V = \pi r^2 h$ showing the contours $V = 0$, π, 2π and 3π for $r \geqslant 0$ and $h \geqslant 0$. Then sketch a perspective view of the surface.

4. Explain why all the contours of the surface $z = f(x^2 + y^2)$ are circles. For each of the following surfaces sketch a contour diagram with at least three labelled contours, and sketch the surface.

(i) $z = x^2 + y^2$

(ii) $z = \dfrac{1}{x^2 + y^2}$

(iii) $z = \sqrt{x^2 + y^2}$

(iv) $z = e^{-(x^2 + y^2)}$

5. Show that all but one of the contours of the surface $z = \dfrac{x + 2y}{x^2 + y^2}$ are circles through the origin O with the point O omitted, and describe the exception. Sketch the contours of this surface.

6. Describe these three-dimensional surfaces.

(i) $z = x^2$

(ii) $z = y^3$

(iii) $z = x^2 + y^3$

7. Sketch the contours of the surface $z = 3x - x^3 - y^2$ for $z = -4, -2, 0, 2, 4$. If $z = 0$ is sea level, describe the geographical feature which might be represented by this surface.

8. With the help of contours and perspective diagrams describe the following surfaces.

(i) $z = \cos(x + y)$

(ii) $z = \cos(x - y)$

(iii) $z = \cos(x^2 - y)$

9. In a simple model of an anticyclone the pressure p at the point (x, y, z) at time t is given by $p = a\{b^2 - (x - ct)^2 - 4(y - ct)^2\}e^{-kz}$, where z is the height above sea level and a, b, c and k are constants.

(i) Where is the point of highest pressure?

(ii) In what direction is the anticyclone moving?

(iii) What shape are the isobars at sea level?

(iv) What is the purpose of the factor e^{-kz}?

10. The *Wind Chill Index* developed from experiments in the Antarctic, gives an indication of how the combination of temperature and wind affects the comfort and safety of a suitably-clothed person outdoors. The unit of wind chill is the rate of loss of heat from the body in $\text{kg cal m}^{-2}\,\text{h}^{-1}$.

Using readings from the diagram, draw a new diagram showing the same information presented as wind chill contours plotted against axes representing temperature (in °C) and wind speed (in k.p.h.).

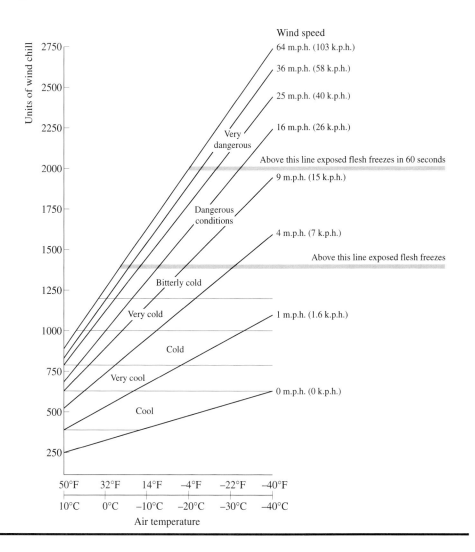

Sections

Another way to give information about a surface is to use the *section* in which a vertical plane cuts the surface. The plane may have any orientation, but for the moment we concentrate on sections parallel to the *x*- or *y*-axis.

A section parallel to the *x*-axis is obtained by fixing the value of *y* (at $y = b$, say). Then the section is the curve $z = f(x, b)$ in the plane $y = b$. Similarly, a section parallel to the *y*-axis is a curve $z = f(a, y)$ in the plane $x = a$ (see figure 3.2).

Multivariable calculus

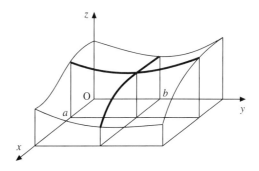

Figure 3.2

EXAMPLE Draw the sections of the surface $z = x^2 + 2y^2 + xy - 7x$ cut by the planes $x = 3$ and $y = 2$.

Solution

$$y = 2 \quad \Rightarrow \quad z = x^2 + 8 + 2x - 7x = x^2 - 5x + 8$$

$$\text{and} \quad x = 3 \quad \Rightarrow \quad z = 9 + 2y^2 + 3y - 21 = 2y^2 + 3y - 12.$$

So the sections are the parabolas $z = x^2 - 5x + 8$ in the plane $y = 2$ and $z = 2y^2 + 3y - 12$ in the plane $x = 3$. These are shown in the diagram together with a perspective sketch showing how they fit together, passing though the point $(3, 2, 2)$ on the surface.

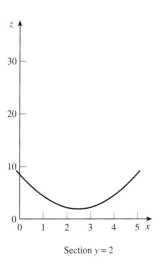

Section $y = 2$

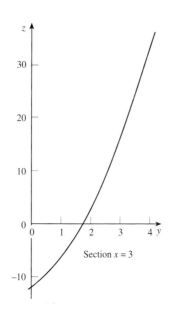

Section $x = 3$

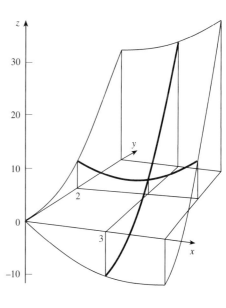

EXAMPLE   The surface $z = \frac{1}{10}(x^2 + 2y^2 + xy - 7x)$ represents a hillside, with the x- and y-axes due east and due north respectively. Footpaths running east-west and north-south cross at the point $(3, 2, 0.2)$. Find the angle each path makes with the horizontal at this point.

Solution

The surface is the same as in the previous example, except for the vertical scaling factor $\frac{1}{10}$.

The east-west footpath is the parabola $z = \frac{1}{10}(x^2 - 5x + 8)$ in the plane $y = 2$.

On this curve $\dfrac{dz}{dx} = \frac{1}{10}(2x - 5) = 0.1$ when $x = 3$, so the path slopes upward to the east at an angle of $\arctan 0.1 \approx 6°$.

On the north-south path $z = \frac{1}{10}(2y^2 + 3y - 12)$ and so $\dfrac{dz}{dy} = \frac{1}{10}(4y + 3) = 1.1$ when $y = 2$, so the path slopes upward to the north at an angle of $\arctan 1.1 \approx 48°$.

Exercise 3B

1. Using the information given in Question 1 of Exercise 3A, draw sections to show:

   (i)   how the length of day at latitude $50°$ N varies throughout the year

   (ii) how the length of day at latitude $80°$ N varies throughout the year

   (iii) how the length of day on October 15th varies with latitude.

2. Sketch the sections of the surface $z = x^2 - y^2$ for which:

   (i) $x = 0, \pm 1, \pm 2$ (ii) $y = 0, \pm 1, \pm 2$.

3. The depth of water, z m, in a wave tank is given at time t seconds by the formula
   $z = h(x, t) = 1.5 + 0.2 \sin(0.5x - t)$, where x m is the distance from one end of the tank.

   (i)   Sketch on the same axes the sections $z = h(x, 3)$ and $z = h(x, 5)$ and explain what these show.

   (ii) Sketch on the same axes the sections $z = h(3, t)$ and $z = h(5, t)$ and explain what these show.

   (iii) How fast do the waves move along the tank?

4. (i)   Find the equations of the sections of the surface $z = x^3 + y^2 - 4xy^2$ cut by the planes $x = 3$ and $y = 1$.

   (ii) Sketch the graphs of these sections.

   (iii) Find the gradient of each of these sections at the point $(3, 1, 16)$.

5. Part of a desert surface can be modelled by the equation $z = 3 + 2\sin(0.4x - 0.3y)$, where the x- and y-axes are due east and due north respectively. Give a general description of the surface. Tracks running east-west and north-south meet at the point where $x = 5$ and $y = 2$. Find the angle each track makes with the horizontal at this point.

6. Given that $f(x, y) = \dfrac{x^3}{y^2 + 1}$ find:

   (i) $f(5, 3)$

   (ii) the equations of the sections of the surface $z = f(x, y)$ cut by the planes $x = 5$ and $y = 3$

   (iii) the gradients of these sections when $x = 5$ and $y = 3$.

7. Find the equations of the sections of the surface $z = (x - 2y)\ln(x^2 + 3y)$ cut by the planes $x = a$ and $y = b$. Find also the gradients of these sections when $x = a$ and $y = b$.

8.

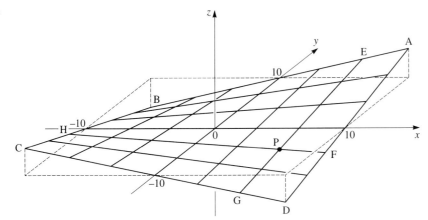

The diagram shows a framework ABCD made by joining the points $A(10, 10, 2)$, $B(-10, 10, -2)$, $C(-10, -10, 2)$, $D(10, -10, -2)$. Each side is divided into the same number of equal segments (six in this diagram) and corresponding points of division are joined by straight lines parallel to the planes $x = 0$ (such as EG) or $y = 0$ (such as FH).

(i) Show that the coordinates of E and G are $(\lambda, 10, \lambda/5)$ and $(\lambda, -10, -\lambda/5)$ respectively, where $-10 \leqslant \lambda \leqslant 10$.

(ii) Find the coordinates of F and H, given that they lie on the plane $y = \mu$, where $-10 \leqslant \mu \leqslant 10$.

(iii) Show that the lines EG and FH intersect, at P say, and find the coordinates of P in terms of λ and μ.

(iv) Deduce that P lies on the surface $z = \dfrac{xy}{50}$.

(v) Show that the contours of this surface are hyperbolas. Are there any exceptions?

(vi) Show that the vertical sections in which the diagonal planes $y = \pm x$ cut this surface are parabolas. Investigate other vertical sections not parallel to the x- or y-axis.

(This surface is called a *hyperbolic paraboloid* because of the properties shown in (v) and (vi). It is an example of a *ruled surface*, i.e. a surface generated by a family of straight lines; in this case there are two families of generators, the sections parallel to the $x = 0$ or $y = 0$ planes. Ruled surfaces can be constructed physically by using cables stretched across a framework as in the diagram above, to support the surface coating. This relatively simple and cheap method is used to form attractively shaped roofs.)

Partial differentiation

The last example (page 108) showed that the gradient at the point $(3, 2, 0.2)$ on the $y = 2$ section of the surface $z = \frac{1}{10}(x^2 + 2y^2 + xy - 7x)$ is 0.1. You can use the same method to find the gradient where $x = a$, on the section where the plane $y = b$ meets the surface $z = f(x, b)$ by:

- putting $y = b$ into the equation to obtain $z = f(x, b)$

- differentiating with respect to x

- putting $x = a$ into the result to find the gradient.

The same can be achieved more simply by:

- differentiating $z = f(x, y)$ with respect to x, treating y as a constant

- putting $x = a$ and $y = b$ into the result.

The process of differentiating a function with respect to one variable while keeping the other variable(s) constant is called *partial differentation*, and the expression for the gradient which this produces is called the *partial derivative*.

Various notations are used for the partial derivative of $z = f(x, y)$ with respect to x. The first two are $\dfrac{\partial z}{\partial x}$ or $\dfrac{\partial f}{\partial x}$. The symbol ∂, which should be distinguished from d or δ in handwriting, was introduced by Carl Jacobi (1804–51): it can be read as 'partial d' or 'curly d', and $\dfrac{\partial z}{\partial x}$ is read as 'partial dz by dx'. The next common notation for the same thing is $f_x(x, y)$, which corresponds to Lagrange's $f'(x)$ for the derivative of a function of one variable; this requires that the function involved should be named (f), but then has the advantage that it can show where the derivative is evaluated. In the example on page 108 the two gradients found can be written as '$\dfrac{\partial z}{\partial x}$ and $\dfrac{\partial z}{\partial y}$ evaluated at $(3, 2)$', but if we introduce $f(x, y) = \frac{1}{10}(x^2 + 2y^2 + xy - 7x)$ then we can use the simpler alternative '$f_x(3, 2)$ and $f_y(3, 2)$'. You may come across other notations too, such as $f'_x(x, y)$ or $f_1(x, y)$, but we shall not use these.

EXAMPLE

Find $\dfrac{\partial z}{\partial x}$ and $\dfrac{\partial z}{\partial y}$ when:

(i) $z = \frac{1}{10}(x^2 + 2y^2 + xy - 7x)$ (ii) $z = x^3 e^{xy^2}$ (iii) $z = h\left(\dfrac{x^3}{y^2}\right)$.

Solution

(i) $\dfrac{\partial z}{\partial x} = \frac{1}{10}(2x + y - 7)$ $\dfrac{\partial z}{\partial y} = \frac{1}{10}(4y + x)$.

(ii) The usual rules of differentiation (product rule, chain rule, etc.) still apply. Keeping y constant:

$$\frac{\partial z}{\partial x} = 3x^2 e^{xy^2} + x^3 y^2 e^{xy^2} = (3 + xy^2)x^2 e^{xy^2}$$

Keeping x constant:

$$\frac{\partial z}{\partial y} = x^3 \times 2xy\, e^{xy^2} = 2x^4 y\, e^{xy^2}.$$

(iii) Put $u = \dfrac{x^3}{y^2}$ so that $z = h(u)$. To differentiate partially with respect to x we use the chain rule, which now takes the form $\dfrac{\partial z}{\partial x} = \dfrac{dz}{du} \times \dfrac{\partial u}{\partial x}$.

$$\frac{\partial z}{\partial x} = h'(u) \times \frac{3x^2}{y^2} = h'\left(\frac{x^3}{y^2}\right) \times \frac{3x^2}{y^2}.$$

Similarly $\dfrac{\partial z}{\partial y} = h'(u) \times \left(-\dfrac{2x^3}{y^3}\right) = -h'\left(\dfrac{x^3}{y^2}\right) \times \dfrac{2x^3}{y^3}$.

Exercise 3C

1. Find $\dfrac{\partial z}{\partial x}$ and $\dfrac{\partial z}{\partial y}$.

   (i) $z = \dfrac{x^2}{y^3}$

   (ii) $z = \arctan\left(\dfrac{y}{x}\right)$

   (iii) $z = x \cos y + y \cos x$

   (iv) $z = x\,e^{\sqrt{xy}}$

2. If $f(x, y) = x^3 + 3xy^2 - \dfrac{y^4}{x}$ find $f_x(1, 3)$ and $f_y(-2, 5)$.

3. The temperature T (in °C) at a point in a room is linked by the function F to the distance x metres from a heater and the time t minutes after the heater has been switched on, so that $T = F(x, t)$. For each of the following statements give the appropriate units and say what the statement means in practical terms.

   (i) $F_x(3, 10) = -0.3$ (ii) $F_t(3, 10) = 0.1$

4. The displacement of a vibrating guitar string 1 metre long from its rest position is given by $d(x, t) = 0.003 \sin(\pi x) \sin(2500t)$, where x m is the distance from one end of the string and t s is the time. Evaluate $d_x(0.4, 2)$ and $d_t(0.4, 2)$ and explain what each of these means in practical terms.

5. If $u = \dfrac{x + y}{1 - xy}$, $v = \arctan x + \arctan y$, prove that

   $$\frac{\partial u}{\partial x}\frac{\partial v}{\partial y} = \frac{\partial u}{\partial y}\frac{\partial v}{\partial x}.$$

6. Van der Waal's equation $RT = \left(p + \dfrac{a}{V^2}\right)(V - b)$ connects the pressure p, volume V and temperature T of a quantity of gas of fixed mass, where a, b and R are constants.

Find $\dfrac{\partial T}{\partial p}, \dfrac{\partial p}{\partial V}$ and $\dfrac{\partial V}{\partial T}$. Hence show that

$$\left(\frac{\partial T}{\partial p}\right)\left(\frac{\partial p}{\partial V}\right)\left(\frac{\partial V}{\partial T}\right) = -1.$$

7. The Cobb-Douglas production model states that $P = aK^bL^{1-b}$, where P is the total yearly production of an economy, K is the total capital investment, L is the total labour force, and a and b are constants with $0 < b < 1$.

   Prove that $K\dfrac{\partial P}{\partial K} + L\dfrac{\partial P}{\partial L} = P$.

8. Prove that the relation $x\dfrac{\partial z}{\partial x} - y\dfrac{\partial z}{\partial y} = z$ is satisfied by every function of the form $z = xF(xy)$, where F is any function which can be differentiated.

9. (i) Prove that if $z = F\left(\dfrac{x}{y}\right)$ then $x\dfrac{\partial z}{\partial x} + y\dfrac{\partial z}{\partial y} = 0$.

   Check this with the function in Question 1 (ii).

   (ii) Prove that if $z = y^n F\left(\dfrac{x}{y}\right)$ then

   $$x\frac{\partial z}{\partial x} + y\frac{\partial z}{\partial y} = nz.$$ Check this with the functions in Questions 2 and 7.

   (This result is Euler's theorem on homogeneous functions, a *homogeneous function of degree n* being a function f such that $f(tx, ty) = t^n f(x, y)$.)

Differentiability

This section points out an important distinction between curves and surfaces.

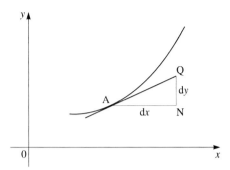

Figure 3.3

Figure 3.3 shows a curve $y = f(x)$ with its tangent at the point A. Another point Q on the tangent can be reached from A by means of an x-step AN followed by a y-step NQ. These steps are called *differentials* and are denoted by dx and dy respectively. Thus dx is an arbitrary (and not necessarily small) change in x and dy is the resulting change in y needed to reach a point on the *tangent* at A. Since the gradient of this tangent is $f'(x)$, the connection between these differentials is $dy = f'(x)\,dx$ (so that dy depends on the two variables x and dx). From this it follows that $\dfrac{dy}{dx} = f'(x)$, which explains the familiar but rather odd Leibniz notation for the derivative.

Activity

(i) If A is the point $(2, 8)$ on the curve $y = x^3$, find dy when dx is (a) 100 (b) -2 (c) 0.01. Give the formula for dy in terms of dx.

(ii) Give the formula for dy in terms of x and dx for the general point A on the curve $y = x^3$.

The tangent at A is the line which fits closest to the curve at A, and so can be used as the *linear approximation* to the curve near to A. This idea, which is the starting point for the discussion of Maclaurin and Taylor approximations in *Pure Mathematics 5*, page 58, can be written in the equivalent forms

$$\delta y \approx \frac{dy}{dx}\,\delta x \ \text{ or } \ \delta y \approx f'(x)\,\delta x \ \text{ or } \ f(x+h) \approx f(x) + h f'(x)$$

where $\delta x \,(= h)$ is a small change in x.

Notice that in moving from A to a neighbouring point P on the curve we use the *approximate* formula $\delta y \approx f'(x)\,\delta x$, which applies only for small values of δx, whereas in moving from A to a point Q on the tangent we use the *exact* formula $dy = f'(x)\,dx$, which is true for all values of dx. Figure 3.4 superimposes both changes, so that $\delta x = dx$, and shows that the error in the approximation is then $\delta y - dy$.

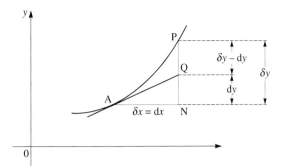

Figure 3.4

The same idea can be expressed in terms of limits instead of approximations:

$$\delta y = (f'(x) + \varepsilon)\,\delta x \ \text{ or } \ f(x + h) = f(x) + h(f'(x) + \varepsilon)$$

where $\varepsilon \to 0$ as $\delta x \to 0$ or $h \to 0$.

You have met examples of functions which have no derivative for certain values of x. At the corresponding points on the graph of the function there is no tangent, and therefore no linear approximation. One simple example is $y = |x|$ at the origin (figure 3.5).

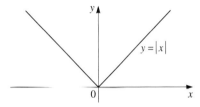

Figure 3.5

For a function of one variable, the properties:

(A) having a derivative when $x = a$

(B) having a tangent to the graph at $(a, f(a))$

(C) having a linear approximation in a neighbourhood of $x = a$

are all equivalent. In the case of a one variable function the statement that $f(x)$ is *differentiable* at $x = a$ tells us that the function has property (C) and (A) and (B).

With functions of two or more variables things are not so simple! Suppose that at the point A on the surface $z = f(x, y)$ we can find the partial derivatives $f_x(x, y)$ and $f_y(x, y)$. Then the sections through A in the x- and y-directions have tangents at A, and there is just one plane through A containing these two tangents. This plane gives satisfactory linear approximations to the surface in the x- and y-directions, but there is no guarantee that the surface stays close to the plane as we head in other directions.

For example, consider the surface $z = f(x, y)$ shown, together with its contours, in figure 3.6. It consists of the horizontal slab $z \leqslant 3$ from which are carved two V-shaped grooves parallel to the x- and y-axes defined by $z = |y|$ and $z = |x|$. Near the origin the surface is something like the adjacent portions of a slab of chocolate.

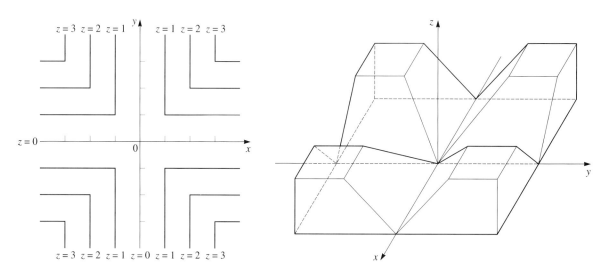

Figure 3.6

It is clear that the sections through the origin are just the x- and y-axes, so both partial derivatives are zero at the origin, and the plane $z = 0$ contains the tangents to these sections. But if you move on the surface heading in any other direction from the origin you have to ascend the face of a groove, moving away from the plane $z = 0$ by an amount proportional to the distance travelled. So the plane $z = 0$ cannot be regarded as a linear approximation to the surface at the origin. Nor can any other plane give a better approximation, since no other plane contains both the x- and y-axes. So this surface has *no* satisfactory approximating plane at the origin, even though both partial derivatives exist there. More formally, it is not possible to find constants p and q such that:

$$f(h, k) = f(0, 0) + h(p + \varepsilon_1) + k(q + \varepsilon_2)$$

where $\varepsilon_1 \to 0$ and $\varepsilon_2 \to 0$ as $\sqrt{h^2 + k^2} \to 0$.

Luckily this rather artificial example is not typical: for most points on most elementary surfaces the plane containing the tangents to the sections in the x- and y-directions can be used as a linear approximation to the entire surface in the neighbourhood of the point. When this happens we say that the function is *differentiable* at that point, and the plane is called the *tangent plane*. The example just given shows that for a function of two variables the properties:

(A′) having both partial derivatives when $x = a$ and $y = b$

(B′) having a tangent plane to the surface at $(a, b, f(a, b))$

(C′) having a linear approximation in the neighbourhood of $x = a$ and $y = b$

are *not* equivalent.

Activity

Insert the correct symbol ($\Rightarrow, \Leftarrow$ or $\Leftrightarrow$) between these properties.

(i) (A′) and (B′)   (ii) (B′) and (C′)   (iii) (C′) and (A′)

Activity

Use contours, sections and, if possible, a computer-generated perspective drawing to describe the surface $z = \sqrt[4]{x^2 y^2}$. In particular, find which (if any) of the properties (A'), (B') and (C') are true at the points:

(i) $(0,0,0)$ (ii) $(1,0,0)$ (iii) $(1,1,1)$.

Tangent planes

A set of conditions sufficient to ensure that a function is differentiable (i.e. has a tangent plane) at a particular point will be given on page 128. Meanwhile you may assume that all the functions you meet in this book are differentiable at all relevant points, unless attention is drawn to the contrary. The practical task of finding the equation of the tangent plane at a point on a given surface is straightforward, as shown in this example.

EXAMPLE

Find the equation of the tangent plane at the point $A(4, 3, 14)$ on the surface $z = x^2 + 2y^2 - 2xy + x$.

Solution

Since $\dfrac{\partial z}{\partial x} = 2x - 2y + 1 = 3$ at A, the gradient of the section parallel to the x-axis at A is 3. This means that, in moving along the tangent to this section at A, z increases by 3 when x increases by 1 and y remains constant. Therefore the vector $\begin{pmatrix} 1 \\ 0 \\ 3 \end{pmatrix}$ is in the direction of this tangent.

Similarly, since $\dfrac{\partial z}{\partial y} = 4y - 2x = 4$ at A, the vector $\begin{pmatrix} 0 \\ 1 \\ 4 \end{pmatrix}$ is in the direction of the tangent at A to the section parallel to the y-axis (see diagram).

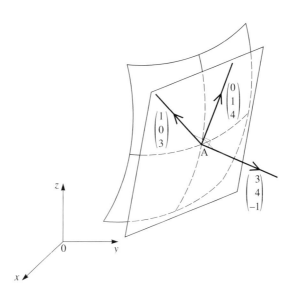

The tangent plane contains both these tangents, and so has normal vector:

$$\mathbf{n} = \begin{pmatrix} 0 \\ 1 \\ 4 \end{pmatrix} \times \begin{pmatrix} 1 \\ 0 \\ 3 \end{pmatrix} = \begin{pmatrix} 3 \\ 4 \\ -1 \end{pmatrix}.$$

The equation of the tangent plane is therefore:

$$3x + 4y - z = 3 \times 4 + 4 \times 3 - 14 = 10$$

or

$$z = 3x + 4y - 10.$$

The general procedure is similar. The equation of the tangent plane of the surface $z = f(x, y)$ at the point $A(a, b, c)$, where $c = f(a, b)$, can be found as follows.

- Find the values of the partial derivatives at the point. Suppose that $\dfrac{\partial f}{\partial x} = p$

  and $\dfrac{\partial f}{\partial y} = q$ when $x = a$ and $y = b$. Then the vectors $\begin{pmatrix} 1 \\ 0 \\ p \end{pmatrix}$ and $\begin{pmatrix} 0 \\ 1 \\ q \end{pmatrix}$ are in

  the directions of the tangents to the sections at A.

- A normal vector to the tangent plane is $\mathbf{n} = \begin{pmatrix} 0 \\ 1 \\ q \end{pmatrix} \times \begin{pmatrix} 1 \\ 0 \\ p \end{pmatrix} = \begin{pmatrix} p \\ q \\ -1 \end{pmatrix}.$

- The equation of the tangent plane is $(\mathbf{r} - \mathbf{a}) \cdot \mathbf{n} = 0$, i.e.

$$\begin{pmatrix} x - a \\ y - b \\ z - c \end{pmatrix} \cdot \begin{pmatrix} p \\ q \\ -1 \end{pmatrix} = 0 \text{ or } z = p(x - a) + q(y - b) + c.$$

This equation can be put into another useful form. The differences $x - a$, $y - b$, $z - c$ are the changes in x, y, z needed to move from A to the general point (x, y, z) on the tangent plane at A. These local coordinates relative to A are called differentials (as on page 112) and denoted by dx, dy, dz respectively. Then it follows from the equation of the tangent plane given above that

$$\begin{pmatrix} dx \\ dy \\ dz \end{pmatrix} \cdot \begin{pmatrix} \partial f / \partial x \\ \partial f / \partial y \\ -1 \end{pmatrix} = 0$$

or

$$dz = \frac{\partial f}{\partial x} dx + \frac{\partial f}{\partial y} dy$$

where the partial derivatives are evaluated at A. This is the result for a function of two variables corresponding to the formula $dy = f'(x) \, dx$ given on page 112. It shows how the change dz needed to remain on the tangent plane is related to the changes dx and dy. Notice that, in general, dz depends on the four variables x, y, dx, dy.

Exercise 3D

1. Find the equation of the normal line and of the tangent plane to the given surface at the given point.

   (i) $z = 2x^2 - 3xy + 4y^2$, $(3, 1, 13)$

   (ii) $z = \dfrac{x^3}{y^2}$, $(4, -8, 1)$

   (iii) $z = \dfrac{20}{x^2 + y^2}$, $(2, 4, 1)$

   (iv) $z = e^x \cos y$, $(1, \pi, -e)$

2. A student who was asked to find the equation of the tangent plane to $z = x^2 y^4$ at $(3, 1, 9)$ gave the answer $z = 2xy^4(x - 3) + 4x^2 y^3(y - 1) + 9$.

   (i) How can you tell at a glance that this is wrong?

   (ii) What mistake did the student make?

   (iii) What is the correct answer?

3. Given that $z = \dfrac{x^3 + y^3}{x - y}$, $x = 10$, $y = 8$, $dx = 2$, $dy = -3$, find dz.

4. Given a particular function $z = f(x, y)$ and particular values of x and y, is it always possible to find values of dx and dy, not both zero, for which $dz = 0$? Interpret your answer geometrically.

5. Find the coordinates of the point on $z = x^2 + 5xy + 2y^2$ where the tangent plane is parallel to $5x - 13y + z = 10$.

6. Show that the equation of the tangent plane to the surface $z = \dfrac{1}{xy}$ at the point $\left(h, k, \dfrac{1}{hk}\right)$ can be written in the form $kx + hy + h^2 k^2 z = 3hk$. Prove that as h and k vary through positive values the tetrahedron formed by this plane and the planes $x = 0$, $y = 0$, $z = 0$ has constant volume.

7. Prove that the surfaces $z = x^2 + y^2 + 50$ and $z = 12x + 16y - x^2 - y^2$ touch, and find the equations of their common normal line and tangent plane.

8. The family of surfaces $S_1, S_2, S_3, \ldots$ is such that S_k has the equation $z = k(x^2 + 3y^2)$ for $k = 1, 2, 3, \ldots$. The line $\begin{pmatrix} x \\ y \\ z \end{pmatrix} = \lambda \begin{pmatrix} a \\ b \\ c \end{pmatrix}$, where a, b, c are constant, meets S_k at the origin and at P_k, and the tangent plane to S_k at P_k is T_k.

   (i) Find the equation of T_k.

   (ii) Show that the planes $T_1, T_2, T_3, \ldots$ are all parallel.

   (iii) Show that the distance of T_k from the origin is inversely proportional to k.

Directional derivatives

EXAMPLE

Find the angle of slope of a footpath leaving the point $(3, 2, 0.2)$ on bearing $072°$, for the hillside $z = \frac{1}{10}(x^2 + 2y^2 + xy - 7x)$ in the example on page 108.

Solution

As on page 108, $\dfrac{\partial z}{\partial x} = 0.1$ and $\dfrac{\partial z}{\partial y} = 1.1$ at $(3, 2, 0.2)$.

Therefore the tangent plane at this point is $dz = 0.1\,dx + 1.1\,dy$.

A horizontal line on a bearing of $072°$ makes an angle of $18°$ with the positive x-axis, so the horizontal unit vector in this direction is $\begin{pmatrix} \cos 18° \\ \sin 18° \end{pmatrix}$.

To move a distance of 1 unit along this horizontal line we therefore need

$$dx = \cos 18° \quad \text{and} \quad dy = \sin 18°$$

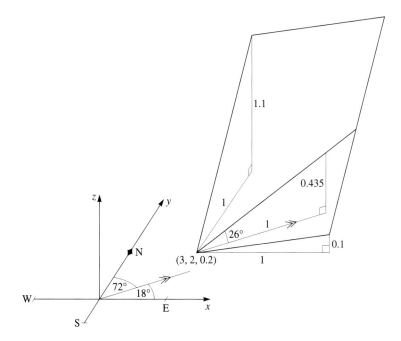

and to get back on to the tangent plane we need

$$dz = 0.1 \times \cos 18° + 1.1 \times \sin 18° \approx 0.435$$

(see the diagram).

So a horizontal shift of 1 unit in the given direction produces a vertical shift of 0.435. The gradient of the footpath is 0.435, and its angle of slope is $\arctan 0.435 \approx 26°$.

The general procedure is similar. Suppose we want to find the gradient on the surface $z = f(x, y)$ at the point $A(a, b, c)$ in the direction defined by the horizontal unit vector $\hat{\mathbf{u}} = \begin{pmatrix} \cos \alpha \\ \sin \alpha \end{pmatrix}$. Putting $dx = \cos \alpha$ and $dy = \sin \alpha$ in the equation of the tangent at A gives $dz = \dfrac{\partial f}{\partial x} \cos \alpha + \dfrac{\partial f}{\partial y} \sin \alpha$, where the partial derivatives are evaluated at A. Since this dz is the vertical change needed to get back to the tangent plane after a *unit* horizontal step, the required gradient is $\dfrac{\partial f}{\partial x} \cos \alpha + \dfrac{\partial f}{\partial y} \sin \alpha$. This scalar quantity is called the *directional derivative* in the direction of $\hat{\mathbf{u}}$.

The vector grad f

The directional derivative we have just found can be written as a scalar product.

$$\frac{\partial f}{\partial x} \cos \alpha + \frac{\partial f}{\partial y} \sin \alpha = \begin{pmatrix} \cos \alpha \\ \sin \alpha \end{pmatrix} \cdot \begin{pmatrix} \partial f/\partial x \\ \partial f/\partial y \end{pmatrix} = \hat{\mathbf{u}} \cdot \begin{pmatrix} \partial f/\partial x \\ \partial f/\partial y \end{pmatrix}.$$

The vector $\begin{pmatrix} \partial f/\partial x \\ \partial f/\partial y \end{pmatrix}$ is called **grad** f; alternative notations for this which you may meet are $\begin{pmatrix} f_x \\ f_y \end{pmatrix}$, $\begin{pmatrix} \partial z/\partial x \\ \partial z/\partial y \end{pmatrix}$, ∇f or ∇z; as usual the partial derivatives are evaluated at the point in question.

NOTE *The symbol ∇ was introduced by Hamilton in 1843; he called it 'nabla' because it is shaped like a Hebrew musical instrument of that name; ∇ is also called 'del'.*

It is natural to wonder how the *two*-dimensional vector **grad** f is related to the *three*-dimensional surface $z = f(x, y)$ from which it arises. Suppose you are on the surface at A and want to walk along the contour line through A. Then you have to head in the direction of a unit vector **û** which makes the directional derivative zero. But:

$$\hat{\mathbf{u}} \cdot \mathbf{grad}\, f = 0 \quad \Rightarrow \quad \hat{\mathbf{u}} \text{ is perpendicular to } \mathbf{grad}\, f$$

(assuming for the moment that **grad** f $\neq$ **0**).

Therefore the direction along the contour through A is perpendicular to **grad** f, or, turning this around, **grad** f is a vector *normal* to the contour through A.

For Discussion

What happens at a point where **grad** f $= 0$?

Now suppose that, instead of wanting to walk on the level, you want to walk in the steepest possible direction. Then you have to choose **û** to maximise the directional derivative. But

$$\hat{\mathbf{u}} \cdot \mathbf{grad}\, f = |\hat{\mathbf{u}}||\mathbf{grad}\, f| \cos \theta$$

where θ is the angle between **û** and **grad** f. Since $|\hat{\mathbf{u}}| = 1$ and $|\mathbf{grad}\, f|$ is fixed, the directional derivative is maximised when $\cos \theta = 1$ (which is when **û** is in the same direction as **grad** f, showing, not surprisingly, that the line of greatest slope is perpendicular to the contour), and the maximum gradient is $|\mathbf{grad}\, f|$.

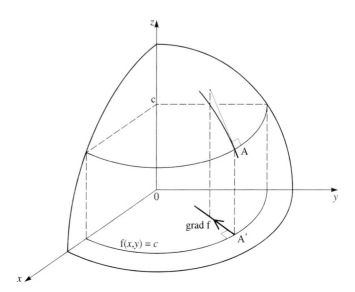

Figure 3.7

To summarise, if $A(a, b, c)$ is a point on the surface $z = f(x, y)$ and the contour with implicit equation $f(x, y) = c$ is drawn in the x–y plane, then the vector **grad** f, drawn starting at the point $A'(a, b)$, is in the x–y plane, is normal to the contour, points in the direction of greatest slope, and has magnitude equal to the greatest gradient at A (see figure 3.7).

Exercise 3E

Where relevant in this exercise, take the x- and y-axes to be due east and due north respectively.

1. For the given expression $f(x, y)$ find **grad** f and sketch the contour through $(3, 2)$ with **grad** f at this point shown as a vector.

   (i) xy

   (ii) $\dfrac{x}{y}$

   (iii) $x^2 + 3y^2$

   (iv) $\dfrac{x}{x^2 + y^2}$

2. Find the directional derivatives at the point $(4, 3, 7)$ on the surface $z = x^2 - y^2$ in the directions of the eight principal points of the compass.

3. The point $P(2, 1, 3.1)$ is on a hill represented by the surface $z = 4 - \dfrac{x^2}{5} - \dfrac{y^2}{10}$.

   Find the gradient and angle of slope of the paths through P in the directions:

   (i)   north-east

   (ii) directly towards the top of the hill.

4. A rectangular metal plate bounded by the lines $x = 0$, $x = 20$, $y = 0$ and $y = 10$ is heated in such a way that its temperature T at the point (x, y) is given by $T = \dfrac{200}{4x^2 + y^2 + 1}$.

   (i)   Where are the hottest and coolest points of the plate?

   (ii) A small bug placed on the plate at $(10, 5)$ moves in the direction in which the temperature decreases as rapidly as possible. Find this direction as a unit vector, and find the rate at which the temperature decreases.

   (iii) Another type of small bug placed on the plate at $(5, 7)$ finds the temperature to its liking, and moves in the direction which keeps the temperature constant. Find this direction as a unit vector.

5. Given that $f_x(a, b) = p$ and $f_y(a, b) = q$, find, in terms of p, q and θ, the gradient of the path on the surface $z = f(x, y)$ which passes through $(a, b, f(a, b))$ on the bearing of $\theta°$.

Implicit functions

These ideas can be used in connection with implicit functions of two variables.

EXAMPLE   Find the equation of the tangent to the ellipse $3x^2 + 5y^2 = 53$ at the point $(4, -1)$.

Solution

Think of the ellipse as the $z = 53$ contour of the surface $z = 3x^2 + 5y^2$.

Then $\mathbf{grad}\, z = \begin{pmatrix} 6x \\ 10y \end{pmatrix} = \begin{pmatrix} 24 \\ -10 \end{pmatrix}$ at $(4, -1)$. This is the direction of the normal,

and so the perpendicular vector $\begin{pmatrix} 10 \\ 24 \end{pmatrix}$ is tangential to the ellipse.

The gradient of $\begin{pmatrix} 10 \\ 24 \end{pmatrix}$ is $\dfrac{24}{10} = 2.4$, so the equation of the tangent is:

$$y - (-1) = 2.4(x - 4)$$

or $y = 2.4x - 10.6$.

The same idea can be used to differentiate the implicit function $f(x, y) = c$.

Since the vector $\mathbf{grad}\, f = \begin{pmatrix} \partial f/\partial x \\ \partial f/\partial y \end{pmatrix}$ is normal to the curve (considered as a

contour), it follows that the perpendicular vector $\begin{pmatrix} -\partial f/\partial y \\ \partial f/\partial x \end{pmatrix}$ is tangential, so

the gradient of the tangent is given by $\dfrac{dy}{dx} = -\dfrac{\dfrac{\partial f}{\partial x}}{\dfrac{\partial f}{\partial y}}$.

Activity

Given that $5x^4 + 4xy^2 - (3x + y)^2 = 10$, find $\dfrac{dy}{dx}$:

(i) by using this result   (ii) by differentiating each term with respect to x.

NOTE

The result $\dfrac{dy}{dx} = -\dfrac{\dfrac{\partial f}{\partial x}}{\dfrac{\partial f}{\partial y}}$ *is best remembered as a rearrangement of*

$df = \dfrac{\partial f}{\partial x}\, dx + \dfrac{\partial f}{\partial y}\, dy$ *in the case* $df = 0$.

Exercise 3F

1. Find $\dfrac{dy}{dx}$ in terms of x and y.

(i) $9x^2 - 7xy - 5y^2 = 27$

(ii) $\sin x + \sinh y = 0$

(iii) $\ln(x^3 + y^3) = 3x^2 + 2y^3$

(iv) $x^y = y^x$

2. Find the equations of the tangent and normal to the curve $x^2 + 2xy + 3y^2 = 22$ at the point $(5, -3)$.

3. Find the equation of the tangent at the point (x_1, y_1) on these conics.

(i) $\dfrac{x^2}{a^2} + \dfrac{y^2}{b^2} = 1$

(ii) $y^2 = 4ax$

(iii) $xy = c^2$

4. The *general conic* C has equation $F(x, y) = 0$, where:

$$F(x, y) = ax^2 + 2hxy + by^2 + 2gx + 2fy + c.$$

(i)   Prove that the equation of the tangent to C at the point (x_1, y_1) is:

$axx_1 + h(xy_1 + x_1y) + byy_1 + g(x + x_1) + f(y + y_1) + c = 0.$

(ii) Explain why the tangents are parallel to the coordinate axes at the points where the lines $F_x(x, y) = 0$ or $F_y(x, y) = 0$ meet C.

(iii) Deduce that if C is an ellipse or a hyperbola then its centre is the point of intersection of the lines $F_x(x, y) = 0$ and $F_y(x, y) = 0$, and hence find the coordinates of the centre.

(iv) By considering how the procedure in (iii) might fail, find the condition for C to be a parabola.

Stationary points

A point on the surface $z = f(x, y)$ at which the tangent plane is horizontal (i.e. parallel to the x–y plane) is called a *stationary point*. At a stationary point both $\dfrac{\partial z}{\partial x}$ and $\dfrac{\partial z}{\partial y}$ are zero, so the search for stationary points starts with solving the equations $\dfrac{\partial z}{\partial x} = 0$ and $\dfrac{\partial z}{\partial y} = 0$ simultaneously.

A stationary point may be the 'top of a hill' or the 'bottom of a hollow' or a 'col' (mountain pass) or something more complicated, as shown in the sketches and shaded contour maps of figure 3.8 (where light or dark shading shows high or low ground respectively).

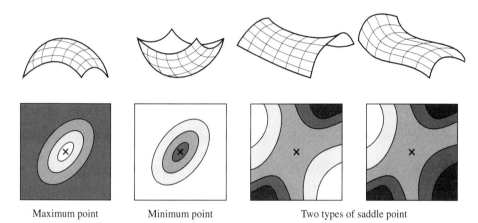

| Maximum point | Minimum point | Two types of saddle point |

Figure 3.8

More formally, a stationary point $(a, b, f(a, b))$ is:

- a *local maximum point* if $f(a + h, b + k) < f(a, b)$ for all sufficiently small h and k (not both zero)

- a *local minimum point* if $f(a + h, b + k) > f(a, b)$ for all sufficiently small h and k (not both zero)

- a *saddle point* if $f(a + h, b + k)$ is less than $f(a, b)$ for some h and k but greater than $f(a, b)$ for others.

There are other possibilities too, such as a point on a ridge (see Question 9 of Exercise 3G).

Activity

By considering contours and/or sections find what sort of stationary point the origin is on these surfaces.

(i) $z = x^2 + y^2$ (ii) $z = x^2 - y^2$ (iii) $z = -x^2 - y^2$ (iv) $z = xy$

There are tests involving second partial derivatives for distinguishing the types of stationary point but (as with the second derivative test for a function of one variable) they sometimes fail, and can be complicated to use. It is often simpler to use the definitions directly, as follows.

EXAMPLE

Investigate the stationary points of $z = x^3 + 6xy - 3y^2$.

Solution

Since $\dfrac{\partial z}{\partial x} = 3x^2 + 6y$ and $\dfrac{\partial z}{\partial y} = 6x - 6y$ we solve the simultaneous equations

$$3x^2 + 6y = 0 \quad \text{and} \quad 6x - 6y = 0$$

giving $y = x$ and $3x^2 + 6x = 0$, so $x = 0$ or -2.

This gives stationary points $(0, 0, 0)$ and $(-2, -2, 4)$.

At $(0, 0, 0)$ the $x = 0$ section is $z = -3y^2$, which might suggest that this is a maximum point. But the $y = 0$ section is $z = x^3$, so $z > 0$ for $x > 0$ and $z < 0$ for $x < 0$. Therefore by moving on the surface in the plane $y = 0$ it is possible either to ascend from the origin or to descend. These show that $(0, 0, 0)$ is a saddle point.

It is easy to show that both the sections $x = -2$ and $y = -2$ have maximum points at $(-2, -2, 4)$, but we still have to deal with sections in other directions. To do this we examine the value of z at the general point in the neighbourhood of the stationary point by putting $x = -2 + h$ and $y = -2 + k$.

Then $z = (-2 + h)^3 + 6(-2 + h)(-2 + k) - 3(-2 + k)^2$

$\qquad = -8 + 12h - 6h^2 + h^3 + 24 - 12k - 12h + 6hk - 12 + 12k - 3k^2$

$\qquad = 4 - 6h^2 + h^3 + 6hk - 3k^2$

$\qquad = 4 - 3(h - k)^2 - h^2(3 - h)$

Notice that the terms in h and k cancel, as they must at a stationary point.

$\qquad < 4$ provided that $h < 3$.

Therefore $(-2, -2, 4)$ is a maximum point.

The surface is shown in the diagram on the next page.

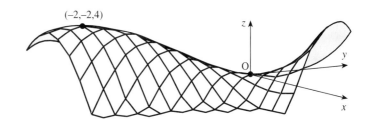

(-2,-2,4)

Exercise 3G

1. Show that $(1, 2, -7)$ is a stationary point of $z = x^2 + 2y^2 - xy - 7y$. Show that if $x = 1 + h$ and $y = 2 + k$ then $z = -7 + (h - \frac{1}{2}k)^2 + \frac{7}{4}k^2$. What can you deduce about the nature of the stationary point?

2. Find the coordinates of the stationary point of $z = x^2 - 4xy - y^2 + 20y - 15$ and determine, stating reasons, whether this is a maximum, minimum or saddle point.

3. Find both of the stationary points of $z = x^2 + y^3 - 12xy$ and determine their nature.

4. Investigate the stationary points of these surfaces.

   (i) $z = (x + y)(xy + 1)$

   (ii) $z = x^2 + y + \dfrac{2}{x} + \dfrac{4}{y}$

   (iii) $z = xy + \ln |x| + 2y^2$

5. Show that
   $f(x, y) = (x^2 + y^2)^2 - 4(x^2 + 2y^2) + 15$ has a stationary value at the origin and find all other points (x, y) at which $f(x, y)$ is stationary.

   By considering small displacements from the origin, $x = ht$ and $y = kt$, where h and k are constants, or otherwise, determine the nature of the stationary value at the origin.   [MEI]

6. Show that $z = Ax^2 + Bxy + Cy^2$ has a stationary point at the origin. Find conditions involving the constants A, B, C which ensure that this is:

   (i)   a maximum point

   (ii) a minimum point

   (iii) a saddle point.

7. (i)   Prove that of all the cuboids with a given volume the one with the smallest surface area is the cube.

   (ii) The base of a closed cuboidal cardboard carton with a given volume has an extra thickness of cardboard added for strength. Ignoring all other overlaps, find the shape of carton which minimises the amount of cardboard used.

8. Find and classify all the stationary points of $z = 2\cos x - y^2$.

   (This is the surface described in the example on page 103.)

9. Investigate the stationary points of $z = e^x(1 - \cos y)$.

10. Prove that the surface $z = \dfrac{1}{x^2 y}$ has no stationary points.

    Find the points on this surface which are closest to the origin, and show that the distance of these points from the origin is $2^{3/4}$.

    (**Hint:** Let the distance from the origin of the point (x, y, z) on this surface be L, find L^2 in terms of y and z and minimise L^2, treating y and z as independent variables.)

11. The diagram shows the cross-section of a proposed water channel. The section is to be an isosceles trapezium with area $20\,\text{m}^2$ and with width $w\,\text{m}$, depth $d\,\text{m}$ and sides inclined at angle θ to the horizontal, as shown.

Show that $w = \dfrac{20}{d} - d\cot\theta$.

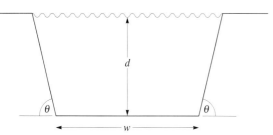

It is known that the average flow velocity of water through the channel is inversely proportional to the wetted perimeter p of the channel (i.e. the perimeter of the trapezium, excluding the top). Therefore the flow is maximised when p is minimised. Find the values of θ, d and w which achieve this.

12. The diagram shows the points $(1,2)$, $(2,5)$, $(4,4)$, $(5,6)$ and the line $y = mx + c$. The broken lines show the vertical deviations of the given points from the line. The *least squares line of best fit* is the line which minimises the sum of the square of these deviations, which we call Q. The squares of the deviations are used so that the positive or negative contributions from points on different sides of the line do not cancel.

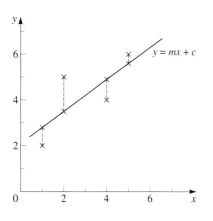

(i) Show that the y-coordinates of the points on the line are $m + c$, $2m + c$, $4m + c$ and $5m + c$ respectively, and hence that:

$$Q = (m + c - 2)^2 + (2m + c - 5)^2$$
$$+ (4m + c - 4)^2 + (5m + c - 6)^2.$$

(ii) Show that:

$$\frac{\partial Q}{\partial m} = 2(m + c - 2) + 4(2m + c - 5)$$
$$+ 8(4m + c - 4) + 10(5m + c - 6).$$

Deduce that:

$$\frac{\partial Q}{\partial m} = 0 \quad \Leftrightarrow \quad 23m + 6c - 29 = 0.$$

(iii) Show that:

$$\frac{\partial Q}{\partial c} = 0 \quad \Leftrightarrow \quad 12m + 4c - 17 = 0.$$

Deduce from this equation that the line of best fit passes through the point $(3, 4.25)$, which is the centroid of the four original points.

(iv) Solve the equations in (ii) and (iii) simultaneously to find the values of m and c which minimize Q, and draw a diagram showing this line, the four original points and their centroid.

(v) By considering the form of the surface for which the equation is given in (i), explain how you can be sure that the stationary point is a minimum point.

Exercise 3G continued

13. This question generalises Question 12.
Suppose there are now n given points (x_i, y_i), $i = 1, 2, \ldots, n$, and that we want to find the values of m and c which minimise the sum of the squares of the vertical deviations of these points from the line $y = mx + c$, i.e. we want to minimise $Q = \sum(mx_i + c - y_i)^2$ (where in this question all the summation is from $i = 1$ to n).

(i) Describe the form of the surface $Q = \sum(mx_i + c - y_i)^2$, and explain why this shows that it has one stationary point, and that this is a minimum.

(ii) Show that at the minimum point
$$m\sum x_i^2 + c\sum x_i - \sum x_i y_i = 0 \quad \text{and}$$
$$m\sum x_i + nc - \sum y_i = 0.$$

(iii) Deduce that the line of best fit passes through the centroid $(\bar{x}, \bar{y})$, where
$$\bar{x} = \frac{\sum x_i}{n}, \bar{y} = \frac{\sum y_i}{n}, \text{ and has gradient}$$
$$\frac{\sum x_i y_i - n\bar{x}\bar{y}}{\sum x_i^2 - n\bar{x}^2}. \text{ Check that the line found in}$$
Question 12 agrees with this.

(In statistics this least squares line of best fit is called the *regression line of y on x*.)

Functions of more than two variables

The behaviour of a function of more than two variables is not easy to visualise. If $w = g(x, y, z)$ it may help to think of w as some physical quantity which varies with the position of the point (x, y, z) (for example, in astrophysics, the density within a star). Quantities which vary with position and time are represented by functions of four variables x, y, z, t, as when a meteorologist deals with the changing distribution of atmospheric pressure, or a designer of microwave ovens is concerned with the temperature distribution in a joint of meat as it cooks.

Here we concentrate on a function of three variables, but using methods which can be applied just as well to more than three. The main result is the natural extension of the two-variable case, $\delta w \approx \dfrac{\partial w}{\partial x}\delta x + \dfrac{\partial w}{\partial y}\delta y + \dfrac{\partial w}{\partial z}\delta z$.

The proof of this, which follows, also gives information about the error in this approximation. The argument is a bit complicated but you should at least try to grasp the main steps, which have been emphasised by the italics.

Let $w = g(x, y, z)$, and let δw be the change in w caused by changes $\delta x, \delta y, \delta z$ in x, y, z. Then:

$$\delta w = g(x + \delta x, y + \delta y, z + \delta z) - g(x, y, z)$$

We have inserted two pairs of extra terms, which cancel.

$$= \{g(x + \delta x, y + \delta y, z + \delta z) - g(x, y + \delta y, z + \delta z)\}$$

$$+ \{g(x, y + \delta y, z + \delta z) - g(x, y, z + \delta z)\}$$

$$+ \{g(x, y, z + \delta z) - g(x, y, z)\}$$

This manipulation has expressed δw as the sum of three differences (in braces { }) in each of which *only one variable changes*. Assuming that *all three partial derivatives exist* in the neighbourhood of $P(x, y, z)$ we can deal with these differences as on page 113:

$$g(x + \delta x, y + \delta y, z + \delta z) - g(x, y + \delta y, z + \delta z) = (g_x(x, y + \delta y, z + \delta z) + \varepsilon_1)\,\delta x$$
where $\varepsilon_1 \to 0$ as $\delta x \to 0$

$$g(x, y + \delta y, z + \delta z) - g(x, y, z + \delta z) = (g_y(x, y, z + \delta z) + \varepsilon_2)\,\delta y$$
where $\varepsilon_2 \to 0$ as $\delta y \to 0$

$$g(x, y, z + \delta z) - g(x, y, z) = (g_z(x, y, z) + \varepsilon_3)\,\delta z$$
where $\varepsilon_3 \to 0$ as $\delta z \to 0$

Notice that g_x and g_y are evaluated at $(x, y + \delta y, z + \delta z)$ and $(x, y, z + \delta z)$ respectively, not at (x, y, z). If we make the further assumption that g_x *and* g_y *are continuous* at (x, y, z) then:

$$g_x(x, y + \delta y, z + \delta z) = g_x(x, y, z) + \eta_1, \quad \text{where } \eta_1 \to 0 \text{ as } \delta y \text{ and } \delta z \to 0$$

$$g_y(x, y, z + \delta z) = g_y(x, y, z) + \eta_2, \quad \text{where } \eta_2 \to 0 \text{ as } \delta z \to 0$$

Putting all this together gives:

$$\delta w = (g_x(x, y, z) + \varepsilon_1 + \eta_1)\,\delta x + (g_y(x, y, z) + \varepsilon_2 + \eta_2)\,\delta y + (g_z(x, y, z) + \varepsilon_3)\,\delta z$$

Since all the partial derivatives are now evaluated at (x, y, z) we can revert to writing them as $\dfrac{\partial w}{\partial x}, \dfrac{\partial w}{\partial y}$ and $\dfrac{\partial w}{\partial z}$. Also we can tidy up all the terms which tend to zero: for if $\delta s = \sqrt{(\delta x)^2 + (\delta y)^2 + (\delta z)^2}$ each of $\delta x, \delta y, \delta z$ must tend to zero as $\delta s \to 0$. Therefore the error terms $(\varepsilon_1 + \eta_1)\,\delta x + (\varepsilon_2 + \eta_2)\,\delta y + \varepsilon_3\,\delta z$ can be replaced by $\varepsilon\,\delta s$, where $\varepsilon \to 0$ as $\delta s \to 0$ (see the activity below). This gives the final version.

$$\delta w = \frac{\partial w}{\partial x}\,\delta x + \frac{\partial w}{\partial y}\,\delta y + \frac{\partial w}{\partial z}\,\delta z + \varepsilon\,\delta s, \quad \text{where } \varepsilon \to 0 \text{ as } \delta s \to 0$$

Activity

Justify the steps in the following argument.

Let $\varepsilon = ((\varepsilon_1 + \eta_1)\,\delta x + (\varepsilon_2 + \eta_2)\,\delta y + \varepsilon_3\,\delta z)/\delta s$.

Then $\quad |\varepsilon| \leqslant |\varepsilon_1 + \eta_1| \left|\dfrac{\delta x}{\delta s}\right| + |\varepsilon_2 + \eta_2| \left|\dfrac{\delta y}{\delta s}\right| + |\varepsilon_3| \left|\dfrac{\delta z}{\delta s}\right|$

$$< |\varepsilon_1 + \eta_1| + |\varepsilon_2 + \eta_2| + |\varepsilon_3|$$

so $\varepsilon \to 0$ as $\delta s \to 0$.

NOTE

(i) The conditions we have used (all partial derivatives exist near P and all but one of these are continuous at P) are sufficient to ensure differentiability at P (see page 115).

(ii) If dx, dy and dz are independent differentials then the total differential of w is defined to be:

$$\mathrm{d}w = \frac{\partial w}{\partial x}\,\mathrm{d}x + \frac{\partial w}{\partial y}\,\mathrm{d}y + \frac{\partial w}{\partial z}\,\mathrm{d}z$$

and we have proved that if the function is differentiable and $\mathrm{d}x = \delta x$, $\mathrm{d}y = \delta y$, $\mathrm{d}z = \delta z$ then $\delta w \approx \mathrm{d}w$ when δs is small.

EXAMPLE

Check this last statement numerically in the case where $w = \dfrac{xy^2}{\sqrt{z}}$, $x = 10$, $y = 3$, $z = 36$, $\delta x = 0.01$, $\delta y = 0.02$ and $\delta z = 0.03$.

Solution

$$\delta w = \frac{10.01 \times 3.02^2}{\sqrt{36.03}} - \frac{10 \times 3^2}{\sqrt{36}} = 0.209\,531$$

Evaluating the partial derivatives at (10, 3, 36):

$$\frac{\partial w}{\partial x} = \frac{y^2}{\sqrt{z}} = \frac{3}{2}, \quad \frac{\partial w}{\partial y} = \frac{2xy}{\sqrt{z}} = 10, \quad \frac{\partial w}{\partial z} = -\frac{1}{2}xy^2 z^{-3/2} = -\frac{5}{24}$$

Therefore $\mathrm{d}w = \frac{3}{2} \times 0.01 + 10 \times 0.02 - \frac{5}{24} \times 0.03 = 0.208\,75$.

So dw gives δw with an error of less than 0.4%.

Approximations and errors

The approximation $\delta w \approx \dfrac{\partial w}{\partial x}\,\delta x + \dfrac{\partial w}{\partial y}\,\delta y + \dfrac{\partial w}{\partial z}\,\delta z$ or its equivalent for two or more than three variables can be used to estimate the effect of errors in a calculation. The availability of powerful calculating aids has reduced the practical need for this, but it still gives a useful way of analysing which errors have the greatest impact.

EXAMPLE

A surveyor calculates the area Δ of a triangular plot by measuring two sides a and b and the included angle C, and then using the formula $\Delta = \frac{1}{2}ab\sin C$. Estimate the percentage error in Δ caused by errors of 1.5% in each of a and b, and an error of $1°$ in C when:

(i) $a = 50\,\text{m}$, $b = 70\,\text{m}$, $C = 75°$ (ii) $a = 50\,\text{m}$, $b = 70\,\text{m}$, $C = 15°$.

Solution

$$\Delta = \tfrac{1}{2}ab\sin C \quad \Rightarrow \quad \delta\Delta \approx \frac{\partial\Delta}{\partial a}\delta a + \frac{\partial\Delta}{\partial b}\delta b + \frac{\partial\Delta}{\partial C}\delta C$$

$$= \tfrac{1}{2}b\sin C\,\delta a + \tfrac{1}{2}a\sin C\,\delta b + \tfrac{1}{2}ab\cos C\,\delta C.$$

Dividing by $\Delta\,(=\tfrac{1}{2}ab\sin C)$ to get the fractional error gives:

$$\frac{\delta\Delta}{\Delta} \approx \frac{\delta a}{a} + \frac{\delta b}{b} + \cot C\,\delta C.$$

The percentage error is $100 \times$ the fractional error, so this shows that the percentage error in Δ is $1.5 + 1.5 + \cot C\,\delta C \times 100$, since the errors in a and b are each 1.5%. (Notice that the actual values of a and b do not appear in this.) Since we have used $\cos C$ as the derivative of $\sin C$ the angle C is here measured in radians, and so $\delta C = \pi/180 \approx 0.0175$. Therefore the percentage error in Δ is approximately $3 + 1.75\cot C$. For the given values this gives:

(i) $3 + 1.75\cot 75° \approx 3.47$ and (ii) $3 + 1.75\cot 15° \approx 9.53$.

Since $\cot C$ is large when C is small this shows that the surveyor needs to be particularly careful to measure C accurately when the corner of the plot being used is very sharp – it would be better to use another corner!

Activity

Estimate the actual errors in the calculated value of Δ for the same values (i) and (ii) in this example.

Exercise 3H

1. Given that $w = kx^p y^q z^r$, where k, p, q, r are positive constants, prove that an error of $\pm1\%$ in each of x, y, z gives a maximum error of approximately $(p + q + r)\%$ in w. What is the corresponding result if q and r are negative?

2. The temperature T at a point (x, y, z) in a rectangular Cartesian coordinate system is given by $T = 10(9x^2 + 10y^2 + 6z^2)^{1/2}$. Find the temperature at the point $(2.00, 1.00, 3.00)$ and use partial differentiation to find an approximation to the difference in temperature δT between this point and the nearby point $(2.03, 0.94, 3.15)$. **[MEI]**

3. A cuboid has volume V and surface area S. Show:

(i) by using partial differentiation

(ii) by a geometrical argument

that if the length of each side increases by the small amount ε then $\delta V \approx \tfrac{1}{2}S\varepsilon$. What is the corresponding result for a sphere?

Exercise 3H continued

4. In a standard physics experiment the value of g (the acceleration due to gravity) is found by measuring the time of swing T of a compound pendulum and using the formula $g = \dfrac{4\pi^2 \ell}{T^2}$. In this formula the length ℓ is given by $\ell = s + \dfrac{k^2}{s}$, where s and k are lengths measured from the pendulum.

If the measurements of s and k are accurate to within 1% find the approximate maximum percentage error in ℓ. If the measurement of T is accurate to within 0.5% find the approximate maximum percentage error in the computed value of g.

5. The following two methods are suggested for calculating the volume V of a right circular cone.

(A) Measure the base radius r and the semi-vertical angle α.

(B) Measure the base radius r and the generator length ℓ.

(i) Find the appropriate formula for V for each method.

(ii) Find an approximate formula for $\dfrac{\delta V}{V}$ in terms of the measured quantities and their errors for each method.

(iii) Explain why method (A) is particularly suitable when $\alpha \approx \dfrac{\pi}{4}$ and method (B) is particularly suitable when $\alpha \approx \arcsin\sqrt{\tfrac{2}{3}}$.

6. The length of the side a of an obtuse-angled triangle is to be calculated using the cosine rule:

$a^2 = b^2 + c^2 - 2bc \cos A$, where A is the obtuse angle.

(i) Find $\dfrac{\partial(a^2)}{\partial b}$, $\dfrac{\partial(a^2)}{\partial c}$ and $\dfrac{\partial(a^2)}{\partial A}$.

(ii) Write down an equation involving the increment in a^2 relating to increments δb in b, δc in c and δA in A.

(iii) Calculate an approximation for the maximum error in a, using a method involving partial differentiation, when the measurements are given as $b = 25 \pm 1.0\,\text{cm}$, $c = 30 \pm 1.5\,\text{cm}$, $A = 120 \pm 2°$.

(iv) The area Δ of the triangle is calculated using the formula $\Delta = \tfrac{1}{2}bc \sin A$. Use methods of partial differentiation to estimate the maximum percentage error in Δ with the measurements given in part (iii). [MEI]

7. The side b of a triangle is found from measurements of the side a and the angles B and C. There may be an error not exceeding h (in either direction) in the measurements of a, and an error not exceeding α in either or both of B and C, where h and α are small. Show that, if $B + C < \dfrac{\pi}{2}$, the greatest error in b is approximately $b\left(\dfrac{h}{a} + \alpha \cot B\right)$. Find the greatest error when $B + C > \dfrac{\pi}{2}$, and explain the difference in the two cases.

The vector grad g

EXAMPLE

The temperature T at the point (x, y, z) inside a solid is given by $T = x^2 + 3xy + 2z^2$. Find:

(i) the average temperature gradient from A$(2, 1, 3)$ to B$(4, 2, 1)$

(ii) the actual temperature gradient at A in the direction towards B.

Solution

(i) Since $T = 28$ at $(2, 1, 3)$ and $T = 42$ at $(4, 2, 1)$, the change in T from A to B is 14. The distance $AB = \sqrt{2^2 + 1^2 + (-2)^2} = 3$, and so the average temperature gradient is $\frac{14}{3}$.

(ii) Since $\mathbf{AB} = \begin{pmatrix} 2 \\ 1 \\ -2 \end{pmatrix}$ and $|\mathbf{AB}| = 3$, the vector of length δs in the direction from A to B is $\delta \mathbf{r} = \begin{pmatrix} 2\delta s/3 \\ \delta s/3 \\ -2\delta s/3 \end{pmatrix}$.

At A the partial derivatives of T are $\dfrac{\partial T}{\partial x} = 2x + 3y = 7$, $\dfrac{\partial T}{\partial y} = 3x = 6$, $\dfrac{\partial T}{\partial z} = 4z = 12$.

Putting all this into $\delta T = \dfrac{\partial T}{\partial x} \delta x + \dfrac{\partial T}{\partial y} \delta y + \dfrac{\partial T}{\partial z} \delta z + \varepsilon \delta s$ gives:

$$\delta T = 7 \times \frac{2\delta s}{3} + 6 \times \frac{\delta s}{3} + 12 \times \left(-\frac{2\delta s}{3} \right) + \varepsilon \, \delta s$$

$$= \left(-\frac{4}{3} + \varepsilon \right) \delta s$$

where $\varepsilon \to 0$ as $\delta s \to 0$.

The limit of $\dfrac{\delta T}{\delta s}$ as $\delta s \to 0$ is $-\frac{4}{3}$, which is the required temperature gradient.

The argument in part (ii) of this solution is typical of the general case.

Let $w = g(x, y, z)$ and let $\delta \mathbf{r} = \begin{pmatrix} \delta x \\ \delta y \\ \delta z \end{pmatrix}$ be a vector of length $|\delta \mathbf{r}| = \delta s$ in a given fixed direction. Then:

$$\delta w = \frac{\partial w}{\partial x} \delta x + \frac{\partial w}{\partial y} \delta y + \frac{\partial w}{\partial z} \delta z + \varepsilon \, \delta s$$

$$\Rightarrow \quad \frac{\delta w}{\delta s} = \frac{\partial w}{\partial x} \frac{\delta x}{\delta s} + \frac{\partial w}{\partial y} \frac{\delta y}{\delta s} + \frac{\partial w}{\partial z} \frac{\delta z}{\delta s} + \varepsilon \qquad \text{①}$$

where $\varepsilon \to 0$ as $\delta s \to 0$.

But $\dfrac{\delta x}{\delta s}, \dfrac{\delta y}{\delta s}$ and $\dfrac{\delta z}{\delta s}$ are the components of the *unit* vector in the given direction, $\hat{\mathbf{u}}$ say.

If we now introduce the vector $\mathbf{grad}\, w = \begin{pmatrix} \partial w/\partial x \\ \partial w/\partial y \\ \partial w/\partial z \end{pmatrix}$ we can write ① as:

$$\frac{\delta w}{\delta s} = \hat{\mathbf{u}} \,.\, \mathbf{grad}\, w + \varepsilon$$

Taking the limit as $\delta s \to 0$ gives the *directional derivative* in the direction of $\hat{\mathbf{u}}$:

$$\frac{dw}{ds} = \hat{\mathbf{u}} \cdot \mathbf{grad}\, w.$$

This corresponds directly with the form of the directional derivative for a function of two variables given on page 119. The only difference is a small change of notation: there we used $\mathbf{grad}\, f$ (rather than $\mathbf{grad}\, z$), and here $\mathbf{grad}\, w$ (rather than $\mathbf{grad}\, g$). To emphasise the common form of the result whatever the number of independent variables, and because it will be more convenient in the next section, from now on we shall refer to the directional derivative in the form

$$\hat{\mathbf{u}} \cdot \mathbf{grad}\, g, \text{ where } \mathbf{grad}\, g = \begin{pmatrix} \partial g/\partial x \\ \partial g/\partial y \\ \partial g/\partial z \end{pmatrix}.$$

As before, $\mathbf{grad}\, g$ may also be written as ∇g.

The greatest possible value of the directional derivative is $|\mathbf{grad}\, g|$; this occurs when $\hat{\mathbf{u}}$ is in the direction of $\mathbf{grad}\, g$. Therefore we can interpret $\mathbf{grad}\, g$ as the vector of which the magnitude and direction give the greatest rate of change of w and the direction in which this occurs.

EXAMPLE

The density w inside a gas cloud is given by $w = \dfrac{88}{\sqrt{x^2 + y^2 + 2z^2}}$. The point A has coordinates $(8, 5, 4)$. Find, at A:

(i)   the density

(ii) the greatest rate of change of density with distance, and the direction in which this occurs

(iii) the rate of change of density with distance in the direction from A to the origin O.

Solution

(i)   For brevity, let $R = \sqrt{x^2 + y^2 + 2z^2}$; then $w = \dfrac{88}{R}$.

At A, $R = \sqrt{64 + 25 + 32} = 11$, and so $w = 8$.

(ii) $\dfrac{\partial w}{\partial x} = -88x(x^2 + y^2 + 2z^2)^{-3/2} = -\dfrac{88x}{R^3}$.

Similarly $\dfrac{\partial w}{\partial y} = -\dfrac{88y}{R^3}$ and $\dfrac{\partial w}{\partial z} = -\dfrac{176z}{R^3}$.

Therefore, at A:

$$\frac{\partial w}{\partial x} = -\frac{64}{121}, \frac{\partial w}{\partial y} = -\frac{40}{121}, \frac{\partial w}{\partial z} = -\frac{64}{121}$$

So $\mathbf{grad}\, w = \begin{pmatrix} -\frac{64}{121} \\ -\frac{40}{121} \\ -\frac{64}{121} \end{pmatrix}$ and $|\mathbf{grad}\, w| = \dfrac{\sqrt{64^2 + 40^2 + 64^2}}{121} = \dfrac{\sqrt{9792}}{121} \approx 0.818.$

Therefore the greatest rate of change of w is 0.818, occurring in the direction

of $\begin{pmatrix} -\frac{64}{121} \\ -\frac{40}{121} \\ -\frac{64}{121} \end{pmatrix}$ or, more simply, $\begin{pmatrix} -8 \\ -5 \\ -8 \end{pmatrix}$.

(iii) $\mathbf{AO} = \begin{pmatrix} -8 \\ -5 \\ -4 \end{pmatrix}$ and $|\mathbf{AO}| = \sqrt{64 + 25 + 16} = \sqrt{105}$, so the unit vector in the

direction from A to O is $\hat{\mathbf{u}} = \dfrac{1}{\sqrt{105}} \begin{pmatrix} -8 \\ -5 \\ -4 \end{pmatrix}$.

The rate of change of density in this direction is:

$$\hat{\mathbf{u}} \cdot \operatorname{grad} w = \frac{1}{\sqrt{105}} \times \frac{1}{121}(8 \times 64 + 5 \times 40 + 4 \times 64) = \frac{968}{121\sqrt{105}} \approx 0.781.$$

Exercise 3I

1. Find the directional derivative of $g(x, y, z)$ at the point P in the direction of $\mathbf{u}$.

   (i) $g(x, y, z) = x^2 - 9y^2 + 4z^2, P = (2, 1, 0),$

   $\mathbf{u} = \mathbf{i} + \mathbf{j} + \mathbf{k}$

   (ii) $g(x, y, z) = \dfrac{xy}{z}, P = (3, 1, -1),$

   $\mathbf{u} = \mathbf{i} - 2\mathbf{j} + 2\mathbf{k}$

   (iii) $g(x, y, z) = e^x \cos y, P = (\ln 3, \pi, 7),$

   $\mathbf{u} = 3\mathbf{i} + 4\mathbf{k}$

2. Find a function $g(x, y, z)$ for which:

   (i) $\operatorname{grad} g = (y + z)\mathbf{i} + (z + x)\mathbf{j} + (x + y)\mathbf{k}$

   (ii) $\operatorname{grad} g = (x\mathbf{i} + y\mathbf{j} + z\mathbf{k})/(x^2 + y^2 + z^2).$

3. Let $\mathbf{r} = x\mathbf{i} + y\mathbf{j} + z\mathbf{k}$. Prove that:

   (i) $w = |\mathbf{r}| \quad \Rightarrow \quad \operatorname{grad} w = \hat{\mathbf{r}}$

   (ii) $w = |\mathbf{r}|^n \quad \Rightarrow \quad \operatorname{grad} w = n|\mathbf{r}|^{n-1}\hat{\mathbf{r}}.$

4. In what directions at a point P does the directional derivative equal $\frac{1}{2}|\operatorname{grad} g|$?

5. The concentration S of sugar in a liquid is given by $S = x^3 + y^2(4 + z^2)$. A sweet-toothed bug is placed in the liquid at $(2, 3, 1)$ and swims in the direction in which the concentration increases fastest.

   (i) Find the unit vector in this direction.

   (ii) If x, y, z are measured in centimetres and the bug swims in this direction at $0.5\,\mathrm{cm\,s^{-1}}$, at what rate does the concentration increase?

6. A space probe is moving along a spiral so that at time t its position vector $\mathbf{r}$ is given by $\mathbf{r} = \cos t\mathbf{i} + \sin t\mathbf{j} + t\mathbf{k}$. It passes a gaseous star centred at $(-6, 0, 0)$ for which the distribution of temperature T is given by $T = \dfrac{1000}{(x + 6)^2 + y^2 + z^2}$.

   (i) Find, in terms of t, the unit vector in the direction of motion of the probe.

   (ii) Hence find the directional derivative of T in terms of t.

   (iii) Show that there are three points along its path at which the temperature experienced by the probe has a stationary value.

   (iv) Find, correct to three significant figures, the global maximum temperature experienced by the probe.

The surface $g(x, y, z) = k$

If $w = g(x, y, z)$ and k is a constant then the set of points (x, y, z) for which $w = k$ forms a surface with implicit equation $g(x, y, z) = k$. Such a surface is the three-dimensional equivalent of a contour or level curve $f(x, y) = c$ of a function of two variables. It is sometimes called a *contour surface* or *level surface* (though in this case 'level' does not mean 'horizontal' or even 'plane'). For example, if $w = x^2 + y^2 + z^2$ then each level surface is a sphere centred at the origin, with equation $x^2 + y^2 + z^2 = k$ (for $k \geqslant 0$).

Suppose that $A(a, b, c)$ is a point of the surface $w = k$, and that $\hat{u}$ is any unit vector tangential to this surface at A. Since w remains constant in the surface, the directional derivative in the direction of $\hat{u}$ is zero. Therefore $\hat{u} \cdot \mathbf{grad}\, g = 0$, and so $\mathbf{grad}\, g$ is perpendicular to $\hat{u}$. Thus $\mathbf{grad}\, g$ is perpendicular to all vectors in the tangent plane at A; in other words, $\mathbf{grad}\, g$ is a normal vector for the tangent plane at A. This makes it very easy to find the equations of the normal line and tangent plane at A:

- the normal line is the line $\mathbf{r} = \mathbf{a} + \lambda\, \mathbf{grad}\, g$
- the tangent plane is the plane $(\mathbf{r} - \mathbf{a}) \cdot \mathbf{grad}\, g = 0$.

EXAMPLE

Show that $A(4, 2, 3)$ is on the surface $2x^2 - 3yz + 4z^2 = 50$, and find the equations of the normal line and the tangent plane at A.

Solution

Let $g(x, y, z) = 2x^2 - 3yz + 4z^2$. Then:

$$g(4, 2, 3) = 2 \times 4^2 - 3 \times 2 \times 3 + 4 \times 3^2 = 32 - 18 + 36 = 50$$

so A is on the surface.

At A, $\dfrac{\partial g}{\partial x} = 4x = 16$, $\dfrac{\partial g}{\partial y} = -3z = -9$, $\dfrac{\partial g}{\partial z} = -3y + 8z = 18$ so $\mathbf{grad}\, g = \begin{pmatrix} 16 \\ -9 \\ 18 \end{pmatrix}$.

The normal line at A is $\begin{pmatrix} x \\ y \\ z \end{pmatrix} = \begin{pmatrix} 4 \\ 2 \\ 3 \end{pmatrix} + \lambda \begin{pmatrix} 16 \\ -9 \\ 18 \end{pmatrix}$.

The tangent plane at A is $\begin{pmatrix} x - 4 \\ y - 2 \\ z - 3 \end{pmatrix} \cdot \begin{pmatrix} 16 \\ -9 \\ 18 \end{pmatrix} = 0$

$$\Leftrightarrow \quad 16(x - 4) - 9(y - 2) + 18(z - 3) = 0$$

$$\Leftrightarrow \quad 16x - 9y + 18z = 100.$$

The equation of the tangent plane can be expressed in another way, using the differentials dx, dy, dz for the steps $x - a$, $y - b$, $z - c$ needed to move from A to the general point (x, y, z) on the tangent plane at A, so that $d\mathbf{r} = \begin{pmatrix} dx \\ dy \\ dz \end{pmatrix}$.

Then the tangent plane is $d\mathbf{r} \cdot \mathbf{grad}\, g = 0$.

The similar version given on page 116 is a special case of this, since the surface $z = f(x, y)$ can be thought of as the level surface $g(x, y, z) = 0$, where $g(x, y, z) = f(x, y) - z$. So:

$$\mathbf{grad}\, g = \begin{pmatrix} \partial f/\partial x \\ \partial f/\partial y \\ -1 \end{pmatrix}.$$

HISTORICAL NOTE   *The ideas of level surface and gradient vector have many applications in physics, in particular in connection with the idea of potential, a scalar quantity which is a function of position. The theory of gravitational potential started with Newton and was developed in the 18th century by Euler, Clairaut, Lagrange, Laplace and Monge (who emphasised the geometrical aspects). George Green (1793–1841, a self-educated son of a miller) used the notion of the potential function in the mathematical treatment of electricity and magnetism, and the experiments of Michael Faraday (1791–1867, a self-educated son of a blacksmith) led to the concept of lines of force (normal to equipotential surfaces). James Clerk Maxwell (1831–71) expressed these ideas in mathematical form in 1864, and from his equations predicted the existence of electro-magnetic waves. These were first actually produced in 1887 by Heinrich Hertz (1857–94), whose experiments led to the present vastly important technology of radio transmission.*

Exercise 3J

1. Find the equation of the tangent plane to the ellipsoid $3x^2 + 5y^2 + z^2 = 39$ at the point $(1, -2, 4)$.

2. Find the equations of the normal line and tangent plane to the surface $x^2 - \dfrac{y}{z^2} = 10$ at:

   (i) $(3, -1, 1)$ (ii) $(-4, 24, -2)$.

3. Find the coordinates of the point(s) on the surface $3x^2 - 2y^2 - z^2 = 4$ at which the tangent plane is parallel to $18x - 14y - 3z = 0$.

4. Prove that the equation of the tangent plane at the point (x_1, y_1, z_1) on the ellipsoid

   $$\frac{x^2}{a^2} + \frac{y^2}{b^2} + \frac{z^2}{c^2} = 1 \text{ can be written in the form}$$

   $$\frac{xx_1}{a^2} + \frac{yy_1}{b^2} + \frac{zz_1}{c^2} = 1.$$

5. (i) Show that
   $$x^2 + y^2 + z^2 + 2ax + 2by + 2cz + d = 0$$
   is the Cartesian equation of a sphere provided that $d < a^2 + b^2 + c^2$, and say what the equation represents when:

   (a) $d = a^2 + b^2 + c^2$ (b) $d > a^2 + b^2 + c^2$.

   (ii) In the case of a sphere, prove that the normal line at every point of the sphere passes through the centre.

6. Prove that the sum of the squares of the intercepts on the x-, y- and z-axes of every tangent plane to the surface $x^{2/3} + y^{2/3} + z^{2/3} = a^{2/3}$ is a constant.

7. Find the acute angle at which the surface $xy^2 + xz^2 = 10$ cuts the surface $z^3 - x - x^2y = 2$ at the point $(2, 1, 2)$. Find also the direction of the curve of intersection at this point.

Exercise 3J continued

8. (i) Given that
$$g(x, y, z) = 3x^2 - 2xy + 2y^2 + z^2 + 4z - 31,$$
find $\dfrac{\partial g}{\partial x}, \dfrac{\partial g}{\partial y}$ and $\dfrac{\partial g}{\partial z}$.

A surface S has equation
$$3x^2 - 2xy + 2y^2 + z^2 + 4z - 31 = 0.$$

(ii) Find the equation of the normal line to S at the point $P(2, 1, 3)$.

(iii) This normal line meets the surface again at the point Q. Find the coordinates of Q.

(iv) Find the two values of k for which $x + z = k$ is a tangent plane to the surface S.   [MEI]

9. The hyperboloid of one sheet

(i) Given that $g(x, y, z) = \dfrac{x^2}{a^2} + \dfrac{y^2}{b^2} - \dfrac{z^2}{c^2}$, show that all the horizontal contours of the surface $g(x, y, z) = 1$ are ellipses, and that all the vertical sections parallel to the planes $x = 0$ or $y = 0$ are hyperbolas. Sketch the surface, which is called a *hyperboloid of one sheet*.

(ii) Show that the equation $g(x, y, z) = 1$ can be written as
$$\left(\frac{y}{b} - \frac{z}{c}\right)\left(\frac{y}{b} + \frac{z}{c}\right) = \left(1 - \frac{x}{a}\right)\left(1 + \frac{x}{a}\right).$$
By putting $\dfrac{\frac{y}{b} - \frac{z}{c}}{1 - \frac{x}{a}} = u$ show that the hyperboloid contains the line of intersection of the two planes
$$\frac{y}{b} - \frac{z}{c} = u\left(1 - \frac{x}{a}\right) \text{ and } u\left(\frac{y}{b} + \frac{z}{c}\right) = 1 + \frac{x}{a}.$$

(As u varies this gives a family of lines (called *generators*) which lie entirely in the surface, and so this is another example of a *ruled surface* (see Question 8 in Exercise 3B).)

(iii) Show similarly that the surface contains a second family of generators, namely the lines of intersection of the pairs of planes
$$\frac{y}{b} + \frac{z}{c} = v\left(1 - \frac{x}{a}\right) \text{ and } v\left(\frac{y}{b} - \frac{z}{c}\right) = 1 + \frac{x}{a}.$$

(iv) Show that the generator in (ii) has
$$\text{direction vector } \mathbf{d} = \begin{pmatrix} 2u/bc \\ (1 - u^2)/ca \\ (1 + u^2)/ab \end{pmatrix}$$
and find, in a similar form, a direction vector $\mathbf{e}$ for the generator in (iii).

(v) Find $\mathbf{d} \times \mathbf{e}$, and show that this is parallel to
$$\begin{pmatrix} (uv - 1)/a \\ (u + v)/b \\ (u - v)/c \end{pmatrix}.$$
Why is this vector normal to the surface at the point where these generators cross?

(vi) By comparing the normal vector in (v) with $\mathbf{grad}\, g$, obtain parametric equations
$$x = a\frac{uv - 1}{uv + 1}, \ y = b\frac{u + v}{uv + 1}, \ z = c\frac{v - u}{uv + 1} \text{ for}$$
this hyperboloid.

(*Hyperbolic cog wheels* make use of this to transmit rotation about one given axis into rotation about another given axis by having two hyperboloids of one sheet which mesh along their generators (figure 3.9).)

Figure 3.9

Investigation

The multivariable chain rule

The diagram shows part of the surface $z = f(x, y)$. Suppose that, instead of being independent, the variables x and y are given in terms of a variable u by $x = p(u)$, $y = q(u)$. Then as u varies the point $P'(x, y, 0)$ moves along the curve $\mathscr{C}'$ in the $z = 0$ plane with parametric equations $x = p(u)$, $y = q(u)$, and the point $P(x, y, z)$ moves along the three-dimensional curve $\mathscr{C}$ in which the verticals through $\mathscr{C}'$ meet the surface. (Imagine a pastry slice in the shape of $\mathscr{C}'$ cutting the surface in $\mathscr{C}$.) Thus z depends only on u, so we may write $z = F(u)$, where $F(u) = f(p(u), q(u))$.

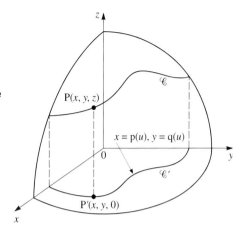

(i) Find $F(u)$ and $F'(u)$ when:

(a) $f(x, y) = x^2 - y^2$, $x = u^3 + 1$, $y = u^3 - 1$

(b) $f(x, y) = (x + y)\, e^y$, $x = \dfrac{1}{u}$, $y = \ln u$

(c) $f(x, y) = \arctan\left(\dfrac{x}{y}\right)$, $x = 2u$, $y = 1 - u^2$.

(ii) It is not necessary to find $F(u)$ explicitly in order to find $F'(u)$. Suppose that changing u by δu causes changes δx and δy in x and y, which in turn cause z to change by δz. Starting with $\delta z = \dfrac{\partial z}{\partial x}\delta x + \dfrac{\partial z}{\partial y}\delta y + \varepsilon \delta s$ (as on page 127),

prove that $\dfrac{dz}{du} = \dfrac{\partial z}{\partial x}\dfrac{dx}{du} + \dfrac{\partial z}{\partial y}\dfrac{dy}{du}$; this is the multivariable chain rule.

(iii) Use the chain rule to obtain $F'(u)$ again for the functions in (i).

(iv) The annual wheat production in a certain location, W units, depends on the mean annual temperature, $T\,°C$, and the mean annual rainfall, $R\,cm$. Agricultural experiments have shown that W decreases by 4.8 units for each $1\,°C$ rise in temperature, and increases by 1.2 units for each $1\,cm$ increase in rainfall. Climatologists judge that, due to global warming, T is increasing by $0.07\,°C$ per year and R is decreasing by $0.16\,cm$ per year. Estimate the change in W from this year to next. How is your calculation related to the multivariable chain rule?

(v) Suppose now that w depends on three variables, x, y, z, each of which depends on two variables, u and v. So $w = f(x, y, z) = F(u, v)$ where $x = p(u, v)$, $y = q(u, v)$, $z = r(u, v)$. Investigate the chain rules for $\dfrac{\partial w}{\partial u}$ and $\dfrac{\partial w}{\partial v}$, and show that they can be written in the form $\mathbf{grad}\, F = \mathbf{M}\,\mathbf{grad}\, f$,

where $\mathbf{M} = \begin{pmatrix} \dfrac{\partial x}{\partial u} & \dfrac{\partial y}{\partial u} & \dfrac{\partial z}{\partial u} \\[2mm] \dfrac{\partial x}{\partial v} & \dfrac{\partial y}{\partial v} & \dfrac{\partial z}{\partial v} \end{pmatrix}$.

KEY POINTS

- Contours $z = c$ and vertical sections $x = a$ or $y = b$ can be used to give information about the surface $z = f(x, y)$.

- The *partial derivative* $\dfrac{\partial z}{\partial x} \left(= \dfrac{\partial f}{\partial x} = f_x(x, y) \right)$ is found by differentiating with respect to x, keeping y constant; this gives the gradient of a vertical section parallel to the x-axis. Similarly $\dfrac{\partial z}{\partial y} \left(= \dfrac{\partial f}{\partial y} = f_y(x, y) \right)$ gives the gradient of a vertical section parallel to the y-axis.

- If the surface $z = f(x, y)$ has a tangent plane at the point (a, b, c) then its equation is $z - c = \dfrac{\partial f}{\partial x}(x - a) + \dfrac{\partial f}{\partial y}(y - b)$, where $\dfrac{\partial f}{\partial x}$ and $\dfrac{\partial f}{\partial y}$ are evaluated at the point (a, b, c).

- The vector $\mathbf{grad}\, f = \begin{pmatrix} \partial f / \partial x \\ \partial f / \partial y \end{pmatrix}$ evaluated at the point A on the surface $z = f(x, y)$ is normal to the contour through A.

- The *directional derivative* $\hat{\mathbf{u}} \cdot \mathbf{grad}\, f$ is the gradient of the surface in the direction of the unit vector $\hat{\mathbf{u}}$.

- The gradient of the implicit function $f(x, y) = c$ is given by
$$\frac{dy}{dx} = -\frac{\dfrac{\partial f}{\partial x}}{\dfrac{\partial f}{\partial y}}.$$

- The stationary points on $z = f(x, y)$ are found by solving $\mathbf{grad}\, f = \mathbf{0}$, i.e. solving $\dfrac{\partial z}{\partial x} = 0$ and $\dfrac{\partial z}{\partial y} = 0$ simultaneously. The nature of the stationary point (maximum, minimum or saddle point) can be found by comparing $f(a + h, b + k)$ with $f(a, b)$.

- The approximation $\delta w \approx \dfrac{\partial w}{\partial x} \delta x + \dfrac{\partial w}{\partial y} \delta y + \dfrac{\partial w}{\partial z} \delta z$ or its equivalent for two or more than three variables can be used to estimate the effect of errors in a calculation.

- The vector $\mathbf{grad}\, g = \begin{pmatrix} \partial g / \partial x \\ \partial g / \partial y \\ \partial g / \partial z \end{pmatrix}$ gives the magnitude and direction of the greatest rate of change of $w = g(x, y, z)$. The directional derivative in the direction of the unit vector $\hat{\mathbf{u}}$ is $\hat{\mathbf{u}} \cdot \mathbf{grad}\, g$.

- For the point A with position vector $\mathbf{a}$ on the surface $g(x, y, z) = k$ the normal line is $\mathbf{r} = \mathbf{a} + \lambda \mathbf{grad}\, g$ and the tangent plane is $(\mathbf{r} - \mathbf{a}) \cdot \mathbf{grad}\, g = 0$, where $\mathbf{grad}\, g$ is evaluated at the point A.

4 Differential geometry

There is nothing in the world except empty, curved space. Matter, charge, electromagnetism and other fields are only manifestations of the curvature of space.

John Archibald Wheeler, 1911–

Envelopes

Activity

(a) Stick a pin into a piece of paper. Place one edge of a ruler against the pin and draw a short, straight line, using the other edge of the ruler, as in diagram (a). Do this repeatedly, turning the ruler slightly around the pin each time. (A firmly held pen or pencil will do instead of a pin.)

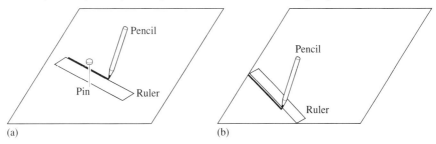

(a)   (b)

(b) Place a short straight edge (e.g. a 15 cm ruler or the edge of a set square) so that its ends are on two perpendicular edges of a piece of paper, and use it to rule a straight line, as in diagram (b). Do this repeatedly, moving the straight edge slightly each time but keeping its ends on these edges of the paper.

Although all you have done in the activity is draw a set of straight lines, if these are close enough together your eye will also see the curve (not actually drawn) which all the lines touch. In (a) this curve is the circle with centre at the pin and radius the width of the ruler. In (b) it is not obvious what curve is produced: it will be identified later.

From plotting points on a graph or using a graphics calculator, you are used to thinking of a curve as a set of points, or as the path of a moving point. This activity shows that a curve can also be produced by a set of lines, or by a moving line. The set of lines is said to form a *family*, and the curve which every member of the family touches is called the *envelope* of the family. (To *envelop* means to 'wrap around'.)

The family may consist of curves rather than straight lines, and the envelope may be more than one curve, or only part of a curve.

EXAMPLE

Describe the envelope of the family of circles:

$$(x - 3\cos\theta)^2 + (y - 3\sin\theta)^2 = 1.$$

Solution

The given circle has radius 1, and its centre $(3\cos\theta, 3\sin\theta)$ lies on the circle with centre O and radius 3 (see the left-hand diagram). It is clear from the right-hand diagram that the envelope consists of the two circles with centre O and radii 2 and 4.

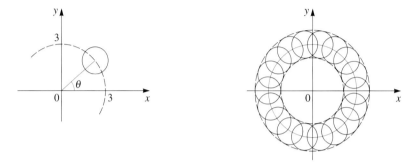

In this example the family of curves can be written in the form $f(x, y, \theta) = 0$, where $f(x, y, \theta) = (x - 3\cos\theta)^2 + (y - 3\sin\theta)^2 - 1$ is a function of the two coordinates x, y and the parameter θ which defines the particular member of the family.

The first step in finding the envelope of a family of curves is to write the equation of the curve similarly, in the form $f(x, y, p) = 0$, where p is a parameter. Thus in part (a) of the activity above, if the width of the ruler is 2.5 cm and the angle ϕ is as in figure 4.1, then the equation of the line is:

$$x\cos\phi + y\sin\phi = 2.5$$

and so the family is $f(x, y, \phi) = x\cos\phi + y\sin\phi - 2.5 = 0$.

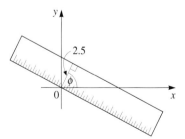

Figure 4.1

Activity

In part (b) of the activity on page 139, take the length of the straight edge to be 15 cm and the edges of the page as coordinate axes. Find the equation of the family:

(i) as $f_1(x, y, \theta) = 0$, where the parameter θ is the angle between the line and the x-axis

(ii) as $f_2(x, y, L) = 0$, where the parameter L is the distance from the origin to the point where the line meets the x-axis.

Figure 4.2 shows some members of the family $f(x, y, p) = 0$ and its envelope. The neighbouring members with equations $f(x, y, p) = 0$ and $f(x, y, p + \delta p) = 0$ meet at A, and touch the envelope at P and Q respectively.

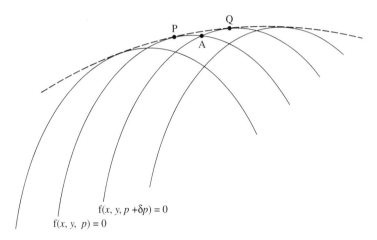

$f(x, y, p + \delta p) = 0$

$f(x, y, p) = 0$

Figure 4.2

The diagram suggests that if these members of the family are close together then A is close to P, or more precisely that P is the limiting position of A as $\delta p \to 0$. This is true for families of 'reasonably smooth' curves (including all those in the following exercise), but it is beyond the scope of this book to define this exactly* and then prove it. So in what follows we shall have to be content to *assume* this basic fact.

The coordinates of A satisfy both $f(x, y, p) = 0$ and $f(x, y, p + \delta p) = 0$, and so they also satisfy:

$$f(x, y, p + \delta p) - f(x, y, p) = 0$$

and therefore
$$\frac{f(x, y, p + \delta p) - f(x, y, p)}{\delta p} = 0.$$

Therefore the coordinates of P (the limiting position of A as $\delta p \to 0$) satisfy:

$$\lim_{\delta p \to 0} \frac{f(x, y, p + \delta p) - f(x, y, p)}{\delta p} = 0$$

i.e.
$$\frac{\partial}{\partial p} f(x, y, p) = 0$$

> The partial derivative with respect to p, as on page 110.

* A good discussion of all this is given in 'What is an envelope?' by J. W. Bruce and P. J. Giblin in The Mathematical Gazette, October 1981, pages 186–192.

So the coordinates of P satisfy both $f(x, y, p) = 0$ and $\dfrac{\partial}{\partial p} f(x, y, p) = 0$.

The equation of the envelope is found by elimination from these two equations. Sometimes it is easy to eliminate p to give the Cartesian equation of the envelope; in other cases it is better to get parametric equations expressing x and y in terms of p.

EXAMPLE

Find the envelope of a line of fixed length a moving with its end points on the coordinate axes (as in part (b) of the activity on page 139).

Solution

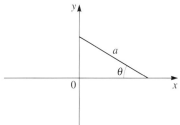

With the notation as in the diagram, the equation of the family is

$$f(x, y, \theta) = x \sec \theta + y \cosec \theta - a = 0.$$

$$\frac{\partial f}{\partial \theta} = x \sec \theta \tan \theta - y \cosec \theta \cot \theta = 0$$

$$\Rightarrow \quad x \sec \theta \tan \theta = y \cosec \theta \cot \theta$$

$$\Rightarrow \quad \frac{x \sin \theta}{\cos^2 \theta} = \frac{y \cos \theta}{\sin^2 \theta}$$

$$\Rightarrow \quad \frac{x}{\cos^3 \theta} = \frac{y}{\sin^3 \theta}$$

so that if $x = \lambda \cos^3 \theta$ then $y = \lambda \sin^3 \theta$.

Substituting these in $f(x, y, \theta) = 0$ gives:

$$\lambda \cos^2 \theta + \lambda \sin^2 \theta - a = 0$$

so that $\lambda = a$ and the parametric equations are $x = a \cos^3 \theta$, $y = a \sin^3 \theta$.

Since $\cos^2 \theta + \sin^2 \theta = 1$ we can eliminate θ to reach the Cartesian equation:

$$x^{2/3} + y^{2/3} = a^{2/3}.$$

The complete envelope (allowing the line to move in each quadrant) is a four-pointed star shape called an *astroid*, as in this diagram.

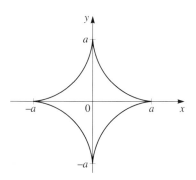

For Discussion

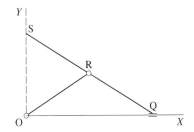

The diagram shows the plan of a door found (in pairs) in many buses. The point Q moves in the groove OX at the top of the door and the point R is joined to the fixed point O by a horizontal link OR which is free to rotate at O and R (there is a similar groove and link at the bottom of the door). The lengths OR, SR and RQ are all equal. When the door is shut QRS lies along OX, with S at O. Explain why the locus of S as the door opens is the straight line OY perpendicular to OX. Deduce that the door sweeps out one quarter of an astroid as it opens.

EXAMPLE Find the envelope of the family of circles which pass through the origin and have their centres on the rectangular hyperbola $xy = c^2$.

Solution

Let the centre be at $\left(ct, \dfrac{c}{t}\right)$. Then the equation of the circle is:

$$(x - ct)^2 + \left(y - \frac{c}{t}\right)^2 = (ct)^2 + \left(\frac{c}{t}\right)^2$$

$$\Leftrightarrow \quad x^2 - 2xct + y^2 - \frac{2yc}{t} = 0$$

> Multiplying throughout by t and rearranging.

so that the equation of the family of circles is:

$$f(x, y, t) = 2xct^2 - (x^2 + y^2)t + 2yc = 0.$$

$$\frac{\partial f}{\partial t} = 4xct - (x^2 + y^2) = 0$$

$$\Rightarrow \quad t = \frac{x^2 + y^2}{4cx}$$

unless $x = 0$, in which case $y = 0$ also.

Substituting for t in $f(x, y, t) = 0$ gives the equation of the envelope:

$$2cx \frac{(x^2 + y^2)^2}{16c^2x^2} - \frac{(x^2 + y^2)^2}{4cx} + 2cy = 0$$

$$\Leftrightarrow \quad (x^2 + y^2)^2 = 16c^2xy$$

(which includes the exceptional point $(0, 0)$ noted above).

The hyperbola, circles and envelope are shown in the following diagram.

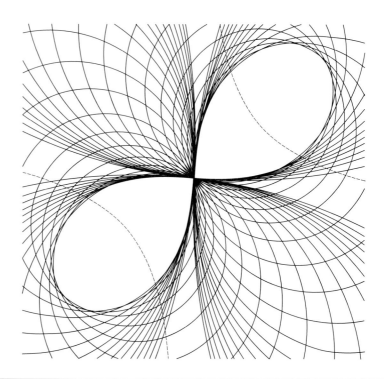

NOTE *The envelope can be identified by changing to polar coordinates.*

$$(x^2 + y^2)^2 = 16c^2xy \quad \Leftrightarrow \quad r^4 = 16c^2r^2 \cos\theta \sin\theta$$

$$\Leftrightarrow \quad r^2 = 8c^2 \sin 2\theta$$

Comparing this with Question 5 of Exercise 2C of **Pure Mathematics 5** *(page 33) shows that the envelope is Bernoulli's lemniscate, but turned so that the line*

$\theta = \dfrac{\pi}{4}$ *is an axis of symmetry. To see this note that*

$$\sin 2\theta = \cos\left(2\theta - \frac{\pi}{2}\right) = \cos 2\left(\theta - \frac{\pi}{4}\right).$$

Exercise 4A

Throughout this exercise a is a constant.

1. Sketch the family of circles which have centres on $y = \frac{1}{2}x$ and touch the x-axis. Show geometrically that the envelope is a pair of straight lines. Find these lines:

   (i) by using $\tan(2\alpha)$, where $\tan\alpha = \frac{1}{2}$

   (ii) by using the equation of the circle with centre $(2p, p)$ and radius p.

2. Find the envelope of the family of lines
$$y = px + \frac{a}{p}.$$

3. Use the standard method to obtain the equations of the envelope in the example on page 140.

4. Find the envelope of the normals of the parabola $y^2 = 4ax$.

5. The diagram shows a common 'curve stitching' activity. The points which are joined are at equal steps along the lines XA and XB, and XA = XB.

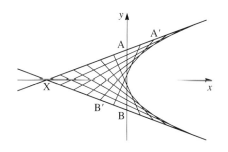

   (i) Taking AB as the y-axis and the perpendicular from X to AB as the x-axis, show that the equations of XA and XB can be written as $y = hx + k$ and $y = -hx - k$, where h and k are constants.

   (ii) Show that the coordinates of corresponding points which are joined, such as A', B', can be taken as $(p, hp + k)$, $(-p, hp - k)$, where p is a parameter. Find the equation of the line joining these points.

   (iii) Deduce that the envelope is a parabola. What happens if the steps on each line continue on AX produced and BX produced?

6. (i) A family of curves is defined by $Ap^2 + Bp + C = 0$, where A, B and C are functions of x and y (so $A = A(x, y)$, etc.). Prove that the envelope is $B^2 = 4AC$.

   (ii) Find the envelope of the family $A\cos\theta + B\sin\theta = C$, where A, B and C are functions of x and y.

7. A particle is thrown from the origin O in the plane of horizontal (x) and vertical (y) axes with initial speed V so that its initial direction of motion makes an angle θ with Ox. After time t its position (x, y) is given by:
$$x = V\cos\theta t, \quad y = V\sin\theta t - \tfrac{1}{2}gt^2.$$
[See page 156 of *Mechanics 1*.]

   (i) Show that its trajectory (flight path) is the parabola:
$$y = x\tan\theta - \frac{gx^2}{2V^2}\sec^2\theta$$
   where g is the acceleration due to gravity, and that this can be written as:
$$y = xp - \frac{gx^2}{2V^2}(1 + p^2)$$
   where $p = \tan\theta$.

   (ii) Show that the envelope of the trajectories obtained as θ varies (with V and g fixed) is also a parabola. Find the coordinates of the focus and vertex of this envelope.

8. Find the envelope of the family of circles which have as their diameters the chords of the circle $x^2 + y^2 = 1$ parallel to the y-axis.

9. Show that the envelope of the family of circles which pass through the origin and have centres on the circle $x^2 + y^2 - 2x = 0$ is the cardioid with polar equation $r = 2(1 + \cos\theta)$.

10. Show that the envelope of the family of ellipses
$$\frac{x^2}{u^2} + \frac{y^2}{v^2} = 1, \text{ where } u^2 + v^2 = a^2, \text{ is a square.}$$

Exercise 4A continued

11. The nephroid

(i) Draw a circle (called the *base circle*) and one of its diameters. Then draw the family of circles which have centres on the base circle and touch this diameter. The curved envelope you obtain is called a *nephroid* (meaning 'kidney-shaped'.) What else is part of the envelope?

(ii) Take the base circle with centre at the origin of coordinates, O, and radius $2a$, and take the x-axis as the diameter. Find the equation of the circle with centre $R(2a \cos \theta, 2a \sin \theta)$ touching the x-axis. Hence find the parametric equations of this nephroid.

(iii) Show that these parametric equations can be written as:

$$x = 3a \cos \theta - a \cos 3\theta$$
$$y = 3a \sin \theta - a \sin 3\theta.$$

(iv) Now consider the circle with centre O and radius $4a$ as one cross-section of a mirror which is part of a circular cylinder (with its axis perpendicular to the plane of the diagram).

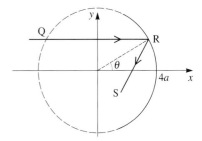

QR is a ray of light parallel to Ox which is reflected along RS. Show that the envelope of RS for parallel rays QR is the same nephroid. Deduce the focal length of the mirror for small θ.

(An envelope of reflected rays is called a *caustic curve*. This particular caustic can be seen sometimes on the surface of a mug of tea held in sunlight.)

Arc length

One practical way to find the length of a curve between two of its points (for example to find the distance along the road between two towns, from a map) is to mark several intermediate points along the arc, join them successively with straight lines to form a polygonal line, and measure the total length of this open polygon to give an approximation to the length of the arc. If you start with a polygon P_1 and construct a new polygon P_2 by inserting extra points along the arc then P_2 will fit better than P_1, and the length of P_2 will be greater than the length of P_1 (see figure 4.3).

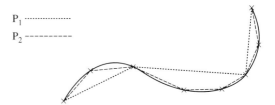

P₁ ······················
P₂ ----------

Figure 4.3

In this way, you can form more and more polygons with successively greater lengths. But since the shortest route between two points is the straight line joining them, the length of any such polygon does not exceed the length of the curve. So you would expect that the lengths of successive approximations would be bounded above (by the length of the curve), and that by putting the intermediate points sufficiently close together you could get an approximation as close as you like to the arc length.

For most curves that occur in practice this approach works, and leads to the calculus method of finding arc length, which is given below. But, contrary to intuition, there are some curves for which there is *no* upper bound to the length of the inscribed polygon between two fixed points. One example is Von Koch's 'snowflake' which is described in *Pure Mathematics 2*, page 49; another is given in Question 11 of Exercise 4B. Essentially what happens with these exceptional curves is that they wiggle so much that no chord, however short, is a good approximation. To rule out this possibility we shall restrict ourselves to curves for which the arc length PQ is nearly the same as the length of the chord PQ whenever the two points P and Q on the curve are close together. To be more precise, we shall *assume* that:

$$\frac{\text{arc PQ}}{\text{chord PQ}} \to 1 \text{ as P} \to \text{Q}.$$

The positive sense along a curve

If the coordinates of a point P on a curve are given in terms of a parameter p then the sense in which P moves along the curve as p increases is called the *positive sense*. The same applies if we are using x or y instead of p as the independent variable. The positive sense on a curve depends on the particular way in which the Cartesian equation or parametric equations are expressed. For example, the equations:

(a) $y = -x^3$ (b) $x = -\sqrt[3]{y}$ (c) $x = p, y = -p^3$ (d) $x = -q, y = q^3$

all give the same curve, using independent variables x, y, p, q respectively. In (a) and (c) the positive sense is from left to right across the page, but in (b) and (d) the opposite sense is positive (see figure 4.4).

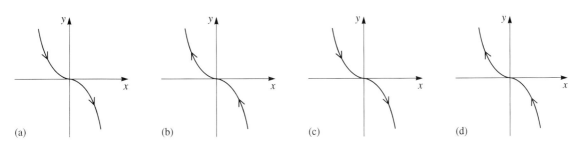

(a)   (b)   (c)   (d)

Figure 4.4

Activity

Draw diagrams to show the positive sense when the unit circle is expressed as:

(a) $x = \cos\theta, y = \sin\theta$ (b) $x = \sin\phi, y = \cos\phi$ (c) $y = \pm\sqrt{1 - x^2}$.

Arc length with Cartesian coordinates

To see how to find the length of an arc, look at part of a general curve, as shown in figure 4.5. There C is a fixed point on the curve, P is the point with parameter p, and s is the arc length from C to P, where s is positive if and only if the motion from C to P is in the positive sense along the curve.

Let P and Q have coordinates (x, y) and $(x + \delta x, y + \delta y)$, corresponding to parametric values p and $p + \delta p$ respectively, and let arc $PQ = \delta s$.

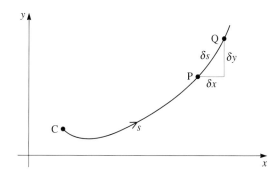

Figure 4.5

The chord length PQ is:

$$|\mathbf{PQ}| = \sqrt{(\delta x)^2 + (\delta y)^2}.$$

Therefore $\dfrac{\delta s}{\delta p} = \dfrac{\delta s}{|\mathbf{PQ}|} \times \dfrac{|\mathbf{PQ}|}{\delta p}$

$$= \frac{\delta s}{|\mathbf{PQ}|} \times \sqrt{\left(\frac{\delta x}{\delta p}\right)^2 + \left(\frac{\delta y}{\delta p}\right)^2}.$$

As $\delta p \to 0$, $\dfrac{\delta s}{\delta p}, \dfrac{\delta x}{\delta p}, \dfrac{\delta y}{\delta p}$ tend to $\dfrac{ds}{dp}, \dfrac{dx}{dp}, \dfrac{dy}{dp}$ respectively, and by the assumption stated above $\dfrac{\delta s}{|\mathbf{PQ}|} \to 1$. So taking limits as $\delta p \to 0$ gives the basic result:

$$\frac{ds}{dp} = \sqrt{\left(\frac{dx}{dp}\right)^2 + \left(\frac{dy}{dp}\right)^2}.$$

From this s can be found by integrating with respect to p.

If the independent variable is x (i.e. the equation of the curve is given in the form $y = f(x)$), then we put $p = x$ in the basic result. Then:

$$\frac{dx}{dp} = \frac{dx}{dx} = 1 \quad \text{and} \quad \frac{dy}{dp} = \frac{dy}{dx}$$

so that

$$\frac{ds}{dx} = \sqrt{1 + \left(\frac{dy}{dx}\right)^2}.$$

Similarly, when the independent variable is y:

$$\frac{ds}{dy} = \sqrt{\left(\frac{dx}{dy}\right)^2 + 1}.$$

All these are easy to remember from this right-angled 'triangle' (see figure 4.6) in which $(\delta s)^2 \approx (\delta x)^2 + (\delta y)^2$ by Pythagoras' theorem. The three results follow in the limit from dividing by $(\delta p)^2$, $(\delta x)^2$ or $(\delta y)^2$ as appropriate, and then taking the positive square root of each side. Notice that the *positive* root is needed in each case, since by definition s increases with the independent variable.

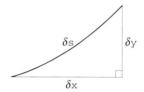

Figure 4.6

EXAMPLE

Find the length of the astroid $x = a\cos^3\theta$, $y = a\sin^3\theta$.

Solution

$$\frac{dx}{d\theta} = -3a\cos^2\theta\sin\theta \qquad \frac{dy}{d\theta} = 3a\sin^2\theta\cos\theta$$

$$\Rightarrow \quad \frac{ds}{d\theta} = \sqrt{9a^2\cos^4\theta\sin^2\theta + 9a^2\sin^4\theta\cos^2\theta}$$

$$= 3a\sqrt{\cos^2\theta\sin^2\theta(\cos^2\theta + \sin^2\theta)}$$

$$= 3a\cos\theta\sin\theta. \qquad \text{①}$$

The values of θ at the four cusps of the curve are as shown in the diagram, so the length of arc in the first quadrant is:

$$\int_0^{\pi/2} 3a\cos\theta\sin\theta\,d\theta = \left[\frac{3a}{2}\sin^2\theta\right]_0^{\pi/2} = \frac{3a}{2}$$

and the length of the complete astroid is $4 \times \dfrac{3a}{2} = 6a.$

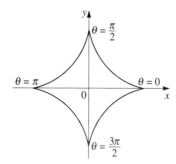

NOTE

If you try to find the whole length in a single integration you get

$$\int_0^{2\pi} 3a \cos\theta \sin\theta \, d\theta = \left[\frac{3a}{2} \sin^2\theta \right]_0^{2\pi} = 0. \text{ This is because } \cos\theta \sin\theta < 0 \text{ in the}$$

second and fourth quadrants, and the positive and negative contributions in the four quadrants have cancelled. When taking the square root at ① (and in similar cases) it is essential to check that you use an expression which is never negative throughout the range of integration.

Arc length with polar coordinates

The method of finding the length of a curve from its polar equation comes from differentiating the relations $x = r\cos\theta$, $y = r\sin\theta$ with respect to θ, remembering that r is a function of θ. Thus:

$$\frac{dx}{d\theta} = \frac{dr}{d\theta}\cos\theta - r\sin\theta \text{ and } \frac{dy}{d\theta} = \frac{dr}{d\theta}\sin\theta + r\cos\theta$$

so that $\left(\dfrac{dx}{d\theta}\right)^2 = \left(\dfrac{dr}{d\theta}\right)^2 \cos^2\theta - 2r\dfrac{dr}{d\theta}\cos\theta\sin\theta + r^2\sin^2\theta$

and $\left(\dfrac{dy}{d\theta}\right)^2 = \left(\dfrac{dr}{d\theta}\right)^2 \sin^2\theta + 2r\dfrac{dr}{d\theta}\cos\theta\sin\theta + r^2\cos^2\theta.$

Adding these and using $\cos^2\theta + \sin^2\theta = 1$ gives:

$$\left(\frac{dx}{d\theta}\right)^2 + \left(\frac{dy}{d\theta}\right)^2 = \left(\frac{dr}{d\theta}\right)^2 + r^2$$

so that $\dfrac{ds}{d\theta} = \sqrt{\left(\dfrac{dr}{d\theta}\right)^2 + r^2}$

from which s is found by integrating with respect to θ.

For Discussion

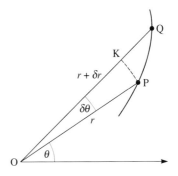

The diagram on the previous page shows a curve through neighbouring points P and Q with polar coordinates (r, θ) and $(r + \delta r, \theta + \delta\theta)$. Identify in the diagram a right-angled 'triangle' with sides δs, δr and $r\,\delta\theta$, and explain how it can be used as a reminder of this result.

EXAMPLE   Find the length of the equiangular spiral $r = 3\,e^{\theta/3}$ from $\theta = 0$ to $\theta = \pi$.

Solution

$$\frac{dr}{d\theta} = e^{\theta/3}$$

$$\Rightarrow \quad \frac{ds}{d\theta} = \sqrt{e^{2\theta/3} + 9\,e^{2\theta/3}} = \sqrt{10}\,e^{\theta/3}$$

$$\Rightarrow \quad s = \int_0^\pi \sqrt{10}\,e^{\theta/3}\,d\theta = \left[3\sqrt{10}\,e^{\theta/3}\right]_0^\pi$$

$$= 3\sqrt{10}(e^{\pi/3} - 1) \approx 17.5$$

Exercise 4B

1. Find the length of the semi-cubical parabola $y^2 = x^3$ from $(0, 0)$ to $(4, 8)$.

   (This curve was the first for which the length was found by calculus methods, by the Dutchman Heinrich van Heuraet, the Englishman William Neil and the Frenchman Pierre de Fermat independently, all between 1658 and 1660.)

2. Find the length of the curve $x = \dfrac{1}{1 + p^2}$, $y = \dfrac{p}{1 + p^2}$ from $p = 0$ to $p = 1$, and draw a sketch of the curve to explain your answer.

3. Find the length of the catenary $y = c\cosh\dfrac{x}{c}$ from $x = 0$ to $x = X$.

4. Show that $x = a\sinh^2 p$, $y = 2a\sinh p$ are parametric equations of the parabola $y^2 = 4ax$, and that the arc length from $p = 0$ to $p = P$ is $a(P + \tfrac{1}{2}\sinh 2P)$.

5. Prove that the length of one complete arch of the cycloid $x = a(\theta - \sin\theta)$, $y = a(1 - \cos\theta)$ is $8a$.

   (This result was first given by Christopher Wren in 1659.)

6. Find the length of the nephroid $x = 3a\cos\theta - a\cos 3\theta$, $y = 3a\sin\theta - a\sin 3\theta$.

7. Prove that the length of the arc of the equiangular spiral $r = a\,e^{k\theta}$ from (r_1, θ_1) to (r_2, θ_2) is proportional to $(r_2 - r_1)$, and find the constant of proportionality in terms of k.

8. Find the length of the cardioid $r = a(1 + \cos\theta)$.

9. For the conic $\dfrac{\ell}{r} = 1 + e\cos\theta$ show that, if e is small, $s \approx \ell(\theta - e\sin\theta)$, where s is measured from where $\theta = 0$.

10. (i) For the ellipse $x = a\cos\theta$, $y = b\sin\theta$ prove that $\dfrac{ds}{d\theta} = a\sqrt{1 - e^2\cos^2\theta}$, where $b^2 = a^2(1 - e^2)$.

    (ii) Prove that the perimeter of this ellipse is exactly the same as the length of one complete wave of the curve $y = ae\cos\dfrac{x}{b}$.

    (iii) Prove that if e is small then the perimeter of this ellipse is approximately $2\pi a(1 - \tfrac{1}{4}e^2)$.

Exercise 4C

1. Find the curved surface area of the solid generated by rotating the curve about the x-axis. Leave your answers in terms of π.

   (i)   the line $4y = 3x$ from $x = 4$ to $x = 8$

   (ii) the circle $x^2 + y^2 = a^2$ from $x = -a$ to $x = a$

   (iii) the catenary $y = c \cosh \dfrac{x}{c}$ from $x = -a$ to $x = a$

   (iv) the parabola $x = ap^2$, $y = 2ap$ from $p = 1$ to $p = 2$

   (v)   one arch of the cycloid $x = a(\theta - \sin\theta)$, $y = a(1 - \cos\theta)$

   (vi) the astroid $x = a\cos^3 p$, $y = a\sin^3 p$

   (vii) one loop of the lemniscate $r^2 = a^2\cos 2\theta$
         $\left(\text{i.e. from } \theta = 0 \text{ to } \theta = \dfrac{\pi}{4}\right)$

   (viii) the cardioid $r = a(1 + \cos\theta)$.

2. Show that the arc of the circle $(x + a\cos\alpha)^2 + y^2 = a^2$ for which $x \geqslant 0$ subtends an angle 2α at the centre of the circle. Find the area of the curved surface generated when this arc is rotated:

   (i)   about the x-axis   (ii) about the y-axis.

3. **Archimedes' tombstone**
   The diagram shows a sphere circumscribed by a cylinder with a vertical axis which touches the sphere at its horizontal equator. Two horizontal planes cut both the sphere and the cylinder. Prove that the portions of the sphere and the cylinder between these planes have equal curved surface areas.

(Archimedes (287–212 BC) was so pleased to discover this that, at his request, a representation of a sphere circumscribed by a cylinder was carved on his tombstone in Sicily, where it was found and restored by the Roman author Cicero about a century later.)

4. Use the 'Archimedes' tombstone' theorem to find:

   (i)   the surface area of a sphere of radius a

   (ii) the surface area of 'the tropics', i.e. the part of the Earth between the circles of latitude $23.47°\,\text{N}$ (the Tropic of Cancer) and $23.47°\,\text{S}$ (the Tropic of Capricorn). Take the Earth to be a sphere of radius $6370\,\text{km}$, and give your answer in km^2 to 3 significant figures.

5. A solid of revolution is generated by rotating the curve $y = \dfrac{1}{x}$ about the x-axis from $x = 1$ to $x = k$, where $k > 1$.

   (i)   Prove that the volume V of this solid is
         $$\pi\left(1 - \frac{1}{k}\right).$$

   (ii) Prove that the curved surface area S is given by $S = \displaystyle\int_1^k \frac{2\pi}{x}\sqrt{1 + \frac{1}{x^4}}\,\mathrm{d}x$. Noting that $1 + \dfrac{1}{x^4} > 1$, deduce that $S > 2\pi\ln k$.

   (This gives a paradox. If an infinitely long hollow vessel of this shape were placed with its axis vertical then a volume π of paint poured into it would completely fill it, but no amount of paint however large would be enough to cover the surface!)

6. In Question 11 of Exercise 4B you proved that the curve $y = f(x)$, where $f(x) = x\cos\left(\dfrac{\pi}{2x}\right)$ for $x \neq 0$, $f(0) = 0$, has infinite length between $x = 0$ and $x = 1$. Investigate whether the surface area of the solid generated by rotating this part of the curve about the x-axis is finite or infinite.

Investigation

The Pappus–Guldin theorems
These two theorems (for volumes (A) and for surfaces areas (B)) were first stated by Pappus of Alexandria in about 320, then rediscovered by Paul Guldin who published them in 1641. They use the idea of the centroid of a plane region or arc: see *Mechanics 3*, page 135.

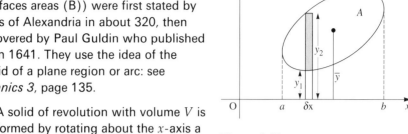

(i)   A solid of revolution with volume V is formed by rotating about the x-axis a region of area A which does not cut the x-axis. The distance of the centroid of the region from the x-axis is $\bar{y}$ (see figure 4.12).

Figure 4.12

Show that $V = \displaystyle\int_a^b \pi(y_2^2 - y_1^2)\,dx$ and that $\bar{y}A = \displaystyle\int_a^b \tfrac{1}{2}(y_1 + y_2)(y_2 - y_1)\,dx$.

(ii)   Deduce that $V = 2\pi\bar{y}A$, and show that this means that

volume of solid of revolution

$\qquad$ = area of region × distance moved by centroid $\qquad\qquad$ (A)

This is the Pappus–Guldin theorem for volumes.

(iii)   An isosceles triangle ABC has AB = AC and BC = $2h$. The length of the perpendicular from A to BC is r. Use the theorem to find the volume of the solid obtained when the triangle is rotated about BC, and deduce the formula for the volume of a right circular cone.

(iv)   A solid of revolution with surface area S is formed by rotating about the x-axis an arc of length s_0 which does not cut the x-axis.

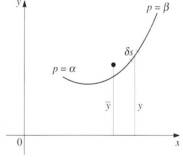

The distance of the centroid of the arc from the x-axis is $\bar{y}$ (see figure 4.13).

Explain why $\bar{y}s_0 = \displaystyle\lim_{\delta s \to 0} \sum y\delta s = \int_\alpha^\beta y\frac{ds}{dp}\,dp,$

where α and β are the initial and final values of the parameter p in terms of which the curve is defined.

Figure 4.13

(v)   Deduce that $S = 2\pi\bar{y}s_0$, and show that this means that

surface area of solid of revolution

$\qquad$ = length of arc × distance moved by centroid $\qquad\qquad$ (B)

This is the Pappus–Guldin theorem for surface areas. The theorem still applies if the arc is a closed loop: in this case (x, y) makes a complete circuit of the loop as p varies from α to β.

(vi)   Find the volume and surface area of the *torus* formed by rotating a circle of radius a about a line at a distance $b\,(> a)$ from its centre.

(vii) Use the theorems to find the position of the centroids of
(a) a semicircular region; (b) a semicircular arc.

(viii) The rectangle in figure 4.14 is
rotated about AB to form a solid.
Use the Pappus–Guldin theorems
to find the volume and surface
area of this solid. Check your
results by other methods.

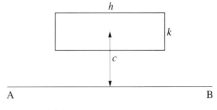

Figure 4.14

Intrinsic equations

With Cartesian or polar coordinates the equation of a particular curve may take
many forms according to the way the coordinate system is set up. For example,
the equation of the same rectangular hyperbola is $x^2 - y^2 = a^2$ when referred to
its axes of symmetry, but $xy = \frac{1}{2}a^2$ when referred to its asymptotes, and it is not
immediately apparent that these two equations give the same curve.

Activity

Show that a unit circle could have polar equation $r = 1$ or $r = 2\sin\theta$ or

$r^2 - 10\sqrt{2}r(\cos\theta + \sin\theta) + 99 = 0$ (if its centre has polar coordinates $\left(10, \frac{\pi}{4}\right)$).

An alternative way to describe a curve is to find an equation connecting its arc
length s with the angle ψ (the Greek letter psi) which its tangent makes with a
fixed direction. It is still necessary to choose:

(A) the point of the curve where $s = 0$ and the direction where $\psi = 0$

(B) the sense along the curve in which s increases (as usual, ψ is measured in
radians in the anticlockwise sense).

The effect of the choices in (A) is minor since they can only change s or ψ by a
constant. The sign of s is more significant, particularly in connection with the
work on curvature in the next section. The relation between s and ψ is called the
intrinsic equation of the curve ('intrinsic' means 'belonging naturally to itself'),
and when s and ψ are used in this way they are called *intrinsic coordinates*.

Activity

Show how to measure s and ψ so that the intrinsic equation of a circle of radius a
is $s = a\psi$.

EXAMPLE

The curve in which a uniform flexible cord or chain hangs when held at its two
ends is called a *catenary*. Find the intrinsic equation of the catenary, and show

that its Cartesian equation can take the form $y = c\cosh\dfrac{x}{c}$.

Solution

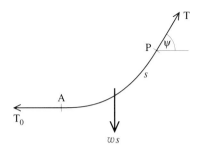

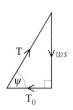

Let the cord have weight w per unit length. Consider the part of the cord from the lowest point A to another point P, where arc AP $= s$ and the tangent at P makes an angle ψ with the horizontal. There are three external forces acting on this portion: the horizontal tension T_0 at A, the tension T at P and the weight ws, as shown in the diagram.

Since the cord is in equilibrium the vectors representing these three forces form a closed triangle, from which:

$$ws = T_0 \tan \psi$$

or

$$s = c \tan \psi$$

where c is the constant $\dfrac{T_0}{w}$. This is the intrinsic equation.

To find the Cartesian equation we use the fact that $\tan \psi = \dfrac{dy}{dx}$ (since ψ is measured from the horizontal), and so, from the intrinsic equation, $\dfrac{dy}{dx} = \dfrac{s}{c}$.

Therefore

$$\frac{ds}{dx} = \sqrt{1 + \left(\frac{dy}{dx}\right)^2} = \sqrt{1 + \frac{s^2}{c^2}} = \frac{\sqrt{c^2 + s^2}}{c}$$

and

$$\frac{dx}{ds} = \frac{c}{\sqrt{c^2 + s^2}}.$$

Integrating with respect to s gives:

$$x = c \operatorname{arsinh} \frac{s}{c} + k$$

where k is a constant. If we choose axes so that the y-axis is the vertical through A then $s = 0$ when $x = 0$, and so $k = 0$.

Therefore $\dfrac{s}{c} = \sinh \dfrac{x}{c}$. But $\dfrac{dy}{dx} = \dfrac{s}{c}$ and so

$$\frac{dy}{dx} = \sinh \frac{x}{c}.$$

Integrating with respect to x gives

$$y = c \cosh \frac{x}{c} + k',$$

where k' is a constant. If we now fix the origin at a distance c below A then $y = c$ when $x = 0$, and so $k' = 0$. With these coordinate axes the Cartesian equation

is $y = c \cosh \dfrac{x}{c}$.

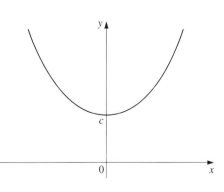

When dealing with links between intrinsic and Cartesian coordinates we shall always measure ψ from the direction of the positive x-axis. Then $\tan \psi = \dfrac{dy}{dx}$ and from this two other useful results follow.

- $\dfrac{ds}{dx} = \sqrt{1 + \left(\dfrac{dy}{dx}\right)^2} = \sqrt{1 + \tan^2 \psi} = \sec \psi$ so that $\dfrac{dx}{ds} = \cos \psi$

- $\dfrac{dy}{ds} = \dfrac{dy}{dx} \times \dfrac{dx}{ds} = \tan \psi \times \cos \psi = \sin \psi.$

These results can be remembered easily by using the 'differential triangle' shown in figure 4.15.

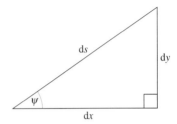

Figure 4.15

The following example illustrates a general method for finding Cartesian or parametric equations of a curve from its intrinsic equation.

EXAMPLE

Find parametric and Cartesian equations for the curve $s = \frac{3}{2}a \sin^2 \psi$, where

$$0 \leqslant \psi \leqslant \frac{\pi}{2}.$$

Solution

$$s = \tfrac{3}{2}a \sin^2 \psi \quad \Rightarrow \quad \frac{ds}{d\psi} = 3a \sin \psi \cos \psi$$

$$\Rightarrow \quad \frac{dx}{d\psi} = \frac{dx}{ds} \times \frac{ds}{d\psi} = \cos \psi \times 3a \sin \psi \cos \psi = 3a \cos^2 \psi \sin \psi$$

and

$$\frac{dy}{d\psi} = \frac{dy}{ds} \times \frac{ds}{d\psi} = \sin \psi \times 3a \sin \psi \cos \psi = 3a \sin^2 \psi \cos \psi.$$

Integrating with respect to ψ gives:

$$x = -a \cos^3 \psi \quad \text{and} \quad y = a \sin^3 \psi$$

provided that $x = -a$ and $y = 0$ when $\psi = 0$.

These are parametric equations for the astroid with Cartesian equation $x^{2/3} + y^{2/3} = a^{2/3}$ (see page 142).

The quarter of the curve from $(-a, 0)$ to $(0, a)$ is described as ψ increases from 0 to $\dfrac{\pi}{2}$.

1. For the curve $y = \ln \sec x$ prove that $\dfrac{ds}{dx} = \sec x$ and that $\psi = x$. Hence find the intrinsic equation.

2. For the semi-cubical parabola $x = 3ap^2$, $y = 2ap^3$ prove that $\tan \psi = p$. Deduce that $\dfrac{ds}{d\psi} = 6a \tan \psi \sec^3 \psi$, and hence find the intrinsic equation.

3. For the nephroid $x = 3a \cos \theta - a \cos 3\theta$, $y = 3a \sin \theta - a \sin 3\theta$ (see Exercise 4A, Question 11) prove that $\psi = 2\theta$ and that the intrinsic equation is $s = 6a(1 - \cos \frac{1}{2} \psi)$.

4. The cycloid is the locus of a point P on the circumference of a circle of radius a as the circle rolls, without slipping, along the x-axis. The diagram shows the circle after it has turned through angle θ; it then touches the x-axis at A, and AB is a diameter.

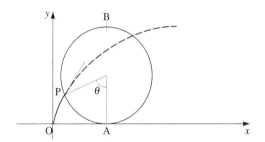

(i) Show from this diagram that the parametric equations are:
$$x = a(\theta - \sin \theta)$$
$$y = a(1 - \cos \theta).$$

(ii) The tangent to the cycloid at P is in the direction of motion of P. Noting that A has zero velocity (since the circle does not slip), deduce that the tangent at P passes through B, and hence that $\psi = \dfrac{\pi}{2} - \dfrac{\theta}{2}$.

(iii) Prove that $\psi = \dfrac{\pi}{2} - \dfrac{\theta}{2}$ by a different method, using calculus.

(iv) Prove that if s is measured from O then the intrinsic equation of one arch is $s = 4a(1 - \sin \psi)$.

(v) Now turn the arch upside-down and put the origin at the minimum point. Prove that with s measured from this new origin the intrinsic equation is now $s = 4a \sin \psi$.

(vi) Find the parametric equations of the cycloid in this new position.

5. If the intrinsic equation of a curve is $s = a\psi$, prove that its Cartesian equation must be $(x - h)^2 + (y - k)^2 = a^2$, where h and k are constants.

6. Given that $s = \ln \tan \dfrac{\psi}{2}$ and that $x = 0$ and $y = \dfrac{\pi}{2}$ when $\psi = \dfrac{\pi}{2}$, find x in terms of y.

7. Use the method of the example on page 160 to prove again that, with a suitable choice of origin:
$$s = c \tan \psi \quad \Rightarrow \quad y = c \cosh \dfrac{x}{c}.$$

8. If $s = a\,e^{k\psi}$ and $z = x + yj$, prove that $\dfrac{dz}{d\psi} = ak\,e^{(k+j)\psi}$. Use this to find the polar equation, and deduce that the curve is an equiangular spiral.

Curvature

If a curve is bending sharply, then the direction of the tangent, measured by ψ, changes considerably in a short distance along the curve, but for a gentle bend there is a smaller change in ψ for the same change in s. To put this more precisely (referring to figure 4.16), if the changes in ψ and s between the points P and Q on the curve are $\delta\psi$ and δs then the *average curvature*

$\dfrac{\delta\psi}{\delta s}$ gives a measure of how much the curve curves between P and Q.

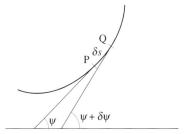

Figure 4.16

Taking the limit as $Q \to P$ leads to the following definition.

The *curvature* at a point P is the rate of change of ψ with respect to s at P.

Curvature is denoted by κ (the Greek letter kappa), and so:

$$\kappa = \frac{d\psi}{ds}.$$

If κ is positive then ψ increases with s, and the curve curves to the left of the positive tangent at P (as in figure 4.16) but if κ is negative then ψ decreases with s and the curve curves to the right. If P is a point of inflection then $\kappa = 0$, but the converse is not true, as will be shown in the example on page 164.

For a straight line ψ is constant, and so $\kappa = 0$ at every point.

For a circle of radius a described anticlockwise from its lowest point (see figure 4.17), the intrinsic equation $s = a\psi$ gives $\dfrac{ds}{d\psi} = a$, and so $\varkappa = \dfrac{1}{a}$.

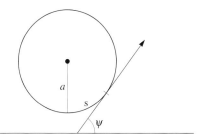

Figure 4.17

As you might expect, this shows that a circle has constant curvature, large when the radius is small and small when the radius is large.

Finding curvature

If the intrinsic equation of a curve is known then finding the curvature is straightforward.

EXAMPLE Find the curvature of the catenary $s = c \tan \psi$ when $\psi = \dfrac{\pi}{4}$ and find where the curvature is greatest.

Solution

$$s = c \tan \psi \quad \Rightarrow \quad \frac{ds}{d\psi} = c \sec^2 \psi$$

$$\Rightarrow \quad \kappa = \frac{d\psi}{ds} = \frac{1}{c} \cos^2 \psi. \qquad \text{Using } \sec \psi = \frac{1}{\cos \psi}$$

$$\psi = \frac{\pi}{4} \quad \Rightarrow \quad \kappa = \frac{1}{2c}.$$

The greatest curvature is $\dfrac{1}{c}$ and this occurs when $\psi = 0$ (at the minimum point).

If the intrinsic equation is not available then it is not quite so simple to find the curvature. Suppose the curve is defined in terms of a parameter p. For brevity we use a prime (') to show differentiation with respect to p, so that x' stands for $\dfrac{dx}{dp}$, x'' for $\dfrac{d^2x}{dp^2}$ and so on.

We know that $s' = \sqrt{x'^2 + y'^2}$ and to find ψ' we differentiate $\tan \psi$ with respect to p:

$$\tan \psi = \frac{dy}{dx} = \frac{y'}{x'} \quad \Rightarrow \quad \sec^2 \psi \psi' = \frac{x'y'' - y'x''}{x'^2}.$$

But

$$\sec^2 \psi = 1 + \tan^2 \psi = 1 + \frac{y'^2}{x'^2} = \frac{x'^2 + y'^2}{x'^2}$$

and so

$$\psi' = \frac{x'y'' - y'x''}{x'^2 + y'^2}.$$

Therefore

$$\kappa = \frac{\psi'}{s'} = \frac{x'y'' - y'x''}{(x'^2 + y'^2)^{3/2}}.$$

Activity

Show that for the parabola $x = ap^2$, $y = 2ap$ this formula gives $\kappa = -\dfrac{1}{2a(1 + p^2)^{3/2}}$.

What is the geometrical significance of the negative sign?

The formula for curvature when y is given in terms of x can be deduced immediately by putting $p = x$. Then $x' = 1$, $x'' = 0$, $y' = \dfrac{dy}{dx}$ and $y'' = \dfrac{d^2y}{dx^2}$.

So

$$\kappa = \frac{\dfrac{d^2y}{dx^2}}{\left\{ 1 + \left(\dfrac{dy}{dx} \right)^2 \right\}^{3/2}}.$$

Activity

Find the formula for κ when x is given in terms of y.

The next example shows that the curvature can be zero even where there is no point of inflection.

EXAMPLE

Find the points of least and greatest curvature on the curve $y = x^4$.

Solution

$$y = x^4 \quad \Rightarrow \quad \frac{dy}{dx} = 4x^3 \quad \text{and} \quad \frac{d^2y}{dx^2} = 12x^2$$

$$\Rightarrow \quad \kappa = \frac{12x^2}{(1 + 16x^6)^{3/2}}.$$

To differentiate κ with respect to x would involve some heavy algebra, which can be avoided to some extent by substituting $u = x^2$ and $v = \dfrac{\kappa^2}{144}$.

Then $\quad v = \dfrac{u^2}{(1 + 16u^3)^3} \quad$ and so $\quad \dfrac{dv}{du} = \dfrac{(1 + 16u^3)^3 \cdot 2u - u^2 \cdot 3(1 + 16u^3)^2 \cdot 48u^2}{(1 + 16u^3)^6}.$

Therefore $\quad \dfrac{dv}{du} = 0 \quad \Leftrightarrow \quad 2u(1 + 16u^3 - 72u^3) = 0$

$$\Leftrightarrow \quad u = 0 \text{ or } 1 - 56u^3 = 0$$

$$\Leftrightarrow \quad u = 0 \text{ or } u = \left(\frac{1}{56}\right)^{1/3}$$

$$\Leftrightarrow \quad x = 0 \text{ or } x = \pm\left(\frac{1}{56}\right)^{1/6} \approx \pm 0.511.$$

The expression for κ is never negative, so the least value of κ is 0, where $x = 0$ (which is at the minimum point of the curve, *not* at a point of inflection). The curvature is greatest when v is greatest, at $x = \pm\left(\dfrac{1}{56}\right)^{1/6}$ where $\kappa \approx 2.151$, as in the diagram.

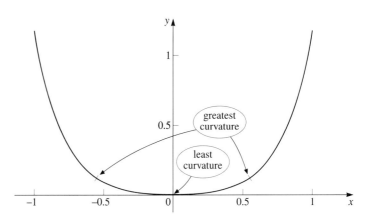

To find the curvature from a polar equation $r = f(\theta)$ it is always possible to treat θ as a parameter, with $x = f(\theta)\cos\theta$ and $y = f(\theta)\sin\theta$, though the working is likely to be complicated. An alternative approach, which is much simpler for some standard curves, is outlined in Question 15 of Exercise 4E.

Exercise 4E

1. Find the curvature of $y = x^3$ at the points $(1,1)$ and $(-2,-8)$.

2. Find the curvature of $y = \ln x$ at $(1,0)$.

3. Prove that the curvature of $y = \sin x$ varies from -1 to 1 during one cycle.

4. The intrinsic equation of an equiangular spiral is $s = a\,e^{k\psi}$. Prove that the curvature is inversely proportional to s.

5. Prove that if $|\psi|$ is small then

$$\kappa \approx \frac{d^2 y}{dx^2}(1 - \tfrac{3}{2}\tan^2\psi).$$ Find the greatest value of $|\psi|$ for which $\dfrac{d^2 y}{dx^2}$ may be used instead of κ with an error of less than 10%.

6. Find the curvature of the astroid:

(i) using the intrinsic equation $s = \tfrac{3}{2}a\sin^2\psi$

(ii) using the parametric equations $x = a\cos^3 p$, $y = a\sin^3 p$.

Explain the connection between your two answers.

7. For the parabola $y^2 = 4ax$ prove that

$$\kappa = -\frac{\sin^3\psi}{2a}.$$

8. For the rectangular hyperbola $xy = c^2$ prove that $\kappa = \dfrac{2c^2}{r^3}$, where $r^2 = x^2 + y^2$.

9. Find κ in terms of θ for the ellipse $x = a\cos\theta$, $y = b\sin\theta$. Check that your answer gives what you would expect when $a = b$.

10. For the catenary $y = c\cosh\dfrac{x}{c}$ prove that $\kappa = \dfrac{c}{y^2}$.

11. Find the greatest curvature on the curve $y = e^x$, and the coordinates of the point where this occurs.

12. If x and y are given in terms of the arc length s, and primes denote differentiation with respect to s, prove that:

(i) $\kappa = -\dfrac{x''}{y'} = \dfrac{y''}{x'}$ (ii) $\kappa^2 = (x'')^2 + (y'')^2$.

Hint: Start by differentiating $x' = \cos\psi$ with respect to s.

13. The Cartesian coordinates of a point P are (x,y). A second pair of axes is obtained by rotating the original pair through the angle α about the origin O. Referring to these axes, the coordinates of P are (X,Y) as shown in the diagram.

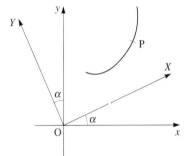

(i) Show that:
$$\begin{pmatrix} X \\ Y \end{pmatrix} = \begin{pmatrix} \cos\alpha & \sin\alpha \\ -\sin\alpha & \cos\alpha \end{pmatrix}\begin{pmatrix} x \\ y \end{pmatrix}.$$

(ii) Now suppose that P is on a curve so that x and y (and therefore X and Y) are given in terms of a parameter p. Show that the same curvature at P is obtained whichever coordinate system is used in its calculation.

(iii) Show in a similar way that a shift of origin does not affect the value of the curvature at P.

(This confirms that curvature is an intrinsic property of the curve, independent of the choice of coordinate axes.)

Exercise 4E continued

14. By working in Cartesian coordinates with θ as a parameter, show that the curvature of the spiral of Archimedes, $r = a\theta$, is $\dfrac{2 + \theta^2}{a(1 + \theta^2)^{3/2}}$.

15. (i) For a curve with polar equation $r = f(\theta)$, the angle between the radius and the positive tangent is denoted by ϕ as shown in the diagram. Prove that $\psi = \phi + \theta$.

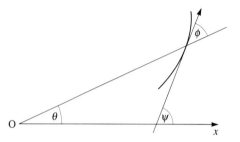

(ii) Deduce from $x = r\cos\theta$, $y = r\sin\theta$ that
$$\tan\psi = \frac{r't + r}{r' - rt}, \text{ where } r' = \frac{dr}{d\theta} \text{ and}$$
$t = \tan\theta$. Hence show that
$$\tan\phi = \tan(\psi - \theta) = \frac{r}{r'} = r\frac{d\theta}{dr}.$$

(iii) Explain how the 'elementary triangle' PKQ shown in the diagram can be used to remember the basic result that
$$\tan\phi = r\frac{d\theta}{dr}.$$

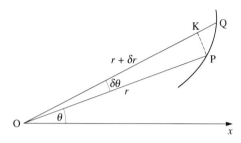

(iv) For the equiangular spiral $r = a\,e^{\theta\cot\alpha}$ where α is constant with $0 < \alpha < \pi$, prove that $\phi = \alpha$. Use $\dfrac{d\psi}{d\theta}$ and $\dfrac{ds}{d\theta}$ to show that
$$\varkappa = \frac{\sin\alpha}{r}.$$

(v) Prove that the curvature of the cardioid $r = a(1 + \cos\theta)$ is $\dfrac{3}{4a}\sec\dfrac{\theta}{2}$.

The circle of curvature

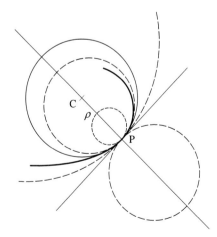

Figure 4.18

Figure 4.18 shows a curve with its tangent and normal at a point P, where the curvature is κ. Every circle through P with centre on the normal touches the curve at P. The member of this family of circles which fits the curve best at P is the one for which the curvature is also κ; this is called the *circle of curvature* at P.

Its centre C is the *centre of curvature*, and its radius ρ (the Greek letter rho) is the *radius of curvature*. Since the curvature of a circle of radius ρ is $\dfrac{1}{\rho}$ it follows that $\dfrac{1}{\rho} = \kappa$, and therefore that $\rho = \dfrac{1}{\kappa} = \dfrac{ds}{d\psi}$.

The radius of curvature can be found directly from the parametric or Cartesian equations.

$$\rho = \frac{(x'^2 + y'^2)^{3/2}}{x'y'' - y'x''} = \frac{\left\{1 + \left(\dfrac{dy}{dx}\right)^2\right\}^{3/2}}{\dfrac{d^2y}{dx^2}}$$

If the curvature is negative then the definition $\rho = \dfrac{1}{\kappa}$ means that ρ is negative too. It may seem strange to talk of a circle with a negative radius (and some texts insist that the radius of curvature should be positive by defining $\rho = \left|\dfrac{1}{\kappa}\right|$), but in fact this turns out to be useful when interpreted correctly.

To show this we introduce two special unit vectors, $\hat{\mathbf{t}}$ and $\hat{\mathbf{n}}$. The *positive tangent* at a point on a curve is the tangent in the direction of the positive sense at that point; the direction of the *positive normal* is $\dfrac{\pi}{2}$ ahead of this in the anticlockwise sense, so that the positive normal makes the angle $\psi + \dfrac{\pi}{2}$ with the positive x-axis (see figure 4.19). The unit vectors in the directions of the positive tangent and positive normal are $\hat{\mathbf{t}}$ and $\hat{\mathbf{n}}$ respectively. In components:

$$\hat{\mathbf{t}} = \begin{pmatrix} \cos\psi \\ \sin\psi \end{pmatrix} \quad \text{and} \quad \hat{\mathbf{n}} = \begin{pmatrix} -\sin\psi \\ \cos\psi \end{pmatrix}.$$

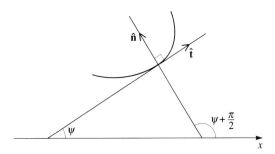

Figure 4.19

If $\kappa > 0$ then the curve bends to the left of the positive tangent and $\hat{\mathbf{n}}$ points towards the concave side of the curve. The centre of curvature C is a distance ρ from P along the positive normal, so that $\mathbf{PC} = \rho\hat{\mathbf{n}}$ (see figure 4.20).

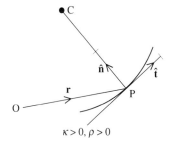

$\kappa > 0, \rho > 0$

Figure 4.20

But if $\kappa < 0$ (in which case ρ is also negative) then $\hat{\mathbf{n}}$ points away from the concave side of the curve, and to get from P to C we go a distance $-\rho$ in the direction of $-\hat{\mathbf{n}}$ (see figure 4.21). Therefore

$$\mathbf{PC} = (-\rho)(-\hat{\mathbf{n}}) = \rho\hat{\mathbf{n}}$$

as before.

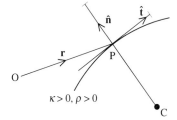

$\kappa > 0, \rho > 0$

Figure 4.21

This shows that, whatever the sign of κ and ρ, the position vector $\mathbf{c}$ of the centre of curvature is given by $\mathbf{c} = \mathbf{r} + \rho\hat{\mathbf{n}}$.

The same result can be expressed in the coordinate form:

$$\xi = x - \rho\sin\psi$$

$$\eta = y + \rho\cos\psi$$

where (ξ, η) is the centre of curvature for the point (x, y). (The symbols ξ and η are the Greek letters xi and eta.)

EXAMPLE

Find $\hat{\mathbf{t}}$, $\hat{\mathbf{n}}$ and $\mathbf{c}$ for the point $(4, 3)$ on the rectangular hyperbola $xy = 12$.

Solution

Taking x as the independent variable, the positive sense along the curve is to the right.

$$y = \frac{12}{x} \quad \Rightarrow \quad \frac{dy}{dx} = -\frac{12}{x^2} = -\tfrac{3}{4} \text{ at } (4, 3).$$

Therefore $\hat{\mathbf{t}} = \begin{pmatrix} \frac{4}{5} \\ -\frac{3}{5} \end{pmatrix}$ and $\hat{\mathbf{n}} = \begin{pmatrix} \frac{3}{5} \\ \frac{4}{5} \end{pmatrix}$.

$$\frac{d^2y}{dx^2} = \frac{24}{x^3} = \frac{3}{8} \text{ at } (4, 3) \quad \Rightarrow \quad \rho = \frac{\{1 + \left(-\frac{3}{4}\right)^2\}^{3/2}}{\frac{3}{8}} = \frac{\{\frac{25}{16}\}^{3/2}}{\frac{3}{8}} = \frac{125}{24}$$

and so $\mathbf{c} = \begin{pmatrix} 4 \\ 3 \end{pmatrix} + \frac{125}{24}\begin{pmatrix} \frac{3}{5} \\ \frac{4}{5} \end{pmatrix} = \begin{pmatrix} 4 \\ 3 \end{pmatrix} + \begin{pmatrix} \frac{25}{8} \\ \frac{25}{6} \end{pmatrix} = \begin{pmatrix} 7\frac{1}{8} \\ 7\frac{1}{6} \end{pmatrix}.$

Exercise 4F

1. Find the coordinates of the centre of curvature of the curve $xy = 16$ at the point $(4, 4)$.

2. Find the vectors $\hat{\mathbf{t}}$, $\hat{\mathbf{n}}$ and $\mathbf{c}$ when $x = \dfrac{\pi}{4}$ on the curve $y = \tan x$.

3. Prove that the coordinates of the centre of curvature are $\left(x - \dfrac{dy}{d\psi}, y + \dfrac{dx}{d\psi} \right)$.

   Hence show that the centre of curvature at the point (X, Y) on $y = \ln \sec x$ is $(X - \tan X, Y + 1)$.
   [**Hint:** See Exercise 4D, Question 1.]

4. For the equiangular spiral $r = k\mathrm{e}^{\theta}$ show that $\psi = \theta + \dfrac{\pi}{4}$. Use the method of Question 3 to find the Cartesian coordinates of the centre of curvature in terms of θ.

5. On a sketch, show the catenary $y = c \cosh \dfrac{x}{c}$ with the tangent and normal at a point P. Prove that if the normal meets the x-axis at D then the centre of curvature C is the reflection of D in the tangent.

6. Let the equation of the circle of curvature at the point $\mathrm{P}\ (3p^2, 2p^3)$ of the curve $27y^2 = 4x^3$ be $x^2 + y^2 + 2ax + 2by + c = 0$. Use the facts that the circle and the curve have triple point contact, i.e.

   (i) both pass through P

   (ii) both have the same gradient at P

   (iii) both have the same value of $\dfrac{d^2y}{dx^2}$ at P

   to find the values of a, b and c.

7. Prove that the centre of curvature of the ellipse $\dfrac{x^2}{a^2} + \dfrac{y^2}{b^2} = 1$ at $(a \cos \theta, b \sin \theta)$ is $\left(\dfrac{a^2 - b^2}{a} \cos^3 \theta, -\dfrac{a^2 - b^2}{b} \sin^3 \theta \right)$. What can be said about the eccentricity of the ellipse if the centre of curvature at $(0, b)$ is:

   (i) inside   (ii) on   (iii) outside

   the ellipse?

The evolute of a curve

As the point P moves along a given curve, the centre of curvature C also moves. The locus of the centre of curvature is called the *evolute* of the curve.

EXAMPLE Find the equation of the evolute of the parabola $y^2 = 4ax$.

Solution

We use the parametric equations $x = ap^2$, $y = 2ap$.

Then $x' = 2ap$, $y' = 2a$, $x'' = 2a$, $y'' = 0$ so that:

$$\rho = \frac{\{(2ap)^2 + (2a)^2\}^{3/2}}{2ap \times 0 - 2a \times 2a} = -2a\{p^2 + 1\}^{3/2}$$

$$\tan \psi = \frac{1}{p} \quad \text{and} \quad \hat{\mathbf{n}} = \begin{pmatrix} -1/\sqrt{p^2 + 1} \\ p/\sqrt{p^2 + 1} \end{pmatrix}.$$

Therefore $\mathbf{c} = \mathbf{r} + \rho\hat{\mathbf{n}}$

$$= \begin{pmatrix} ap^2 \\ 2ap \end{pmatrix} - 2a\{p^2+1\}^{3/2} \begin{pmatrix} -1/\sqrt{p^2+1} \\ p/\sqrt{p^2+1} \end{pmatrix}$$

$$= \begin{pmatrix} ap^2 \\ 2ap \end{pmatrix} - 2a(p^2+1) \begin{pmatrix} -1 \\ p \end{pmatrix}$$

$$= \begin{pmatrix} 2a + 3ap^2 \\ -2ap^3 \end{pmatrix}.$$

The parametric equations of the evolute are:

$$x = 2a + 3ap^2, \quad y = -2ap^3$$

from which the Cartesian equation is $27ay^2 = 4(x-2a)^3$.

The evolute is a semi-cubical parabola with its cusp at $(2a, 0)$, as shown in the diagram.

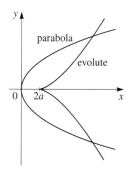

The evolute as the envelope of the normals

An alternative view of the evolute comes from differentiating the relation $\mathbf{c} = \mathbf{r} + \rho\hat{\mathbf{n}}$ with respect to the arc length s:

$$\frac{d\mathbf{c}}{ds} = \frac{d\mathbf{r}}{ds} + \frac{d\rho}{ds}\hat{\mathbf{n}} + \rho\frac{d\hat{\mathbf{n}}}{ds}.$$

The right-hand side simplifies considerably, since:

$$\frac{d\mathbf{r}}{ds} = \begin{pmatrix} dx/ds \\ dy/ds \end{pmatrix} = \begin{pmatrix} \cos\psi \\ \sin\psi \end{pmatrix} = \hat{\mathbf{t}}$$

and

$$\rho\frac{d\hat{\mathbf{n}}}{ds} = \frac{ds}{d\psi}\frac{d\hat{\mathbf{n}}}{ds} = \frac{d\hat{\mathbf{n}}}{d\psi} = \begin{pmatrix} -\cos\psi \\ -\sin\psi \end{pmatrix} = -\hat{\mathbf{t}}.$$

Therefore $\dfrac{d\mathbf{c}}{ds} = \dfrac{d\rho}{ds}\hat{\mathbf{n}}.$

In this equation the left-hand side is a vector in the direction of the tangent to the evolute at C, and the right-hand side is a vector in the direction of the normal PC. Therefore this normal touches the evolute at C, and so the evolute is the envelope of the normals. Finding this envelope is often the simplest way to find the evolute.

EXAMPLE Use the envelope of the normals to find the evolute of the parabola $y^2 = 4ax$ again.

Solution

As before:

$$x = ap^2, \quad y = 2ap \quad \Rightarrow \quad \frac{dy}{dx} = \frac{1}{p}$$

so the gradient of the normal is $-p$ and the equation of the normal is:

$$y - 2ap = -p(x - ap^2)$$

or $$f(x, y, p) = px + y - 2ap - ap^3 = 0$$

$$\frac{\partial f}{\partial p} = x - 2a - 3ap^2 = 0 \quad \Rightarrow \quad x = 2a + 3ap^2$$

Substituting this in $f(x, y, p) = 0$ gives:

$$2ap + 3ap^3 + y - 2ap - ap^3 = 0 \quad \Rightarrow \quad y = -2ap^3$$

so we get the same parametric equations as on page 170, with much less effort (see diagram).

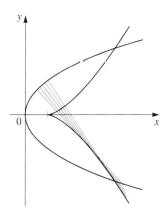

For Discussion

What happens to the evolute:

(i) when there is a cusp on the curve

(ii) when there is a point of inflection on the curve

(iii) when the curvature is greatest or least?

Exercise 4G

1. Find parametric equations for the evolute of the rectangular hyperbola $x = ct$, $y = \dfrac{c}{t}$.

2. Show that the equation of the normal of the standard ellipse at $(a\cos\theta, b\sin\theta)$ can be written in the form $ax\sec\theta - by\csc\theta - a^2 + b^2 = 0$. Use the envelope of the normals to find parametric equations for the evolute. [Compare this with Exercise 4F, Question 7.] Show the ellipse and its evolute in a sketch, indicating how the centre of curvature moves round the evolute as θ increases from 0 to 2π.

3. For the nephroid $x = 3a\cos\theta - a\cos 3\theta$, $y = 3a\sin\theta - a\sin 3\theta$, use the fact that $\psi = 2\theta$ (see Exercise 4D, Question 3) to show that the evolute has parametric equations $x = \frac{1}{2}a(3\cos\theta + \cos 3\theta)$, $y = \frac{1}{2}a(3\sin\theta + \sin 3\theta)$. Show that the evolute is another nephroid, half the size of the original and turned through a right angle. Draw a sketch of the original nephroid and its evolute.

4. Show that the equation of the normal to the astroid $x = a\cos^3\theta$, $y = a\sin^3\theta$ can be put into the form $x\cos\theta - y\sin\theta = a(\cos^2\theta - \sin^2\theta)$. Find the coordinates of the points U and V at which this normal meets the lines $y = x$ and $y = -x$, and show that $UV = 2a$ for all θ. Deduce that the evolute of the astroid is another astroid twice as large, and show both astroids in a sketch.

5. The equiangular spiral with polar equation $r = ak^\theta$ has the remarkable property that turning the curve about O is equivalent to enlarging it from the centre O: to see this, note that adding β to θ multiplies r by k^β. Deduce from this that, for every point P on the curve, the angle between OP and the tangent at P is constant (hence the name *equiangular*). (See also Exercise 4E, Question 15(iv).)

The diagram shows the normal PN meeting at N the line through O perpendicular to OP. The constant angle between OP and the tangent is α.

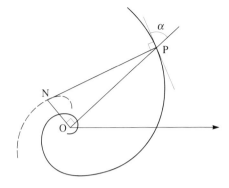

(i) Explain why the locus of N as P varies is a spiral (shown dotted) which is similar to, and therefore congruent to, the original spiral.

(ii) Show that PN touches this new spiral.

(iii) Deduce that the new spiral is the evolute of the original spiral, and that N is the centre of curvature at P.

(iv) Is it possible for a curve to be its own evolute?

6. Prove that the evolute of the cycloid $x = a(\theta - \sin\theta)$, $y = a(1 - \cos\theta)$ has parametric equations $x = a(\theta + \sin\theta)$, $y = -a(1 - \cos\theta)$. Prove that the evolute is a congruent cycloid, and draw a sketch to show how the two cycloids are related.

7. A point P moves so that its position vector at time t is $\mathbf{r}$.

(i) By differentiating $\mathbf{r}$ with respect to t, prove that its velocity is $v\hat{\mathbf{t}}$, where $v = \dfrac{ds}{dt}$.

(ii) Prove that its acceleration is $\dfrac{dv}{dt}\hat{\mathbf{t}} + \dfrac{v^2}{\rho}\hat{\mathbf{n}}$.

(This splits the acceleration into *tangential* and *normal* components; these are the same as if P were moving round the circle of curvature.)

8. The envelope of the normals again

(i) Let the coordinates of a point (x, y) on a curve be given parametrically in terms of ψ, and let (ξ, η) be any point on the normal to the curve at (x, y). Show that
$$\eta - y + \cot \psi (\xi - x) = 0.$$

(ii) Let $f(\xi, \eta, \psi) = \eta - y + \cot \psi (\xi - x)$, so that $f(\xi, \eta, \psi) = 0$ is the equation of the family of normals, with parameter ψ.

Show that $\dfrac{\partial f}{\partial \psi}$ can be written in the form
$$-\rho \sin \psi - \operatorname{cosec}^2 \psi (\xi - x) - \cot \psi \, . \, \rho \cos \psi$$
and deduce that
$$f = 0 \text{ and } \frac{\partial f}{\partial \psi} = 0$$
$$\Rightarrow \quad \xi - x = -\rho \sin \psi \text{ and } \eta - y = \rho \cos \psi.$$

(iii) Deduce that the envelope of the normals has parametric equations:
$$\xi = x - \rho \sin \psi, \quad \eta = y + \rho \cos \psi$$
or
$$\mathbf{c} = \mathbf{r} + \rho \hat{\mathbf{n}}.$$

Involutes

The equation $\dfrac{d\mathbf{c}}{ds} = \dfrac{d\rho}{ds} \hat{\mathbf{n}}$ on page 170 gives another interesting result. You have seen that $\dfrac{d\mathbf{r}}{ds} = \hat{\mathbf{t}}$, which means that differentiating the position vector of a point on a curve with respect to the arc length of the curve gives the tangential unit vector. Now suppose that the arc length of the evolute is σ (the Greek letter sigma), measured from a fixed point of the evolute in the positive sense along the evolute (i.e. the sense in which C moves as s increases). Then the result for the evolute corresponding to $\dfrac{d\mathbf{r}}{ds} = \hat{\mathbf{t}}$ for the original curve is $\dfrac{d\mathbf{c}}{d\sigma} = \hat{\mathbf{n}}$, since $\hat{\mathbf{n}}$ is the tangential unit vector for the evolute.

So $\hat{\mathbf{n}} = \dfrac{d\mathbf{c}}{d\sigma}$

$\qquad = \dfrac{d\mathbf{c}}{ds} \times \dfrac{ds}{d\sigma}$

$\qquad = \dfrac{d\rho}{ds} \hat{\mathbf{n}} \times \dfrac{ds}{d\sigma} \left(\text{since } \dfrac{d\mathbf{c}}{ds} = \dfrac{d\rho}{ds} \hat{\mathbf{n}} \right)$

$\qquad = \dfrac{d\rho}{d\sigma} \hat{\mathbf{n}}.$

Therefore $\dfrac{d\rho}{d\sigma} = 1$, which means that ρ and σ differ by a constant.

So if C moves from C_1 to C_2 as P moves from P_1 to P_2 then arc $C_1 C_2 = \sigma_2 - \sigma_1 = \rho_2 - \rho_1$ (see figure 4.22).

For Discussion

From these examples, pick out cases where:

(i)   the operation cannot be applied at all in the set mentioned

(ii) the operation can be applied to general elements of the set, but not to some particular elements.

Most of these operations take *two* elements of the set and combine them in some way to give a definite result; such a rule of combination is called a *binary operation*. Many binary operations are denoted by conventional symbols, such as + for addition (though this could mean addition of numbers or vectors or matrices according to the context) and ÷ or / for division. Other symbols such as * or • can be used to show binary operations, so that $x * y$ or $x • y$ means the result of combining x and y. The elements to be combined do not have to be distinct: it is usually possible to combine an element with itself, as with $x * x$ or $x • x$.

Other operations (for example, taking the square root of a number or finding the inverse of a matrix) act on a single element; these are called *unary operations*.

Activity

(i)   Identify two other unary operations from the examples above.

(ii) Give two examples of *ternary* operations (which combine three elements).

Here we concentrate on some properties which a particular binary operation may or may not possess. Suppose that • is a binary operation on elements $x, y, z, \ldots$ of the set S.

1. Closure

If the result of the operation is always in the original set S, i.e. $x • y \in S$ for all $x, y \in S$, then the operation is said to be *closed*. For example, within the set of positive integers, multiplication is closed (since the product of two positive integers is always a positive integer) but division is not (since for example $2 ÷ 3$ is not a positive integer).

NOTE   *In set notation, $\in$ means 'belongs to' or 'is a member of'.*

2. Commutativity

If changing the order of the elements has no effect on the result, i.e. if $y • x = x • y$ for all $x, y \in S$, then the operation is said to be *commutative*. For example, if S is the set of three-dimensional vectors, then the scalar product is commutative ($\mathbf{a . b} = \mathbf{b . a}$ for all $\mathbf{a, b} \in S$), but the vector product is not (since $\mathbf{a} \times \mathbf{b}$ and $\mathbf{b} \times \mathbf{a}$ have opposite directions for all non-parallel $\mathbf{a}$ and $\mathbf{b}$).

3. Associativity

If the operation is closed then the result of combining x and y can be combined with another element, z say, to give $(x \bullet y) \bullet z$. Alternatively we can form $x \bullet (y \bullet z)$, where the bracketing now shows that x is combined with the result of combining y and z. If the bracketing has no effect, i.e. if $(x \bullet y) \bullet z = x \bullet (y \bullet z)$ for all $x, y, z \in S$, then the operation is said to be *associative*. For example, addition of real numbers is associative, but subtraction is not (since for example $(10 - 6) - 2 \neq 10 - (6 - 2)$). More is said about associativity on page 186.

Activity

From the examples listed above, select a set and a binary operation which is:

(i)   non-closed, commutative

(ii) closed, non-commutative, associative

(iii) closed, commutative, non-associative

(iv) closed, non-commutative, non-associative.

Modular arithmetic

Within the set of integers, two numbers are said to be *congruent modulo m* if the difference between them is a multiple of m. For example, since $77 - 29 = 48 = 12 \times 4$, 77 and 29 are congruent modulo 12; this can be abbreviated to $77 = 29 \pmod{12}$.

Activity

(i) Give three other integers, one of them negative, which are congruent to 77 modulo 12.

(ii) Give three other positive integers m such that $77 = 29 \pmod{m}$.

If the remainder when a is divided by m is b, then $a = b + hm$ for some integer h, and so $a = b \pmod{m}$. The only possible remainders are $0, 1, 2, \ldots, m - 1$, and so every integer is congruent to one of these. *Modular arithmetic* (with modulus m) is carried out on the set $\{0, 1, 2, \ldots, m - 1\}$ using the binary operations $+_m$ and $\times_m$, which are like ordinary addition and multiplication respectively, except that if the sum or product falls outside the set then the answer is the remainder when this is divided by m.

Abstract algebra

EXAMPLE Find $8 +_{12} 11$ and $8 \times_{12} 11$.

Solution

The modulus is 12.

$8 + 11 = 19 = 12 \times 1 + \underline{7}$, so that $8 +_{12} 11 = 7$

$8 \times 11 = 88 = 12 \times 7 + \underline{4}$, so that $8 \times_{12} 11 = 4$.

The theory of congruences and modular arithmetic has far-reaching consequences, but here we use it only to provide examples in the development of the main topic, groups.

Exercise 5A

1. Write out:

   (i)   the addition table for $\{0, 1, 2, 3, 4\}$ under $+_5$

   (ii) the multiplication table for $\{0, 1, 2, 3, 4, 5, 6\}$ under $\times_7$.

2. Answer these questions, and explain the connections with modular arithmetic.

   (i)   If today is Monday, what day of the week will it be after 100 days?

   (ii) If a train departs at 21.37 and the journey takes 5 hours 40 minutes, at what time (on the 24-hour clock) does the train arrive?

   (iii) Is 48 902 918 529 263 a perfect square?

3. Prove or disprove each of these statements.

   (i) $x + 2 = y + 4 \pmod{12}$
         $\Rightarrow x = y + 2 \pmod{12}$

   (ii) $2x = 4y \pmod{12} \Rightarrow x = 2y \pmod{12}$.

4. (i) Prove that:
        $a = b \pmod{m}$ and $c = d \pmod{m}$
        $\Rightarrow a + c = b + d \pmod{m}$.

        [**Hint:** If $a = b \pmod{m}$ and $c = d \pmod{m}$ then there are integers h and k such that $a = b + hm$ and $c = d + km$.]

   (ii) Prove that:
        $a = b \pmod{m}$ and $c = d \pmod{m}$
        $\Rightarrow ac = bd \pmod{m}$.

   (iii) Investigate whether the converses of (i) and (ii) are true.

5. The integer x written as '$a_n a_{n-1} \ldots a_2 a_1 a_0$', has units digit a_0, tens digit a_1, hundreds digit a_2, and so on. Prove that:

   $$x = a_n + a_{n-1} + \cdots + a_2 + a_1 + a_0 \pmod 9.$$

   Deduce that x is divisible by 9 if and only if the digit sum $a_n + a_{n-1} + \cdots + a_2 + a_1 + a_0$ is divisible by 9. State and prove a similar test for divisibility by 3.

6. Prove that $10^k = (-1)^k \pmod{11}$ for all integers $k \geqslant 0$.

   Using the notation of Question 5, deduce that x is divisible by 11 if and only if the alternating digit sum $\sum_{r=0}^{n} (-1)^r a_r$ is divisible by 11.

7. Prove that each of the operations of $+_m$ and $\times_m$ on the set $\{0, 1, 2, \ldots, m-1\}$ is closed, commutative and associative.

Groups

Here are four examples for you to work through; keep your results so that you can refer to them as the chapter develops.

Example A
Write out the addition table for $\{0, 1, 2, 3, 4, 5\}$ under $+_6$ (addition modulo 6).

Example B
Draw an equilateral triangle on a piece of card, and cut it out. Then draw the three axes u, v and w on the card, and put distinguishing marks on the corners of the triangle, as in figure 5.1, with the same marks in corresponding positions on the back of the triangle too. The triangle can be fitted into the hole in the card in various ways, and can be moved by the following transformations:

P: rotation 120° anticlockwise about the axis through the centre perpendicular to the plane of the card

Q: rotation 120° clockwise about the same axis

U: half-turn about axis u

V: half-turn about axis v

W: half-turn about axis w.

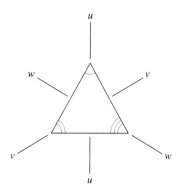

Figure 5.1

Applying P and then U gives: $\triangle \overset{P}{\longmapsto} \triangle \overset{U}{\longmapsto} \triangle$

$\overset{P}{\longmapsto}$ means 'is mapped by P to'

But $\triangle \overset{V}{\longmapsto} \triangle$, so P followed by U is equivalent to V, and

we write $UP = V$. Notice that, as is usually the case when combining mappings, the first transformation P is written on the right of the second transformation U; it is important to be clear about this, since the same transformations applied in the other order have a different effect: $PU \neq UP$.

Activity

Which transformation is equivalent to PU? Try combining other pairs of transformations, and record your results. You will find that some combinations are equivalent to the 'stay put' identity transformation I.

The following table gives the results of combining these six transformations.

First transformation

Second transformation	followed by	I	P	Q	U	V	W
	I	I	P	Q	U	V	W
	P	P	Q	I	W	U	V
	Q	Q	I	P	V	W	U
	U	U	V	W	I	P	Q
	V	V	W	U	Q	I	P
	W	W	U	V	P	Q	I

Notice that the entry for UP is written in row U and column P.

Example C

A *permutation* of a finite set of objects is a mapping of the set onto itself. For example, if the objects are ♣, ◇, ♡, ♠ then one possible permutation is the mapping ♣ ↦ ◇, ◇ ↦ ♠, ♡ ↦ ♡, ♠ ↦ ♣, which can be written more compactly as:

$$\begin{pmatrix} ♣ & ◇ & ♡ & ♠ \\ ◇ & ♠ & ♡ & ♣ \end{pmatrix}.$$

With three objects, called 1, 2, 3, say, there are $3! = 6$ permutations, as follows.

$$i = \begin{pmatrix} 1 & 2 & 3 \\ 1 & 2 & 3 \end{pmatrix} \qquad p = \begin{pmatrix} 1 & 2 & 3 \\ 2 & 3 & 1 \end{pmatrix} \qquad q = \begin{pmatrix} 1 & 2 & 3 \\ 3 & 1 & 2 \end{pmatrix}$$

$$u = \begin{pmatrix} 1 & 2 & 3 \\ 1 & 3 & 2 \end{pmatrix} \qquad v = \begin{pmatrix} 1 & 2 & 3 \\ 3 & 2 & 1 \end{pmatrix} \qquad w = \begin{pmatrix} 1 & 2 & 3 \\ 2 & 1 & 3 \end{pmatrix}$$

Permutations can be combined in the same way as any other functions. For example, under the permutation up (meaning p followed by u), 1 goes to 2 which goes to 3 (i.e. $1 \mapsto 2 \mapsto 3$) and similarly $2 \mapsto 3 \mapsto 2$ and $3 \mapsto 1 \mapsto 1$, so that:

$$up = \begin{pmatrix} 1 & 2 & 3 \\ 3 & 2 & 1 \end{pmatrix} = v.$$

Activity

Show that $pu = w$. Make a table like the one in Example B to show all the results of combining two of these permutations.

Example *D*

Let $\omega = \cos\left(\dfrac{2\pi}{3}\right) + j\sin\left(\dfrac{2\pi}{3}\right)$, one of the complex cube roots of unity.

Activity

(i)   Find the set of all the complex numbers which can be obtained by multiplying together any of the numbers ω, $-\omega$, and previous products.

(ii) Make a multiplication table for this set of numbers.

Comparing the examples

Activity

List the common features shared by the four examples, *A*, *B*, *C* and *D*.

Now compare your list with the following items.

- In each example there is a set of elements: numbers (mod 6), transformations, permutations or complex numbers.

- Each set has six elements.

- In each example there is a binary operation, combining any two elements of the set. In Example *A* the binary operation is addition modulo 6, in Examples *B* and *C* the binary operation is the composition of successive transformations or permutations, and in Example *D* it is multiplication. The same binary operation can be used to combine an element with itself.

- In each example the binary operation is *closed* in the set.

- The results of combining elements can be shown in a table, such as the one given in Example *B*, which is called a *Cayley table*.

- Each Cayley table is a *Latin square*, which means that every element occurs just once in each row or column of the body of the table.

- In each example there is just one element in the set which, when combined with any element of the set, leaves that element unchanged: this element is called the *identity* or *neutral element*. In Examples *A*, *B*, *C* and *D* the identities are, respectively, the number 0, the 'stay put' transformation *I*, the permutation $i = \begin{pmatrix} 1 & 2 & 3 \\ 1 & 2 & 3 \end{pmatrix}$ and the number 1.

- In each example for each element in the set there is just one element (called the *inverse*) which combines with it to produce the identity. Sometimes an element is its own inverse, in which case it is called a *self-inverse* element.

- Each binary operation is *associative*.

Associativity

Most of the points listed above are easy to verify, but associativity cannot be seen immediately from a Cayley table.

The obvious way to test for associativity is to use particular elements. For example, using the Cayley table in Example B gives $U(QV) = UW = Q$ and $(UQ)V = WV = Q$, so that $U(QV) = (UQ)V$. The problem with this approach is that to prove associativity you have to check all possible combinations, in this case $6^3 = 216$ of them! Happily this is not necessary, because it is easy to show that the operation of composition of mappings (in this case transformations) is always associative.

To see this, suppose that X, Y and Z are mappings, and that $\bullet$ denotes composition of mappings. Let a be any element in the domain of Z, and suppose that Z maps a to b, Y maps b to c, and X maps c to d, as in figure 5.2.

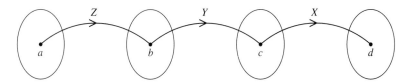

Figure 5.2

So we have a set of (unary) operations (the mappings $X, Y, \ldots$) which are the elements combined by a (binary) operation (composition of mappings $\bullet$), and in order to prove associativity we have to 'look inside' the mappings to see what happens to the elements $(a, b, \ldots)$ upon which they act.

The mapping $(X \bullet Y) \bullet Z$ means "Z followed by the composition of Y and X", under which a maps to b which then maps straight to d, as shown in figure 5.3.

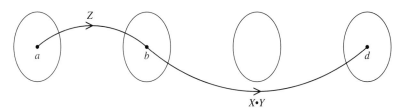

Figure 5.3

Similarly $X \bullet (Y \bullet Z)$ means "the composition of Z and Y, followed by X", under which a maps straight to c which then maps to d, as in figure 5.4.

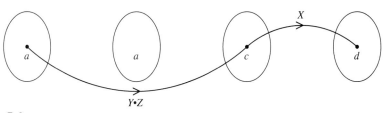

Figure 5.4

The 'by-passing' indicated by the bracketing has no effect on the overall result, which in each case is to map a to d. Therefore $(X \bullet Y) \bullet Z = X \bullet (Y \bullet Z)$. This proves that any binary operation which can be interpreted as the composition of mappings is associative.

This argument establishes associativity in Examples B and C. In Examples A and D there is no problem since modular addition and multiplication of complex numbers are both associative operations.

Activity

Which of the binary operations $+$, $-$, $\div$, $\wedge$ applied to real numbers (where $x \wedge y$ means x^y) are associative?

Exercise 5B

1. The *absolute difference* between two numbers x and y is written as $x \sim y$, so that $x \sim y = |x - y|$.

   (i)   Make a Cayley table showing the results of applying the binary operation $\sim$ to the set $S = \{0, 1, 2, 3, 4\}$.

   (ii) Is S closed under $\sim$?

   (iii) Is the operation $\sim$ associative?

   (iv) Is there an identity or neutral element? If so, what is it?

   (v)   Does each element have an inverse? If so, what are they?

2. A binary operation, $\circ$, is defined on the set R of real numbers by:

   $$a \circ b = \begin{cases} \text{maximum of } a \text{ and } b & \text{if } a \neq b \\ a & \text{if } a = b \end{cases}$$

   State whether $\circ$ is (i) closed, (ii) associative, giving your reasons.

   Show that R does not contain an identity element for this operation.

   Suggest a set S on which this operation can be defined which does contain an identity. Does every element of your set S have an inverse in S? [MEI]

3. The *union* of two sets A and B is the set $A \cup B$ of elements belonging to A or B or both. The *intersection* $A \cap B$ is the set of elements belonging to both A and B. Investigate whether the operations $\cup$ and $\cap$ are associative.

4. (i) The operation $\bullet$ is defined on the set of complex numbers by:

   $$m \bullet n = (m^2 + n^2)^{1/2}.$$

   Investigate closure, associativity, and the existence of an identity and inverses.

   (ii) How do your answers to (i) differ when the operation is applied to the set of non-negative real numbers?

   (iii) If the operation is applied to the set Z^+ of positive integers show that $m \bullet n \in Z^+$ if $n = \frac{1}{2}(m^2 - 1)$ and state the condition this imposes on m. Find the relation between the integers p, q and r if $m = p + q$, $n = q + r$ and $m \bullet n = p + q + r$. [MEI, adapted]

5. Prove that the operation $\bullet$ defined on the real numbers by:

   $$x \bullet y = \frac{xy + k^2}{x + y} \quad (x + y \neq 0)$$

   is associative for all values of the constant k. Show that there are in general no solutions to the equation $x \bullet y = y$, and discuss any exceptional cases.

Exercise 5B continued

6. Suppose that f is a function with an inverse and that • is an associative binary operation. A new binary operation $\Diamond$ is defined by
$x \Diamond y = f^{-1}(f(x) \bullet f(y))$. Prove that $\Diamond$ is associative. Find $x \Diamond y$ in the following particular cases:

(i) $f(x) = x + 3$, $\bullet = \times$

(ii) $f(x) = \dfrac{2x - 1}{x - 1}$, $\bullet = +$

(iii) $f(x) = \arctan x$, $\bullet = +$.

7. The operation $\circ$ is defined on the number pairs $A = (a_1, a_2)$ and $B = (b_1, b_2)$ so that $A \circ B = (a_1 + b_2, a_2 + b_1)$ and $A = B$ if and only if $a_1 = b_1$ and $a_2 = b_2$. Find whether the operation is associative.

Find P such that $A \circ P = A$, and Q such that $Q \circ A = A$, and determine whether I, J can be found such that, for all A, $A \circ I = A$ and $J \circ A = A$.

Given that $A^* = (-a_2, -a_1)$, discuss the following:

$$B \circ A = C \circ A$$
$$\Rightarrow \quad B \circ A \circ A^* = C \circ A \circ A^*$$
$$\Rightarrow \quad B \circ (0, 0) = C \circ (0, 0)$$
$$\Rightarrow \quad B = C. \qquad \text{[MEI]}$$

Group axioms

Your collection of the four examples A, B, C and D shows in a small way that there are many similarities between the various systems formed by particular sets of mathematical objects and their operations. This interested mathematicians increasingly during the nineteenth century, when the emphasis moved from concentrating on particular systems (such as the algebra of complex numbers, or matrix algebra, or vector algebra, or the algebra of transformations) to *abstract algebra*. In abstract algebra we assume certain basic statements (called *axioms*) about unspecified objects and operations, and then see what can be deduced from these. One abstract theorem can then be interpreted in many ways by taking particular objects and operations which satisfy the axioms. Thus abstract methods have great generality and often reveal connections between the properties of apparently quite separate systems.

Of course, much depends on the choice of axioms: one of the simplest but most productive abstract systems is the *group*, which is defined as follows.

A group(S, •) is a non-empty set S with a binary operation • such that

(C) • is *closed* in S, i.e. $a \bullet b \in S$ for all $a, b \in S$

(A) • is *associative*, i.e. $(a \bullet b) \bullet c = a \bullet (b \bullet c)$ for all $a, b, c \in S$

(N) there is an *identity* or *neutral* element $e \in S$ such that:

$$a \bullet e = e \bullet a = a \text{ for all } a \in S$$

(I)   each $a \in S$ has an *inverse* element $a^{-1} \in S$ such that:

$$a \bullet a^{-1} = a^{-1} \bullet a = e.$$

Activity

Check that each of the examples A, B, C and D satisfies all four group axioms.

Note that the group axioms do not require • to be commutative, and in the case of Examples B and C it is not. If the binary operation • is commutative, i.e. if $a \bullet b = b \bullet a$ for all $a, b \in S$, then the group is said to be commutative or *Abelian*, named after one of the pioneers of group theory, the Norwegian Niels Abel (1802–29). Examples A and D are Abelian groups, since modular addition and the multiplication of complex numbers are commutative operations.

NOTE *One way of remembering this and the four group axioms (C), (A), (N), (I) is to think of the biblical brothers Cain and Abel. Of course, the initials only work if you use 'neutral' rather than 'identity', though for most purposes the latter seems to be more common.*

The order of a group

In each of the examples A, B, C and D there are six distinct elements in the set. You will meet many examples of groups where the set S has a finite number of elements (not necessarily six). These are called *finite* groups and the number of elements of S is called the *order* of the group. There are also many *infinite* groups, where S has infinitely many distinct elements: simple examples are the set of integers under the operation of addition (for which the neutral element is 0 and the inverse of a is $-a$), or the set of all rotations about a fixed centre under the operation of combination of transformations.

Activity

Explain why:

(i) the set of rational numbers forms a group under addition but not under multiplication

(ii) the set of positive real numbers forms a group under multiplication but not under addition.

Notation

From now on we shall often leave out the symbol •, or whatever has been used to show the binary operation, so that $a \bullet b$ will be written as ab. This fits with the usual way of showing the multiplication of numbers or matrices or the composition of mappings, and is considerably quicker to write. But you must remember to distinguish between ab and ba unless you specifically know that the operation is commutative.

Indices can be used to show repeated combination of the same element, for example

$$a^2 = a \bullet a \text{ and } a^3 = a \bullet (a \bullet a) = (a \bullet a) \bullet a.$$

If the operation is addition (of numbers, vectors, matrices, or in modular arithmetic) we shall continue to use $+$, and denote the inverse of a by $-a$.

EXAMPLE

Let a, b and c be the functions defined (for all x except 0 and 1) by:

$$a(x) = x \qquad b(x) = 1 - x \qquad c(x) = \frac{1}{x}.$$

Show that the set of functions formed by combining any two of these (including repetitions) forms a finite non-Abelian group under the operation of composition of functions. Give the Cayley table of this group.

Solution

Stage 1

Clearly a is the identity function, and $bb(x) = cc(x) = x$, so that $bb = cc = a$.

Also $bc(x) = b(c(x)) = 1 - \frac{1}{x} = \frac{x-1}{x} = d(x)$, say, so that $bc = d$,

whereas $cb(x) = c(b(x)) = \frac{1}{1-x} = e(x)$, say, so that $cb = e$.

The fact that $bc \neq cb$ shows that the operation is not commutative.

Next we find cd:

$$cd(x) = \frac{x}{x-1} = f(x), \text{ say, so that } cd = f.$$

Stage 2

We could find dc and the other entries by substitution in a similar way, but it is quicker to make use of what is already known (including the fact that composition of functions is associative):

$$dc = (bc)c = b(cc) = ba = b$$

Several other results can be found in a similar way. In what follows we make frequent use of associativity, but condense the working by leaving out the details of the bracketing.

$$bd = bbc = ac = c \qquad ce = ccb = ab = b \qquad cf = ccd = ad = d$$
$$de = bccb = bab = bb = a \qquad df = dcd = bd = c \qquad eb = cbb = ca = c$$
$$ed = cbbc = cac = cc = a \qquad fc = cdc = cb = e \qquad ff = cdcd = cbd = cc = a.$$

Stage 3

To make further progress we need one more substitution:

$$ee(x) = \frac{1}{1 - \dfrac{1}{1-x}} = \frac{1-x}{1-x-1} = \frac{x-1}{x} = d(x), \text{ so that } ee = d.$$

Stage 4

Everything else follows from this.

$$dd = dee = ae = e \quad be = bdd = cd = f \quad bf = bbe = ae = e$$
$$db = bcb = be = f \quad ec = ddc = db = f \quad ef = ddf = dc = b$$
$$fb = beb = bc = d \quad fd = bed = ba = b \quad fe = fdd = bd = c.$$

This shows that the set of six functions is closed under the operation of composition of functions. The remaining axiom (existence of inverses) is easily checked from the Cayley table, so this is a non-Abelian group of order 6. (In the table the entries obtained at stages 1, 2, 3 and 4 are shown by the type faces k, k, **k**, **k** respectively).

	First function					
followed by	a	b	c	d	e	f
a	a	b	c	d	e	f
b	b	a	d	c	**f**	**e**
c	c	e	a	f	b	d
d	d	**f**	b	**e**	a	c
e	e	c	**f**	a	**d**	**b**
f	f	**d**	e	**b**	**c**	a

Second function (row label) / *First function* (column label)

Immediate consequences of the axioms

Other properties common to all groups can be deduced from the group axioms; here are a few important simple ones. Notice the rather formal style of argument which is needed to show how the property links back to the axioms.

(a) The identity element is unique

Proof

Suppose that a group has two identity elements, e and f.

Then $ef = e$ (since f is an identity)

and $ef = f$ (since e is an identity).

Therefore $e = f$.

So there is only one identity.

(b) Each element has a unique inverse

Proof

Suppose that e is the identity, and that an element a has two inverses, p and q, so that $ap = pa = e$ and $aq = qa = e$.

Then $p = pe$ (e is the identity)

$= p(aq)$ ($aq = e$)

$= (pa)q$ (associativity)

$= eq$ ($pa = e$)

$= q$ (e is the identity).

Therefore the inverse of a is unique. A consequence of this result is that there is no ambiguity in using the notation a^{-1} for the inverse of a.

(c) The cancellation laws

Suppose that a, b, h are elements of a group such that $ha = hb$. Then we can 'cancel h on the left' to obtain $a = b$.
Similarly $ah = bh \implies a = b$ ('cancelling on the right').

Proof

$$ha = hb$$

$\implies h^{-1}(ha) = h^{-1}(hb)$ (using the unique inverse of h)

$\implies (h^{-1}h)a = (h^{-1}h)b$ (associativity)

$\implies ea = eb$ ($h^{-1}h = e$)

$\implies a = b$ (e is the identity).

Activity

(i)   Prove similarly that $ah = bh \implies a = b$.

(ii) Produce an example to show that $ha = bh$ does not imply that $a = b$.

(d) The solution of $ax = b$

If a and b are elements of a group then the equation $ax = b$ has the unique solution $x = a^{-1}b$.

Proof

$$ax = b$$

$\iff a^{-1}(ax) = a^{-1}b$ (a^{-1} exists since each element has an inverse)

$\iff (a^{-1}a)x = a^{-1}b$ (associativity)

$\iff ex = a^{-1}b$ ($a^{-1}a = e$)

$\iff x = a^{-1}b$ (e is the identity).

Note that $a^{-1}b$ is an element of the group, by the closure axiom.

Activity

Solve the following equations.

(i) $xa = b$ (ii) $axb = c$

HISTORICAL NOTE *The idea of an abstract algebraic system developed gradually throughout the nineteenth century. Galois, Abel and Cauchy worked with particular groups, and Arthur Cayley wrote about the notion of an abstract group as early as 1849, attracting no attention at the time (see Kline's comment at the start of the chapter), though he is remembered by his name being given to the Cayley table, which shows how the elements of a finite group combine. The first axiomatic definitions of a group were given by Heinrich Weber and Walther von Dyck in independent papers which appeared in the same journal in 1882. The subject grew rapidly, and many mathematicians in the early twentieth century came to believe that all worthwhile mathematics would ultimately be included in the theory of groups: in 1921 Poincaré wrote, '... the theory of groups is, so to say, the whole of mathematics divested of its matter and reduced to pure form.'*

Exercise 5C

1. Write out the Cayley table for combining the symmetry transformations of a square, using I for the identity transformation, R, H, S for rotations 90° anticlockwise, 180° and 90° clockwise about O respectively, X, Y, A, B for half-turns about axes x, y, a, b respectively.

Check that all the group axioms are satisfied.

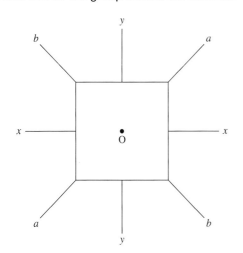

2. Prove that the Cayley table of a finite group has the Latin square property, i.e. each element occurs just once in each row and in each column. Give an example of a 4×4 Latin square which is not a group table, stating which axioms are not satisfied.

3. Prove that if a and b are elements of a group then $(ab)^{-1} = b^{-1}a^{-1}$.

4. In the Cayley table of a group with identity e the following rectangles occur:

Express x and y in terms of a and b.

5. Which of the following sets of 2×2 matrices form a group under the operation of matrix multiplication? Which of these groups are Abelian?

(i)   those with rational elements

(ii) those with integer elements and unit determinant

(iii) those of the form $\begin{pmatrix} \cos\theta & -\sin\theta \\ \sin\theta & \cos\theta \end{pmatrix}$

(iv) those of the form $\begin{pmatrix} z & -w^* \\ w & z^* \end{pmatrix}$

where z and w are complex numbers, not both zero.

Exercise 5C continued

6. If A and B are sets, the *symmetric difference* $A \triangle B$ is defined as:

$$A \triangle B = (A \cap B') \cup (A' \cap B)$$

as shown in this Venn diagram.

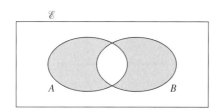

(i) Prove that the set of all subsets of the set $\mathscr{E}$ forms an Abelian group under the operation $\triangle$.

(ii) Given that $\mathscr{E} = \{1, 2, 3, 4, 5, 6, 7, 8, 9\}$, $A = \{1, 2, 3\}$ and $B = \{2, 4, 6, 8\}$, solve the equation $A \triangle X = B$.

7. Prove that if $a^2 = e$ (the identity) for every element a of a group then the group is Abelian. Give an example of such a group.

8. (i) Show that the sets $\{1, 4, 7, 13\}$ and $\{3, 6, 9, 12\}$ form Abelian groups under multiplication modulo 15, but $\{2, 5, 8, 14\}$ does not.

(ii) Write out the multiplication table for $\{1, 2, 3, 4, 5, 6, 7\}$ under multiplication modulo 8, and explain why this is not a group. Find subsets of this set which, under the same operation, give groups of order 2 and 4.

9. G is a group and $D(G)$ is the set of ordered pairs of elements (g, ε) where g is an element of G and $\varepsilon = \pm 1$. Show that $D(G)$ together with the binary operation:

$$(g, \varepsilon) \bullet (h, \delta) = (gh^{\varepsilon}, \varepsilon\delta)$$

is a group if and only if G is Abelian. (Here (h, δ) denotes another element of $D(G)$.)

Find the condition on G which makes $D(G)$ an Abelian group and show that, if this condition is satisfied, then $D(D(G))$ is also an Abelian group. [MEI]

10. The Lorentz group

(i) In the special theory of relativity as applied to motion along a straight line a central problem is to find the linear transformations

$$\begin{pmatrix} x' \\ t' \end{pmatrix} = \begin{pmatrix} p & q \\ r & s \end{pmatrix} \begin{pmatrix} x \\ t \end{pmatrix}$$

for which $x'^2 - c^2 t'^2 = x^2 - c^2 t^2$, where c is the speed of light.

Show that this leads to the conditions $p^2 - c^2 r^2 = 1$, $pq = c^2 rs$ and $c^2 s^2 - q^2 = c^2$. Verify that these are all satisfied if:

$$p = \frac{1}{\alpha}, \quad q = -\frac{v}{\alpha},$$

$$r = -\frac{v}{c^2 \alpha}, \quad s = \frac{1}{\alpha},$$

where v is any number with $|v| < c$ and

$$\alpha = \sqrt{1 - \frac{v^2}{c^2}}.$$

(ii) Let $\mathbf{L}(v) = \begin{pmatrix} \dfrac{1}{\alpha} & -\dfrac{v}{\alpha} \\ -\dfrac{v}{c^2\alpha} & \dfrac{1}{\alpha} \end{pmatrix}$. Show that

$(\mathbf{L}(v))^{-1} = \mathbf{L}(-v)$ and that

$\mathbf{L}(v_1)\mathbf{L}(v_2) = \mathbf{L}(v_3)$ where $v_3 = \dfrac{v_1 + v_2}{1 + \dfrac{v_1 v_2}{c^2}}$.

(This gives the rule for adding velocities in the special theory of relativity.)

(iii) Show that the set of matrices $\mathbf{L}(v)$ with $|v| < c$ forms an Abelian group under matrix multiplication; this is called the Lorentz group.

Isomorphism

You may have already noticed that the Cayley tables of Examples B and C on pages 183–184 are very similar. In fact you can change one into the other merely by replacing capital letters by small letters, or vice versa: both tables show exactly the same pattern of elements. When this happens we say that the mapping:

$$\text{capital letter} \leftrightarrow \text{small letter}$$

preserves the structure of the two groups. The mapping is called an *isomorphism*, and the groups are said to be *isomorphic* (from the Greek *isos* meaning 'equal' and *morphe* meaning 'form' or 'shape'). Isomorphic groups are essentially the same, differing only in the notation used to describe them, and so any property of a particular group can immediately be translated into the corresponding property of any other isomorphic group.

Activity

Show that the vertices of the equilateral triangle in Example B and their original positions on the surrounding card can be numbered 1, 2, 3 in such a way that each transformation in Example B produces the corresponding permutation in Example C.

To put this more formally, two groups $(S, \bullet)$ and $(T, \Diamond)$ are *isomorphic* if there is a one-to-one mapping (called an *isomorphism*) which associates each of the elements $a, b, c, \ldots$ of S with one of the elements $x, y, z, \ldots$ of T in such a way that, if a maps to x and b maps to y, then the result of combining a and b under $\bullet$ maps to the result of combining x and y under $\Diamond$, i.e.

$$a \leftrightarrow x \text{ and } b \leftrightarrow y \text{ (where } a, b \in S \text{ and } x, y \subset T) \quad \Rightarrow \quad a \bullet b \leftrightarrow x \Diamond y.$$

This is shown diagrammatically in figure 5.5.

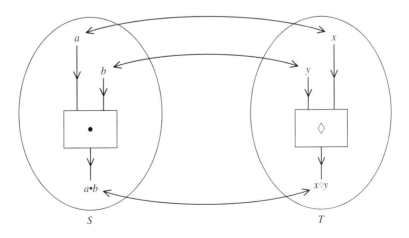

Figure 5.5

How can we find whether two groups are isomorphic? Sometimes there may be an obvious isomorphism, or an obvious way to construct one (as in the activity above). Otherwise the search can be aided by the following simple theorems.

(a) Two isomorphic groups are either both finite and of the same order or both infinite

Proof

If one group is finite and of order n then so is the other, since there is a one-to-one mapping between them; the only alternative is that both groups are infinite.

(b) The identity elements of isomorphic groups are mapped to each other by the isomorphism

Proof

Suppose that $a \leftrightarrow x$ and $e \leftrightarrow u$, where e is the identity of $(S, \bullet)$. Then $a \bullet e \leftrightarrow x \Diamond u$. But $a \bullet e = a$ so $a \bullet e \leftrightarrow x$. Therefore $x \Diamond u = x$, and so u is the identity of $(T, \Diamond)$.

(c) The inverses of corresponding elements are mapped to each other by the isomorphism

Proof

Suppose that $a \leftrightarrow x$ and $a^{-1} \leftrightarrow y$. Then $a \bullet a^{-1} \leftrightarrow x \Diamond y$.

But $a \bullet a^{-1} = e \leftrightarrow u$. Therefore $x \Diamond y = u$, and so $y = x^{-1}$ (the inverse of x in $(T, \Diamond)$).

Activity

(i) An element a is *self-inverse* if $a^{-1} = a$. Prove that two isomorphic groups must have equal numbers of self-inverse elements.

(ii) Are the groups in Examples B and D isomorphic?

EXAMPLE

Investigate whether there is an isomorphism between the group B and the group of functions given in the example on page 190 (which we shall call group E).

Solution

The Cayley tables for groups B and E are shown below.

	I	P	Q	U	V	W
I	I	P	Q	U	V	W
P	P	Q	I	W	U	V
Q	Q	I	P	V	W	U
U	U	V	W	I	P	Q
V	V	W	U	Q	I	P
W	W	U	V	P	Q	I

Group B

	a	b	c	d	e	f
a	a	b	c	d	e	f
b	b	a	d	c	f	e
c	c	e	a	f	b	d
d	d	f	b	e	a	c
e	e	c	f	a	d	b
f	f	d	e	b	c	a

Group E

The identities of the two groups are I and a, and so in any isomorphism these elements must correspond. The leading diagonals show that in each group there are four self-inverse elements (including the identities). The elements which are not self-inverse are P and Q in B and d and e in E, and these too must correspond in an isomorphism, giving two possible pairings. Suppose for example that $P \leftrightarrow d$; then it follows that $Q \leftrightarrow e$, but we still have to link the remaining self-inverse elements, U, V, W and b, c, f. Choosing one pairing arbitrarily, suppose that $U \leftrightarrow f$. Then if the mapping is an isomorphism, $V = UP \leftrightarrow fd = b$ and $W = UQ \leftrightarrow fe = c$. So the mapping given by the table:

I	P	Q	U	V	W
a	d	e	f	b	c

may be an isomorphism. To decide whether it is, rewrite the Cayley table for group E with the elements arranged in this order.

	a	d	e	f	b	c
a	a	d	e	f	b	c
d	d	e	a	c	f	b
e	e	a	d	b	c	f
f	f	b	c	a	d	e
b	b	c	f	e	a	d
c	c	f	b	d	e	a

It is now easy to check that the pattern of entries here is the same as the pattern in the Cayley table for group B, so that these two groups are isomorphic.

Activity

Two of the steps of the solution of this example involved making arbitrary choices. Work through the solution again, making different choices at these two points, and show that you still get an isomorphism. How many different isomorphisms between these two groups are there?

Exercise 5D

1. Find two different isomorphisms between the group of integers under addition and the group of multiples of 5 under addition.

2. Investigate whether there is an isomorphism between group A and group D in the examples on pages 183–185.

3. Show that the following groups are all isomorphic:

   (i)   the set of complex numbers $a + bj$, where a and b are integers, under addition

   (ii) the set of rational numbers of the form $5^m 7^n$, where m and n are integers, under multiplication

   (iii) the set of translations which map an infinite lattice grid of rectangles to itself, under composition of transformations.

4. Prove that the group of symmetry transformations of a non-square rectangle is isomorphic to the group of symmetry transformations of a non-square rhombus:

   (i)   by defining the transformations and writing out the Cayley table for each group

   (ii) by joining the midpoints of the edges of the rectangle.

5. (i)   Prove that all groups of order 2 are isomorphic.

   (ii) Show that there is only one way of completing this Latin square.

$$
\begin{array}{ccc}
e & a & b \\
a & \ldots & \ldots \\
b & \ldots & \ldots
\end{array}
$$

   Deduce that all groups of order 3 are isomorphic. Give a specific example of one such group.

6. Prove that it is impossible for an Abelian group to be isomorphic to a non-Abelian group.

7. Prove that the group of positive real numbers under multiplication is isomorphic to the group of all real numbers under addition. [**Hint:** Use logarithms.]

8. Prove that the group of matrices $\begin{pmatrix} a & 0 \\ 0 & a \end{pmatrix}$ under matrix addition and the group of matrices $\begin{pmatrix} 1 & 0 \\ b & 1 \end{pmatrix}$ under matrix multiplication, where a and b are real numbers, are isomorphic.

9. Show that each of the following sets of four elements with the stated operation forms a group, and sort these six groups into isomorphic sets.

   (i) $1, -1, j, -j$; multiplication of complex numbers

   (ii) functions $e(x) = x$, $f(x) = -\dfrac{1}{x}$, $g(x) = \dfrac{x-1}{x+1}$, $h(x) = \dfrac{1+x}{1-x}$; composition of functions

   (iii) $1, 2, 3, 4$; multiplication modulo 5

   (iv) $1, 3, 5, 7$; multiplication modulo 8

   (v) $\begin{pmatrix} 1 & 0 \\ 0 & 1 \end{pmatrix}$, $\begin{pmatrix} 1 & 0 \\ 0 & -1 \end{pmatrix}$, $\begin{pmatrix} -1 & 0 \\ 0 & 1 \end{pmatrix}$, $\begin{pmatrix} -1 & 0 \\ 0 & -1 \end{pmatrix}$; matrix multiplication

   (vi) $\begin{pmatrix} 1 & 0 \\ 0 & 1 \end{pmatrix}$, $\begin{pmatrix} 0 & 1 \\ -1 & 0 \end{pmatrix}$, $\begin{pmatrix} -1 & 0 \\ 0 & -1 \end{pmatrix}$, $\begin{pmatrix} 0 & -1 \\ 1 & 0 \end{pmatrix}$; matrix multiplication.

10. Suppose that a group of order 4 has the distinct elements e, a, b, c, where e is the identity.

   (i)   Explain why ab cannot equal a or b.

   (ii) Deduce that there are four possibilities:

     (a) $ab = ba = e$ (b) $ab = e, ba = c$

     (c) $ab = c, ba = e$ (d) $ab = ba = c$.

   (iii) By trying to complete the Cayley table, show that cases (b) and (c) lead to contradictions, that (a) gives the Cayley table for one group, and that (d) gives the Cayley tables for three groups.

   (iv) Show that the groups in (iii) are isomorphic to those in Question 9.

(This shows that there essentially only two groups of order 4: the *cyclic group* (with two self-inverse elements) and the *Klein 4-Group* (with four self-inverse elements), named after Felix Klein (1849–1925).)

Subgroups

The symmetry transformations of the equilateral triangle are of two types, rotations (including the identity transformation) and reflections. In Example B on page 183 these are I, P, Q and U, V, W respectively. From the Cayley table for this group we can see that the set of rotations is closed, contains the identity, and contains the inverse of each rotation. Therefore $\{I, P, Q\}$ forms a group under the same operation (composition of transformations); this is called a *subgroup*.

	I	P	Q	U	V	W
I	I	P	Q	U	V	W
P	P	Q	I	W	U	V
Q	Q	I	P	V	W	U
U	U	V	W	I	P	Q
V	V	W	U	Q	I	P
W	W	U	V	P	Q	I

The set of reflections $\{U, V, W\}$ is not a subgroup, since it does not contain the identity, but the subsets $\{I, U\}$, $\{I, V\}$ and $\{I, W\}$ are other subgroups.

Activity

Find whether $\{I, U, V, W\}$ is a subgroup.

The formal definition is as follows.

> A *subgroup* of a group $(S, \bullet)$ is a non-empty subset of S which forms a group under the same binary operation $\bullet$.

Every group has its *trivial subgroup* $\{e\}$, the subset consisting of just the identity. Since it is usual to count the set S as a subset of itself, the whole group is technically also a subgroup. Sometimes we want to exclude these two extremes, so we use the terms *proper subgroup* for any subgroup which is not the whole group, and *non-trivial subgroup* for any subgroup other than $\{e\}$.

EXAMPLE

Let S_4 be the group of permutations of four elements. Find a subgroup of S_4 which is isomorphic to the group G formed by the set $\{1, 2, 3, 4\}$ under $\times_5$ (multiplication mod 5).

Solution

The Cayley table for G is:

$\times_5$	1	2	3	4	permutation
1	1	2	3	4	e
2	2	4	1	3	a
3	3	1	4	2	b
4	4	3	2	1	c

Each row of this table is a permutation of the set $\{1, 2, 3, 4\}$. Label these permutations as shown, so that:

$$e = \begin{pmatrix} 1 & 2 & 3 & 4 \\ 1 & 2 & 3 & 4 \end{pmatrix} \quad a = \begin{pmatrix} 1 & 2 & 3 & 4 \\ 2 & 4 & 1 & 3 \end{pmatrix}$$

$$b = \begin{pmatrix} 1 & 2 & 3 & 4 \\ 3 & 1 & 4 & 2 \end{pmatrix} \quad c = \begin{pmatrix} 1 & 2 & 3 & 4 \\ 4 & 3 & 2 & 1 \end{pmatrix}.$$

It is easy to check that the set $\{e, a, b, c\}$ (which contains four of the 24 permutations in S_4) forms a subgroup isomorphic to G, with the following Cayley table.

	e	a	b	c
e	e	a	b	c
a	a	c	e	b
b	b	e	c	a
c	c	b	a	e

Direct comparison with the Cayley table for G shows that the subgroup $\{e, a, b, c\}$ of S_4 is isomorphic to G.

Exactly the same method is used in the proof of the following general result connecting any finite group with a subgroup of permutations.

Cayley's theorem

Let S_n be the group of permutations of n elements, and let G be any group of order n. Then G is isomorphic to a subgroup of S_n.

Proof

Let the elements of G be $a_1, a_2, \ldots, a_n$. By the Latin square property, each row of the Cayley table of G is a permutation of these n elements: the row corresponding to a_i is:

$$(a_i a_1 \quad a_i a_2 \quad \cdots \quad a_i a_n).$$

Let p_i be the permutation $\begin{pmatrix} a_1 & a_2 & \cdots & a_n \\ a_i a_1 & a_i a_2 & \cdots & a_i a_n \end{pmatrix}$, so that p_i maps a_k to $a_i a_k$, which we can write as $p_i(a_k) = a_i a_k$. Then the mapping $a_i \leftrightarrow p_i$ gives a one-to-one correspondence linking the elements of G with a subset of S_n.

Now let $\bullet$ be the operation of combination of permutations, so that $p_j \bullet p_i$ means p_i followed by p_j. Then:

$$(p_j \bullet p_i)(a_k) = p_j(p_i(a_k))$$

$$= p_j(a_i a_k)$$

$$= a_j(a_i a_k)$$

$$= (a_j a_i)a_k \qquad \text{(using associativity)}.$$

So under the permutation $p_j \bullet p_i$ the general element a_k maps to $(a_j a_i) a_k$, and therefore:

$$a_j a_i \quad \leftrightarrow \quad p_j \bullet p_i.$$

This shows that the mapping $a_i \leftrightarrow p_i$ is an isomorphism. Therefore G is isomorphic to a subgroup of S_n.

This completes the proof of Cayley's theorem.

Exercise 5E

1. Find all the proper subgroups of the symmetry group of a square. (Use the notation of Exercise 5C, Question 1.)

2. (i) List the elements of a subgroup of S_4 (the group of permutations of 4 elements) which is isomorphic to the symmetry group of a non-square rectangle.

   (ii) List the elements of a subgroup of S_8 which is isomorphic to the symmetry group of a square.

3. Prove that if H and K are two subgroups of a group G then $H \cap K$ is also a subgroup. Give an example to show that $H \cup K$ need not be a subgroup.

4. Show that the symmetry group of a regular n-sided polygon has $2n$ elements, comprising n rotations (including the identity) and n reflections. (When dealing with the reflections, take separately the cases when n is odd and when n is even.) This is called the *dihedral* group, D_n. Show that if k is any factor of $2n$ then D_n has a subgroup of order k.

5. (i) Show that successive half-turns about the centres (a, b) and (c, d) are equivalent to the translation $2\begin{pmatrix} c - a \\ d - b \end{pmatrix}$.

   (ii) Show that the set of rotations and translations of a plane under the composition of transformations forms a group.

   (iii) Show that the set of translations is a subgroup and explain why the set of rotations is not a subgroup. Find two subgroups of which all of the elements are rotations.

6. Let G be a group and let H be any non-empty subset of the elements of G. Prove that H is a subgroup of G if and only if $xy^{-1} \in H$ for all $x, y \in H$.

   The centre C of a group G is defined to be the set of all elements of G which commute with every element of G (that is, $c \in C$ if and only if $gc = cg$ for all $g \in G$). Prove that C is a subgroup of G.

   Show that the set of matrices of the form
   $$\begin{pmatrix} 1 & p & q \\ 0 & 1 & r \\ 0 & 0 & 1 \end{pmatrix}, \text{ where } p, q, r \text{ are rational}$$
   numbers, forms a group under matrix multiplication. Find the centre of this group. [SMP]

7. The table shows a Latin square for $\{E, A, B, C, D\}$ under the operation $\bullet$, together with the permutations e, a, b, c, d defined by the rows of this square.

$\bullet$	E	A	B	C	D	permutation
E	E	A	B	C	D	e
A	A	C	D	B	E	a
B	B	D	C	E	A	b
C	C	E	A	D	B	c
D	D	B	E	A	C	d

Show that the set of permutations is not closed, and deduce that $\{E, A, B, C, D\}$ does not form a group under $\bullet$.

Lagrange's theorem

The symmetry group of the equilateral triangle (i.e. Example B from page 183) is a finite group of order 6, and has subgroups of order 1 ($\{I\}$), 2 (three of them, $\{I, U\}$, $\{I, V\}$, $\{I, W\}$), 3 ($\{I, P, Q\}$), and 6 (group B itself). The order of each subgroup is a factor of the order of the group.

Activity

Use the answer to Question 1 of Exercise 5E to check that the order of each subgroup of the symmetry group of a square is a factor of the order of the group.

The purpose of this section is to prove that what happens in these cases must always happen: the order of each subgroup of a finite group is a factor of the order of the group. This result is known as *Lagrange's theorem*.

As a tool for proving this (and an important idea in its own right) we first introduce the idea of a *coset*. Suppose that $H = \{h_1, h_2, h_3, \ldots, h_m\}$ is a subgroup of a group $(S, \bullet)$, and that x is any element of S. Then the set of elements $\{x \bullet h_1, x \bullet h_2, x \bullet h_3, \ldots, x \bullet h_m\}$ is called the *left coset* of H by x, and is denoted by xH.

For example, if $S = \{I, P, Q, U, V, W\}$ and $H = \{I, P, Q\}$ as in Example B then

$$PH = \{PI, PP, PQ\} = \{P, Q, I\} = H \text{ and } UH = \{UI, UP, UQ\} = \{U, V, W\}.$$

Activity

(i) Find the other four left cosets IH, QH, VH and WH. How many distinct left cosets of H are there? What is the union of all these cosets?

(ii) Let $K = \{I, U\}$. Show that there are only three distinct left cosets of K, each containing two elements, and that the union of these is the whole set S.

NOTE *The right coset Hx is defined similarly as $\{h_1 \bullet x, h_2 \bullet x, h_3 \bullet x, \ldots, h_m \bullet x\}$. If the group is Abelian then $xH = Hx$, but this need not be true in other cases: for example, in the activity above, $PK = \{P, W\}$ but $KP = \{P, V\}$. The arguments that follow use left cosets throughout, even when the word 'left' is omitted.*

The behaviour in this activity is typical of the general case: the left cosets of a subgroup $H = \{h_1, h_2, h_3, \ldots, h_m\}$ of a group $(S, \bullet)$ have the following properties.

Property (1)

Every coset has m distinct elements, where m is the order of H.

Proof

In the coset $xH = \{x \bullet h_1, x \bullet h_2, x \bullet h_3, \ldots, x \bullet h_m\}$ all the elements are distinct, since $x \bullet h_i = x \bullet h_j \Rightarrow h_i = h_j$ by the left cancellation rule.

Property (2)

If $y \notin xH$ then xH and yH are disjoint (i.e. have no common element).

Proof

Suppose that xH and yH do have a common element, a say.

Then $a = x \bullet h_i = y \bullet h_j$, and so $y = x \bullet h_i \bullet h_j^{-1}$.

But since H is a subgroup $h_j^{-1} \in H$ and so (by closure) $h_i \bullet h_j^{-1} \in H$.

Therefore $x \bullet h_i \bullet h_j^{-1} \in xH$, i.e. $y \in xH$, which is a contradiction.

Therefore there can be no common element.

We can now prove the main result.

Lagrange's theorem

The order of any subgroup of a finite group is a factor of the order of the group.

Proof

Let H be a subgroup of a finite group $(S, \bullet)$, where H has order m and S has order n.

If $H = S$ then $m = n$ and the theorem is true.

If H is a proper subgroup then choose an element of S which is not in H: let this element be x, say. From property (1) the coset xH has m elements, and from property (2) H and xH are disjoint.

Therefore $H \cup xH$ has $2m$ elements.

If $H \cup xH = S$ then $n = 2m$ and the theorem is true.

If $H \cup xH \neq S$ then choose an element of S which is not in $H \cup xH$: let this element be y, say. From property (1) the coset yH has m elements and from property (2) H, xH and yH are all disjoint.

Therefore $H \cup xH \cup yH$ has $3m$ elements.

If $H \cup xH \cup yH = S$ then $n = 3m$ and the theorem is true.

If $H \cup xH \cup yH \neq S$ then proceed as before, choosing an element of S not in the union of the cosets used so far, and forming its left coset. At each stage this brings m extra elements into the union of cosets, so the total number of elements remains a multiple of m. This process must terminate, since S is finite, so at some stage the union of cosets equals S. If this occurs when there are k cosets (including H itself) then $n = km$, which proves the theorem.

HISTORICAL NOTE

During his (unsuccessful) search for a formula for the roots of a quintic equation Joseph Louis de Lagrange (1736–1813) found a result in the theory of equations which is a special case of the theorem which now bears his name. Although this was before the concept of a group was recognised, Lagrange's method of proving his result was essentially the same as the proof above, using the equivalents of cosets to partition the set whose elements he wanted to count.

Exercise 5F

1. Find the left cosets XK, AK and RK, where K is the subgroup $\{I, A, H, B\}$ of the symmetry group of a square (as in Question 1 of Exercise 5E).

2. Let C_n be the subgroup of rotations of the dihedral group D_n (as in Question 4 of Exercise 5E), and let $R \in D_n$ be any reflection. Prove that $D_n = C_n \cup RC_n$.

3. Let G be a group in which every element x satisfies the equation $x^2 = e$ (where e is the identity element of G). Show that G is Abelian. If, in addition, G is finite and contains more than two elements, show that G contains a subgroup H of order four, and deduce that the number of elements of G is of the form $4k$, where k is an integer.

   By considering cosets of the form gH (where g denotes an element of G), prove that the product of all the elements of G is equal to e. [MEI]

4. A subgroup H of a finite group G is said to be a *normal subgroup* if, for each element $h \in H$ and each element $g \in G$, the element $g^{-1}hg \in H$. Prove that:

   (i) G and $\{e\}$ (where e is the identity element of G) are normal subgroups of G

   (ii) if G is a commutative group, then every subgroup of G is a normal subgroup

   (iii) H is a normal subgroup if and only if every left coset of H is a right coset of H (that is, $gH = Hg$, for all $g \in G$)

   (iv) if the order of G is twice the order of H, then H is a normal subgroup of G. Give (and justify) an example of a finite group G and a subgroup H which is *not* a normal subgroup. [MEI]

Investigation

Change ringing

The object of change ringing a peal of n bells is to ring all the $n!$ possible permutations of the bells without any omissions or repetitions. Clearly it is quite a task to organise this, and to ring the permutations in a sequence which makes it relatively easy for a particular ringer to remember what to do. This investigation shows one method for a peal of four bells, for which there are $4! = 24$ permutations.

(i) For an ordered set of four objects, let A be the operation of interchanging the first and second and interchanging the third and fourth objects, and let B be the operation of interchanging the second and third objects, as shown diagrammatically below.

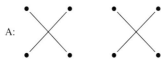

Show that $A: (1 \quad 2 \quad 3 \quad 4) \rightarrow (2 \quad 1 \quad 4 \quad 3)$ and
$B: (2 \quad 1 \quad 4 \quad 3) \rightarrow (2 \quad 4 \quad 1 \quad 3)$
as shown in the following diagram.

Copy and complete the diagram to show that the sequence A, B, A, B, A, B, A, B produces a loop of eight permutations.

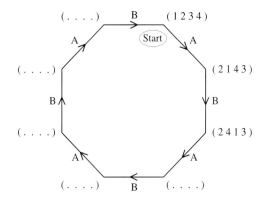

(ii) Using, for example, (2 1 4 3) as shorthand for the permutation
$\begin{pmatrix} 1 & 2 & 3 & 4 \\ 2 & 1 & 4 & 3 \end{pmatrix}$, show that the eight permutations in (i) form a subgroup H
of the permutation group S_4.

(iii) Now suppose that the final operation B in (i) is replaced by the operation C
which interchanges the third and fourth objects, as in this diagram.

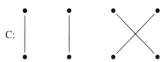

This leads to (1 3 4 2) instead of (1 2 3 4), thus stopping the loop
from closing. Applying the sequence A, B, A, B, A, B, A, B again now gives
another loop of eight permutations, which is a coset of H.

(iv) Again replacing the final B by C gives a third loop containing the final eight
permutations; these form a second coset of H. And if the last B in this loop
is replaced by C we get back to (1 2 3 4), the starting point. This is
shown in the diagram below, which you should copy and complete.

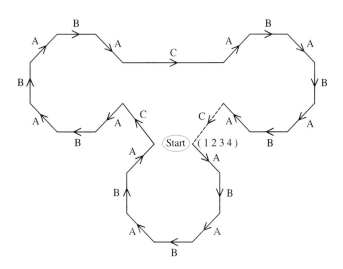

(v) Now that you have found this sequence of all 24 permutations you can see how each individual bell ringer can know where to ring in each permutation. This is shown for the ringers of bells 1 and 2 in the diagram below. Ringer 1 has the easiest task, moving steadily through the order of ringing from first to last and then from last to first ('hunting'). Ringer 2 does mostly the same, but with a variant each time C is used ('dodging').

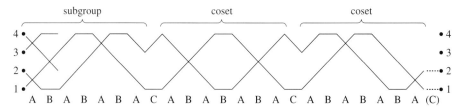

Copy the diagram and complete the patterns for ringers 3 and 4. (It is actually easier to draw the whole diagram from scratch, since it consists of the patterns for A, B and C (above) turned upright for convenience and put together in the correct order.)

Cyclic groups

Activity

See what happens when you work out successive powers of each element of the group $\{1, 2, 4, 8\}$ under $\times_{15}$.

Let x be an element of a finite group $(S, \bullet)$, and consider the set of powers of x:

$$P = \{x, x^2, x^3, \ldots\}$$

where $x^r = x \bullet x \bullet x \bullet \cdots \bullet x$ with r factors.

By closure, all the elements of P belong to S, which is finite. Therefore P must be finite, and so at some stage in the list of powers of x there must be an element which has occurred earlier in the list. Suppose that the first time this happens the earlier element is x^j and the later element is x^{j+k}, so that $x^{j+k} = x^j$. Multiplying both sides of this by $(x^{-1})^j$ gives $x^k = e$, where e is the identity element. Therefore P contains e.

The smallest positive power k such that $x^k = e$ is called the *order* (or *period*) of the element x. This shows that every element of a finite group has a finite order.

Activity

Referring yet again to the examples on pages 183–185, find the order of

(i) 5 in A (ii) U in B (iii) q in C (iv) $-\omega$ in D.

If the element x of $(S, \bullet)$ has order k then $P = \{e, x, x^2, \ldots, x^{k-1}\}$. To see this, note that:

(i) any higher index r can be written in the form $r = ak + b$, where $0 \leqslant b \leqslant k - 1$, so that $x^r = (x^k)^a x^b = x^b$ (since $x^k = e$), which is already listed

(ii) the elements listed are all distinct, since if two were equal then k would not be the *smallest* positive power for which $x^k = e$.

The set P is closed under the group operation since if the result of combining two elements is x^r with $r \geqslant k$ the procedure in (i) gives $x^r = x^{r-k}$, which is in P. Moreover, the inverse of each element of P is also in P, since the inverse of x^b $(0 < b \leqslant k - 1)$ is x^{k-b}. Therefore P is a subgroup of S; this is called the subgroup *generated* by x.

Activity

List the elements of the subgroups of Example D generated by:

(i) 1   (ii) -1 (iii) ω (iv) $-\omega$.

The subgroup generated by x has order k, where k is the order of the element x. Therefore, by Lagrange's theorem, the order of each element of a group is a factor of the order of the group.

A group generated by a single element is called a *cyclic* group. The cyclic group of order n is denoted by C_n, and is isomorphic to the group generated by the rotation of a plane through $2\pi/n$ about a fixed point. Therefore cyclic groups of all orders do exist.

Activity

Are there any cyclic groups among the examples A, B, C and D?

Finally, if the order of a finite group is a *prime* number p then the only possible orders for the elements of the group are 1 and p. Only the identity element e has order 1, so every other element must have order p, and must generate the whole of the group. The group is therefore isomorphic to C_p, the cyclic group of order p. This shows that all groups of prime order are cyclic.

EXAMPLE Prove that there are only two non-isomorphic groups of order 4.

Solution

One method has already been outlined in Question 10 of Exercise 5D, but we can now give a more economical solution.

In a group G of order 4 each element, except for the identity e, must have order 2 or 4. There are two possibilities.

(1) If G contains an element a of order 4 then the elements e, a, a^2, a^3 are distinct, and so G is C_4, the cyclic group of order 4.

(2) If G contains no element of order 4 then each element, except for e, has order 2. Suppose e, a, b are distinct elements of G, with $a^2 = b^2 = e$.

The element ab does not equal any of these, since:

$$ab = e = a^2 \Rightarrow b = a, \quad ab = a \Rightarrow b = e, \quad ab = b \Rightarrow a = e$$

all of which contradict the supposition that e, a, b are distinct.

Therefore ab is the fourth element of G. Similarly:

$$ba = e = a^2 \Rightarrow b = a, \quad ba = a \Rightarrow b = e, \quad ba = b \Rightarrow a = e$$

all of which are contradictory. Therefore $ba = ab$. It is now easy to complete the Cayley table (using, for example, $b(ab) = b(ba) = b^2a = ea = a$):

	e	a	b	ab
e	e	a	b	ab
a	a	e	ab	b
b	b	ab	e	a
ab	ab	b	a	e

which shows that G is V, the Klein 4-Group.

Exercise 5G

1. List the orders of all the elements of D_4, the symmetry group of a square, and find all the cyclic subgroups.

2. Prove that in a group the elements ab and ba have the same order.

3. Prove by induction that, in a group, $(b^{-1}ab)^n = b^{-1}a^nb$. Deduce that $b^{-1}ab$ has the same order as a.

4. (i) Show that C_5 (the cyclic group of order 5) can be generated by every one of its elements except for the identity. Show that the corresponding statement about C_6 is not true.

(ii) Investigate similar statements about C_7 and C_8.

(iii) Given that a is a generator of C_n, find a necessary and sufficient condition for a^k to generate C_n.

5. The Big Wheel at a fairground has 20 seats spaced equally around the rim. At the end of the ride the machinery automatically advances the wheel through successive angles of $\dfrac{\pi}{10}$ radians so that passengers may dismount in turn. On one occasion, to the dismay of all the passengers except one, the machinery develops a fault at the end of the ride and begins to advance the wheel through successive angles of $\dfrac{n\pi}{10}$, where n is an integer. The remaining passenger, a mathematician, notices that $n = 9$ and patiently awaits his turn to dismount. Explain his confidence. What other values of n would leave him unperturbed?   [MEI]

6. Let a and b be distinct elements of a finite group G. Show that the set of elements obtained by taking all possible products of a finite number of a's and b's, including repetitions (for example aba^2b^3a), is a subgroup H of G.

   (We say that a and b are generators of H.)

7. With the notation of Question 1 of Exercise 5C, for the group D_4:

   (i)   by expressing H, S, Y, A, B in terms of R and/or X show that R and X are generators of the group

   (ii) show that $R^4 = X^2 = I$ and $RX = XR^3$ (these are called *defining relations*).

8. Show that D_n is generated by two elements R and X with $R^n = X^2 = I$ and $RX = XR^{n-1}$.

9. Let $p = 2^k + 1$ be a prime number, and let G be the group of integers $1, 2, \ldots, p - 1$, with multiplication defined modulo p.

   (i)   Show that if $0 < m < k$ then $0 < 2^m - 1 < p$ and deduce that $2^m \neq 1 \pmod{p}$.

   (ii) Show that if $k < m < 2k$ then $2^m = 1 \pmod{p} \Rightarrow 2^{2k-m} = -1 \pmod{p}$ and deduce that $2^m \neq 1 \pmod{p}$.

   (iii) Use (i) and (ii) to show that the order of the element 2 in G is $2k$.

   (iv) Deduce that k is a power of 2.

10. A set G consists of 1 and all positive integers having no squared factors. Prove that G is a group with respect to the law of composition:

$$a \bullet b = \frac{m}{d}$$

   where m is the least common multiple of a and b, and d is the greatest common divisor.

   Prove that the subgroup generated by the distinct elements $a_1, a_2, \ldots, a_k$ has order 2^t, $1 \leqslant t \leqslant k$ and that $t = k$ if $a_1, a_2, \ldots, a_k$ are coprime in pairs. (The pair a, b is *coprime* if and only if $d = 1$.)   [MEI]

Fields

The real number system is an example of a more complicated algebraic structure with a set $\mathbb{R}$ of real numbers and two basic binary operations, addition ($+$) and multiplication ($\times$). This system has the following properties:

   (i) $(\mathbb{R}, +)$ is an Abelian group, with identity 0

   (ii) $(\mathbb{R}\backslash\{0\}, \times)$ is an Abelian group, with identity 1

$\mathbb{R}\backslash\{0\}$ means the set of non-zero real numbers.

   (iii) multiplication is *distributive* over addition; that is:

$$a \times (b + c) = (a \times b) + (a \times c) \text{ for all } a, b, c \in \mathbb{R}.$$

Abstract algebra

The other two operations, subtraction ($-$) and division ($\div$), are defined by:

$$a - b = a + (-b), \text{ where } -b \text{ is the additive inverse of } b$$

$$a \div b = a \times b^{-1}, \text{ where } b^{-1} \text{ is the multiplicative inverse of } b.$$

Note that $-$ and $\div$ are neither commutative nor associative.

For Discussion

Are there any other distributive properties involving $+$, $-$, $\times$, $\div$?

There are other sets which have these properties, and we say that a set S with two binary operations which has all the properties (i), (ii) and (iii) forms a *field* $(S, +, \times)$.

The product $a \times b$ is often written as ab, and the usual convention that multiplication takes precedence over addition saves writing some brackets: thus the distributive property can be written as $a(b + c) = ab + ac$.

Activity

Check that the set of rational numbers under addition and multiplication forms a field, and explain why the set of irrational numbers does not.

The rational numbers and the real numbers both have all the field properties, and yet some statements (for example "there is a number x such that $x^2 = 2$") are true for real numbers but false for rational numbers. So the field properties are not sufficient to determine the behaviour of a set of numbers completely. However, all the standard algebraic processes of simplifying, expanding, factorising and solving equations (but not extracting roots or using inequalities) can be derived from the field axioms.

EXAMPLE

Prove that, in every field, $a \times 0 = 0$.

Solution

$$
\begin{array}{lll}
& 1 + 0 = 1 & \text{(0 is neutral)} \\
\Rightarrow & a \times (1 + 0) = a \times 1 & \\
\Rightarrow & a \times 1 + a \times 0 = a \times 1 & \text{(distributivity)} \\
\Rightarrow & a + a \times 0 = a & \text{(1 is neutral)} \\
\Rightarrow & a \times 0 + a = a & \text{(commutativity)} \\
\Rightarrow & (a \times 0 + a) + (-a) = a + (-a) & \\
\Rightarrow & a \times 0 + (a + (-a)) = a + (-a) & \text{(associativity)} \\
\Rightarrow & a \times 0 + 0 = 0 & \text{(use of inverse)} \\
\Rightarrow & a \times 0 = 0 & \text{(0 is neutral).}
\end{array}
$$

Activity

Prove that, in every field, $ab = 0 \Rightarrow a = 0$ or $b = 0$.

NOTE *This is the fundamental result needed for the method of solving equations by factorising.*

The field of complex numbers

It is easy to check that the set of enlargement matrices $\begin{pmatrix} x & 0 \\ 0 & x \end{pmatrix}$, where $x \in \mathbb{R}$, forms a field under the operations of matrix addition and matrix multiplication with the matrices $\mathbf{O} = \begin{pmatrix} 0 & 0 \\ 0 & 0 \end{pmatrix}$ and $\mathbf{I} = \begin{pmatrix} 1 & 0 \\ 0 & 1 \end{pmatrix}$ as the identities. Moreover, the mapping $x \leftrightarrow \begin{pmatrix} x & 0 \\ 0 & x \end{pmatrix}$ is an isomorphism between the field of real numbers and the field of enlargement matrices, preserving the structure of both addition and multiplication.

Activity

Check this.

This idea can be extended by considering matrices of the form $\begin{pmatrix} x & -y \\ y & x \end{pmatrix}$, where $x, y \in \mathbb{R}$; these are called *spiral* matrices.

Activity

Account for this name by describing the geometrical transformation represented by this matrix and check that the set of spiral matrices forms a field under matrix addition and matrix multiplication.

This field has the set of enlargement matrices as a *subfield* (i.e. subset which is itself a field under the same operations), isomorphic to the real numbers. But it also has a crucial property which the real numbers lack, for if $\mathbf{J}$ denotes the particular spiral matrix $\begin{pmatrix} 0 & -1 \\ 1 & 0 \end{pmatrix}$ then $\mathbf{J}^2 = -\mathbf{I}$.

Activity

Check this. What transformation does $\mathbf{J}$ represent?

Every spiral matrix $\mathbf{Z} = \begin{pmatrix} x & -y \\ y & x \end{pmatrix}$ can be written in the form $\mathbf{Z} = x\mathbf{I} + y\mathbf{J}$, and such expressions can be manipulated by 'ordinary' algebra, using $\mathbf{I}^2 = -\mathbf{J}^2 = \mathbf{I}, \mathbf{IJ} = \mathbf{JI} = \mathbf{J}$.

As a further simplification we can write 1 for **I** and j for **J** to obtain expressions of the form $z = x + yj$, where $x, y \in \mathbb{R}$, which behave as if they were real numbers with the extra property that $j^2 = -1$. This is exactly the *Bold Hypothesis* with which complex numbers were introduced on page 57 of *Pure Mathematics 4*, but now we can be sure that the complex numbers form a field (isomorphic to the field of spiral matrices). This resolves the longstanding (though probably forgotten) question of whether our Bold Hypothesis might lead to a contradiction.

Exercise 5H

1. Prove that, in every field, if $a \neq 0$ then:

$$ax + b = c \quad \Leftrightarrow \quad x = (c - b) \div a,$$

making clear which field property is used at each step.

2. Prove that the usual 'rules of signs':

$$(-a)b = a(-b) = -(ab), \ (-a)(-b) = ab$$

hold in every field.

3. Prove that the set of numbers $p + q\sqrt{2}$, where p and q are rational, forms a field under ordinary addition and multiplication.

4. Investigate the circumstances in which the set $\{0, 1, 2, \ldots, m - 1\}$ forms a field under the operations of addition and multiplication modulo m.

5. The set Z consists of all integers (positive, zero and negative). The operations of 'addition', $\oplus$, and 'multiplication', $\otimes$, are defined on Z as follows:

$$m \oplus n = m + n - 1 \quad m \otimes n = m + n - mn.$$

Show that the system $(Z, \oplus, \otimes)$ is closed, commutative, and associative with respect to each operation, and that 'multiplication' is distributive over 'addition'. Show also that each operation has an identity element in Z, and find them. Determine whether each element of Z has an inverse with respect to each operation. [MEI]

6. Use the isomorphism between spiral matrices and complex numbers to write the matrix equation:

$$\begin{pmatrix} 2 & -1 \\ 1 & 2 \end{pmatrix} Z + \begin{pmatrix} 4 & 5 \\ -5 & 4 \end{pmatrix} = \begin{pmatrix} 1 & -3 \\ 3 & 1 \end{pmatrix} Z$$

in terms of complex numbers, and hence find the matrix **Z**.

7. Use complex numbers to find two matrices **Z** for which:

$$\mathbf{Z}^2 + \begin{pmatrix} -9 & -2 \\ 2 & -9 \end{pmatrix} \mathbf{Z} + \begin{pmatrix} 23 & 7 \\ -7 & 23 \end{pmatrix} = \begin{pmatrix} 0 & 0 \\ 0 & 0 \end{pmatrix}.$$

8. Find $\mathbf{P}^6$ if $\mathbf{P} = \begin{pmatrix} \frac{\sqrt{3}}{2} & -\frac{1}{2} \\ \frac{1}{2} & \frac{\sqrt{3}}{2} \end{pmatrix}$.

9. (i) What operation on complex numbers corresponds to transposing (i.e. interchanging rows and columns) spiral matrices?

(ii) Establish the result $zz^* = |z|^2$ by using matrices.

(iii) What property of determinants corresponds to the fact that $|z_1 z_2| = |z_1| |z_2|$?

10. (i) Show that the set of matrices

$$\mathbf{Q} = \begin{pmatrix} z & -w^* \\ w & z^* \end{pmatrix}$$

where z and w are complex numbers, satisfies all the field axioms under matrix addition and multiplication, except that multiplication is not commutative.

(ii) Show that if $z = a + bj$ and $w = c + dj$ then $\mathbf{Q} = a\mathbf{I} + b\mathbf{E} + c\mathbf{F} + d\mathbf{G}$, where:

$$\mathbf{I} = \begin{pmatrix} 1 & 0 \\ 0 & 1 \end{pmatrix} \quad \mathbf{E} = \begin{pmatrix} j & 0 \\ 0 & -j \end{pmatrix}$$

$$\mathbf{F} = \begin{pmatrix} 0 & -1 \\ 1 & 0 \end{pmatrix} \quad \mathbf{G} = \begin{pmatrix} 0 & j \\ j & 0 \end{pmatrix}$$

(iii) Show that $\mathbf{E}^2 = \mathbf{F}^2 = \mathbf{G}^2 = -\mathbf{I}$, $\mathbf{EF} = -\mathbf{FE} = -\mathbf{G}$, $\mathbf{FG} = -\mathbf{GF} = -\mathbf{E}$, $\mathbf{GE} = -\mathbf{EG} = -\mathbf{F}$.

Show also that the same relations can be written more compactly as $\mathbf{E}^2 = \mathbf{F}^2 = \mathbf{G}^2 = -\mathbf{EFG} = -\mathbf{I}$.

(iv) By writing 1, e, f and g in place of $\mathbf{I}$, $\mathbf{E}$, $\mathbf{F}$ and $\mathbf{G}$ show that the system in (i) is isomorphic to the system of expressions of the form $q = a + be + cf + dg$, where a, b, c and d are real and the elements e, f and g act like real numbers with the additional properties:

$$e^2 = f^2 = g^2 = -1$$

$$ef = -fe = -g,\ fg = -gf = -e,$$

$$ge = -eg = -f.$$

('Super-complex' numbers of this form are called *quaternions*. Invented by William Rowan Hamilton in 1843, quaternions are the earliest example of a non-commutative algebraic system. They were studied extensively in the nineteenth century (usually being written in the form $a + bi + cj + dk$) and were found to have many applications which would now be handled by vectors.)

Vector spaces

The array $\begin{bmatrix} 4 & 9 & 2 \\ 3 & 5 & 7 \\ 8 & 1 & 6 \end{bmatrix}$ is perhaps the oldest known *magic square*, a square array of numbers such that the sum of the entries in any row, column or diagonal is identical. The elements of this magic square are the consecutive integers 1 to 9, but for our purposes the elements of a magic square may be any real numbers provided that each row, each column and each diagonal has the same sum. With this definition, the array $\begin{bmatrix} 5 & -2 & 3 \\ 0 & 2 & 4 \\ 1 & 6 & -1 \end{bmatrix}$ is also a magic square. Adding corresponding entries of these two magic squares gives another magic square, $\begin{bmatrix} 9 & 7 & 5 \\ 3 & 7 & 11 \\ 9 & 7 & 5 \end{bmatrix}$.

Activity

Let V be the set of all 3×3 magic squares, and define two binary operations $\oplus$ and $\otimes$ as follows:

$\oplus$ denotes the addition of two magic squares term-by-term as illustrated above

$\otimes$ denotes multiplying a magic square by a real number so that the product $\lambda \otimes \mathbf{M}$ is the array obtained by multiplying each element of magic square $\mathbf{M}$ by real number λ.

(i) Show that: (a) $(V, \oplus)$ is an Abelian group, with identity $\begin{bmatrix} 0 & 0 & 0 \\ 0 & 0 & 0 \\ 0 & 0 & 0 \end{bmatrix}$

(b) if X is a member of V then $\lambda \otimes X$ is also a member of V

(c) $\otimes$ is distributive over $\oplus$;

i.e. $\lambda \otimes (\mathbf{M} \oplus \mathbf{N}) = (\lambda \otimes \mathbf{M}) \oplus (\lambda \otimes \mathbf{N})$

where $\mathbf{M}$ and $\mathbf{N}$ are any 3×3 magic squares, and λ is any real number.

(ii) Given that $\mathbf{L} = \begin{bmatrix} 1 & 1 & 1 \\ 1 & 1 & 1 \\ 1 & 1 & 1 \end{bmatrix}$, $\mathbf{M} = \begin{bmatrix} 1 & -1 & 0 \\ -1 & 0 & 1 \\ 0 & 1 & -1 \end{bmatrix}$, $\mathbf{N} = \begin{bmatrix} 0 & -1 & 1 \\ 1 & 0 & -1 \\ -1 & 1 & 0 \end{bmatrix}$

show that $\begin{bmatrix} 4 & 9 & 2 \\ 3 & 5 & 7 \\ 8 & 1 & 6 \end{bmatrix}$ can be expressed as $(\lambda \otimes \mathbf{L}) \oplus (\mu \otimes \mathbf{M}) \oplus (\nu \otimes \mathbf{N})$ by

putting $\lambda = 5$, $\mu = -1$, $\nu = -3$.

(The activity on page 222 shows that any 3×3 magic square can be expressed in this form.)

Activity

Let V be the set of all arithmetic sequences, and define two binary operations $\oplus$ and $\otimes$ as follows:

$\oplus$ denotes the addition of two sequences term by term so that if $S \equiv 3, 5, 7, 9, \ldots$ and $T \equiv 9, 14, 19, 24, \ldots$ then $S \oplus T \equiv 12, 19, 26, 33, \ldots$.

$\otimes$ denotes multiplying a sequence by a real number so that the product $\lambda \otimes S$ is the sequence obtained by multiplying each term of sequence S by real number λ.

(i) Show that: (a) $(V, \oplus)$ is an Abelian group

(b) if X is a member of V then $\lambda \otimes X$ is also a member of V

(c) $\otimes$ is distributive over $\oplus$.

(ii) Write down any arithmetic sequence and express it in the form $(\lambda \otimes A) \oplus (\mu \otimes D)$, where λ, μ, are real numbers, and A and D are the sequences $1, 1, 1, 1, \ldots$ and $0, 1, 2, 3, \ldots$ respectively.

The set V of vectors of the form $\begin{pmatrix} x \\ y \end{pmatrix}$ where x, y are real numbers, together with the usual vector addition, exhibits similar features.

(a) $(V, +)$ is an Abelian group.

(b) V is closed with respect to multiplication by a scalar (i.e. $\mathbf{v} \in V \Rightarrow \lambda \mathbf{v} \in V$, where λ is real).

(c) Multiplication by a scalar is distributive over vector addition.

(d) Each member of V can be expressed in the form $\lambda \mathbf{i} + \mu \mathbf{j}$, where $\mathbf{i} = \begin{pmatrix} 1 \\ 0 \end{pmatrix}$, $\mathbf{j} = \begin{pmatrix} 0 \\ 1 \end{pmatrix}$.

These same properties have also been observed in several other contexts, including:

- n-dimensional vectors
- the solutions of differential equations such as $\dfrac{d^2 y}{dx^2} - 5\dfrac{dy}{dx} + 6y = 0$
- polynomials of degree not exceeding (say) 3

- the solutions of matrix equations such as $\begin{pmatrix} 6 & 2 \\ 3 & 1 \end{pmatrix} \begin{pmatrix} x \\ y \end{pmatrix} = \begin{pmatrix} 0 \\ 0 \end{pmatrix}$

- functions which are continuous for $0 \leqslant x \leqslant 1$.

These common properties introduce us to the axioms of another abstract system. Let F be the field $(\{\lambda, \mu, \nu, \ldots\}; +, \times)$. Members of F are usually called *scalars*. Let V be the set of elements $\{\mathbf{a}, \mathbf{b}, \mathbf{c}, \ldots\}$, usually called *vectors*. Let $\oplus$ be a binary operation between members of V, usually called addition. Let $\otimes$ be a binary operation between a member of F and a member of V, usually called multiplication. Then $(V, \oplus, \otimes)$ is a *vector space over the field F* if $(V, \oplus)$ is an Abelian group and the product of scalar λ and vector $\mathbf{a}$ is $\lambda \otimes \mathbf{a}$, such that, for all scalars λ and μ in F and all vectors $\mathbf{a}$ and $\mathbf{b}$ in V the following hold.

(C')	$\lambda \otimes \mathbf{a}$ is a member of V.	V is closed with respect to multiplication by a scalar.
(A')	$(\lambda\mu) \otimes \mathbf{a} = \lambda \otimes (\mu \otimes \mathbf{a})$	This resembles the associative property for multiplication, but $\lambda\mu$ is the product of two scalars, while the other multiplication involves multiplying a vector by a scalar.
(N')	When $\lambda = 1$, $\lambda \otimes \mathbf{a} = \mathbf{a}$.	This resembles the neutral element property of a group.
(D)	$\lambda \otimes (\mathbf{a} \oplus \mathbf{b}) = (\lambda \otimes \mathbf{a}) \oplus (\lambda \otimes \mathbf{b})$	Multiplication by a scalar is distributive over (vector) addition.
(D')	$(\lambda + \mu) \otimes \mathbf{a} = (\lambda \otimes \mathbf{a}) \oplus (\mu \otimes \mathbf{a})$	This resembles a distributive property but the additions are different binary operations.

It is quite usual to suppress $\otimes$, the sign we have coined for multiplying a vector by a scalar. It is also quite usual to use $+$ for the addition of two vectors, where we have used $\oplus$; even in D', where $+$ will have two different meanings, this need not cause confusion. Strictly a vector space consists of the set V, together with two operations, which we have denoted $\oplus$ and $\otimes$. Provided there is no loss of clarity, we shall, however, often refer to a vector space V without specifically mentioning the two operations. You may assume that the field in use is the field of real numbers unless stated otherwise.

To check that a particular system is a vector space you will have to check that the system satisfies a total of ten axioms.

The 'vectors' and their 'addition' rule form a group.	– four axioms: C, A, N, I
That group is commutative (i.e. Abelian).	– one axiom
The 'scalars' combine with the 'vectors' as described above.	– five axioms: C', A', N', D, D'

EXAMPLE

The set V consists of all points on a plane π which contains the origin O.

Define addition of points by $P \oplus Q \equiv (x_P + x_Q, y_P + y_Q, z_P + z_Q)$, where $P \equiv (x_P, y_P, z_P)$, etc.

Define multiplication of a point by a scalar by $\lambda P \equiv (\lambda x_P, \lambda y_P, \lambda z_P)$.

Show that V is a vector space over the real numbers.

This vector space consists of points!

Solution

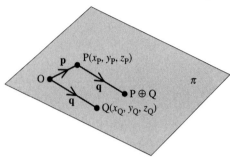

We need to check that each of the ten axioms is satisfied.

The origin O is a point in plane π and the definitions given are equivalent to:

$P \oplus Q$ is the point with position vector $\mathbf{p} + \mathbf{q}$

λP is the point with position vector $\lambda \mathbf{p}$.

As usual, the position vector of P is $\mathbf{p}$.

C: Since the origin is in plane π:

P and Q are members of V $\Rightarrow$ $\mathbf{p}$ and $\mathbf{q}$ are vectors in plane π

$\Rightarrow$ $\mathbf{p} + \mathbf{q}$ is a vector in plane π

$\Rightarrow$ $P \oplus Q$ is a point in plane π

$\Rightarrow$ V is closed under $\oplus$.

A: $(P \oplus Q) \oplus R$ has position vector $(\mathbf{p} + \mathbf{q}) + \mathbf{r}$.

$P \oplus (Q \oplus R)$ has position vector $\mathbf{p} + (\mathbf{q} + \mathbf{r})$.

Since vector addition is associative, these points are identical, proving the associative property.

N: The origin O is a member of V and $\mathbf{p} + \mathbf{0} = \mathbf{p}$ $\Rightarrow$ $P \oplus O = P$ showing that the origin O is the neutral (or identity) element.

I: The inverse of P is the point with position vector $-\mathbf{p}$, which is a point on π and therefore a member of V.

Commutative: Since $\mathbf{p} + \mathbf{q} = \mathbf{q} + \mathbf{p}$, $\oplus$ is commutative, completing the proof that $(V, \oplus)$ is an Abelian group.

C': Since λP is the point with position vector $\lambda \mathbf{p}$, and π contains the origin, P is on $\pi \Rightarrow \lambda P$ is on $\pi \Rightarrow V$ is closed with respect to multiplication by a scalar.

A': $(\lambda\mu)$P has position vector $(\lambda\mu)\mathbf{p} = \lambda(\mu\mathbf{p})$ which represents $\lambda(\mu P)$, as required.

N': The scalar 1 is a real number and 1P is clearly P, as necessary.

D: $\lambda(P \oplus Q)$ has position vector $\lambda(\mathbf{p} + \mathbf{q}) = \lambda\mathbf{p} + \lambda\mathbf{q}$ which represents $\lambda P \oplus \lambda Q$, showing that multiplication by a scalar is distributive over $\oplus$.

D': $(\lambda + \mu)$P has position vector $(\lambda + \mu)\mathbf{p} = \lambda\mathbf{p} + \mu\mathbf{p}$ which represents $\lambda P \oplus \mu P$, completing the proof that V is a vector space over the real numbers.

(An alternative form of solution would show algebraically that each axiom is satisfied.)

The symbol $\mathbb{R}$ is often used for the set of real numbers, and it is convenient to use:

$\mathbb{R}^2$ for the set of ordered pairs of real numbers, i.e. $\mathbb{R}^2 = \{(x, y) : x, y \in \mathbb{R}\}$

$\mathbb{R}^3$ for the set of ordered triples (x, y, z) of real numbers

$\mathbb{R}^n$ for the set of ordered n-tuples of real numbers.

These sets $\mathbb{R}$, $\mathbb{R}^2$, $\mathbb{R}^3$ and $\mathbb{R}^n$ may be interpreted as points of a line, a plane, three-dimensional space, or n-dimensional space respectively, referred to fixed coordinate axes.

Exercise 5I

1. (i) Complete the proof (started on page 213) that the set V of 3×3 magic squares (with $\oplus$ and $\otimes$) forms a vector space over $\mathbb{R}$. (As you have already shown that $(V, \oplus)$ is an Abelian group and have dealt with C' and D, you only need to establish the truth of A', N' and D'.)

   (ii) Complete the proof (started on page 214) that the set V of arithmetic sequences (with $\oplus$ and $\otimes$) forms a vector space over $\mathbb{R}$. (Again you only need to justify A', N' and D'.)

2. Let V be the set of positive real numbers. Define the operations $\oplus$ and $\otimes$ as follows:

   $x \oplus y = xy$ for all x and y in V;

   $\lambda \otimes x = x^\lambda$ for all x in V and all real numbers λ.

   Prove that V is a vector space, and state which element of V corresponds to the zero vector.

3. Let V be the set $\mathbb{R}^2$. Define the operations $\oplus$ and $\otimes$ as follows.

   $(x_1, y_1) \oplus (x_2, y_2) = (x_1 + x_2 + 1, y_1 + y_2 + 1)$

   for all $(x_1, y_1), (x_2, y_2)$ in V

   $\lambda \otimes (x, y) = (\lambda x, \lambda y)$ for all (x, y) in V

   and all real numbers λ.

   Prove that V is not a vector space. Which axioms are not satisfied?

4. The set V and the operations $\oplus$ and $\otimes$ are defined in the table; show that $(V, \oplus, \otimes)$ is a vector space over $\mathbb{R}$.

	V	$\mathbf{a}$	$\mathbf{b}$	$\mathbf{a} \oplus \mathbf{b}$	$\lambda \otimes \mathbf{a}$
(i)	Solutions of the matrix equation: $$\begin{pmatrix} 6 & 2 \\ 3 & 1 \end{pmatrix}\begin{pmatrix} x \\ y \end{pmatrix} = \begin{pmatrix} 0 \\ 0 \end{pmatrix}$$	$\begin{pmatrix} x_1 \\ y_1 \end{pmatrix}$	$\begin{pmatrix} x_2 \\ y_2 \end{pmatrix}$	$\begin{pmatrix} x_1 + x_2 \\ y_1 + y_2 \end{pmatrix}$	$\begin{pmatrix} \lambda x_1 \\ \lambda y_1 \end{pmatrix}$
(ii)	Solutions of the differential equation: $$\frac{d^2y}{dx^2} - 5\frac{dy}{dx} + 6y = 0$$	$y = f(x)$	$y = g(x)$	$y = f(x) + g(x)$	$y = \lambda f(x)$
(iii)	The zero polynomial and all polynomials of degree $\leqslant 3$	$f(x)$	$g(x)$	$f(x) + g(x)$	$\lambda f(x)$
(iv)	Functions which are continuous for $0 \leqslant x \leqslant 1$	$f(x)$	$g(x)$	$f(x) + g(x)$	$\lambda f(x)$
(v)	Solutions of the pair of equations: $6x - 3y - 4z = 0$ $x - y - z = 0$	(p, q, r)	(s, t, u)	$(p+s, q+t, r+u)$	$(\lambda p, \lambda q, \lambda r)$

5. The set V and the operations $\oplus$ and $\otimes$ are defined in the table; decide whether $(V, \oplus, \otimes)$ is a vector space over $\mathbb{R}$, and justify your answer.

	V	$\mathbf{a}$	$\mathbf{b}$	$\mathbf{a} \oplus \mathbf{b}$	$\lambda \otimes \mathbf{a}$
(i)	Vectors of the form $\begin{pmatrix} x \\ y \end{pmatrix}$ where x, y are real numbers	$\begin{pmatrix} x_1 \\ y_1 \end{pmatrix}$	$\begin{pmatrix} x_2 \\ y_2 \end{pmatrix}$	$\begin{pmatrix} x_1 + x_2 \\ y_1 + y_2 \end{pmatrix}$	$\begin{pmatrix} \lambda y_1 \\ \lambda x_1 \end{pmatrix}$
(ii)	Solutions of the differential equation: $$\frac{d^2y}{dx^2} - 5\frac{dy}{dx} + 6y = x$$	$y = f(x)$	$y = g(x)$	$y = f(x) + g(x)$	$y = \lambda f(x)$
(iii)	All complex numbers	z_1	z_2	$z_1 + z_2$	λz_1
(iv)	All 2×3 matrices	$\mathbf{M}$	$\mathbf{N}$	$\mathbf{M} + \mathbf{N}$	$\lambda \mathbf{M}$
(v)	Solutions of the pair of equations: $4x - 3y + 2z = 4$ $x - y + z = 2$	(p, q, r)	(s, t, u)	$(p+s, q+t, r+u)$	$(\lambda p, \lambda q, \lambda r)$

6. Let V be the set of all vectors of the form $\begin{pmatrix} x \\ y \end{pmatrix}$ where x and y are real numbers. Define addition ($\oplus$) of vectors in the usual way. For all λ define:

$$\lambda \bullet \begin{pmatrix} x \\ y \end{pmatrix} = \begin{pmatrix} 0 \\ 0 \end{pmatrix}, \qquad \lambda \otimes \begin{pmatrix} x \\ y \end{pmatrix} = \begin{pmatrix} \lambda x \\ \lambda y \end{pmatrix}$$

and $\lambda \circ \begin{pmatrix} x \\ y \end{pmatrix} = \begin{pmatrix} \lambda x/2 \\ \lambda y/2 \end{pmatrix}$.

(i) Show that $(V, \oplus, \bullet)$ satisfies all the axioms of a vector space except N'.

(ii) Let F be the field $\{0, 2, 4\}$ where addition and multiplication are performed modulo 6. Show that:

(a) $(V, \oplus, \otimes)$ is not a vector space over F

(b) $(V, \oplus, \circ)$ is a vector space over F.

7. The set V consists of all geometric sequences. Let $\oplus$ denote the 'addition' of two sequences so that if (a_n) and (b_n) are sequences and $c_n = a_n b_n$ for all n we write $(c_n) = (a_n) \oplus (b_n)$. Multiplying a sequence (a_n) by the scalar λ is accomplished by multiplying each term of (a_n) by λ. Decide whether V is a vector space over the real numbers.

8. This question is about sequences that satisfy the formula $u_{n+2} = 3u_{n+1} - 2u_n$ ①

(i) Choose any values for u_0 and u_1 (preferably not both the same) and find the first five terms of the sequence (u_n).

(ii) Repeat (i) for a different choice of values of u_0 and u_1.

(iii) Show that adding (term by term) the two sequences generated in (i) and (ii) gives rise to a new sequence which also satisfies ①.

(iv) Show that multiplying each term of one of your sequences by a constant gives another sequence which satisfies ①.

(v) Show that the set of all sequences that satisfy ①, with addition and multiplication by a scalar as defined in (iii) and (iv), form a vector space over the real numbers.

Immediate consequences of the axioms of a vector space

As with groups, properties that can be proved for an abstract vector space apply equally to all vector spaces. We now look at some of these.

Firstly, since the vector space $(V, \oplus, \otimes)$ contains the Abelian group $(V, \oplus)$ the properties already established for groups continue to apply when we are dealing with the single operation $\oplus$. In the list below, **a** and **b** are members of V.

(a) The neutral element for $\oplus$ is unique

It is often called the *null* or *zero vector* and is frequently denoted by **0**, though we sometimes use **z** to aid clarity.

(b) Each element has a unique inverse with respect to $\oplus$

The inverse of **a** is usually denoted $-\mathbf{a}$.

(c) The cancellation laws apply

Thus:
$$\mathbf{a} \oplus \mathbf{x} = \mathbf{a} \oplus \mathbf{y} \quad \Rightarrow \quad \mathbf{x} = \mathbf{y}$$
$$\mathbf{x} \oplus \mathbf{a} = \mathbf{y} \oplus \mathbf{a} \quad \Rightarrow \quad \mathbf{x} = \mathbf{y}$$

and, since $(V, \oplus)$ is Abelian, $\quad \mathbf{a} \oplus \mathbf{x} = \mathbf{y} \oplus \mathbf{a} \quad \Rightarrow \quad \mathbf{x} = \mathbf{y}$.

(d) The equation $\mathbf{a} \oplus \mathbf{x} = \mathbf{b}$ has a unique solution

That unique solution may be written as $\mathbf{x} = (-\mathbf{a}) \oplus \mathbf{b}$ or $\mathbf{x} = \mathbf{b} \oplus (-\mathbf{a})$.

But the vector space $(V, \oplus, \otimes)$ also involves the second binary operation, $\otimes$. Though the symbol for multiplication of a vector by a scalar is often suppressed, the operation leads to the following additional properties. They may seem obvious, but they require proof as they have not been specified in the axioms.

(e) The product of any scalar and the null vector is the null vector

Proof

Let $\mathbf{z}$ be the null vector and λ any scalar.

Then
$$\mathbf{z} \oplus \mathbf{z} = \mathbf{z} \qquad \text{(by N)}$$
$$\Rightarrow \qquad \lambda \otimes (\mathbf{z} \oplus \mathbf{z}) = \lambda \otimes \mathbf{z} \qquad \text{(multiplying both sides by } \lambda\text{)}$$
$$\Rightarrow \qquad (\lambda \otimes \mathbf{z}) \oplus (\lambda \otimes \mathbf{z}) = (\lambda \otimes \mathbf{z}) \oplus \mathbf{z} \qquad \text{(by D and N)}$$
$$\Rightarrow \qquad \lambda \otimes \mathbf{z} = \mathbf{z} \qquad \text{(cancelling } \lambda \otimes \mathbf{z} \text{ on both sides)}$$

(f) The product of the zero element of the field and any vector is the null vector

Proof

Let 0 be the zero element of the field, $\mathbf{a}$ any vector, and $\mathbf{z}$ the null vector.

Then
$$0 + 0 = 0 \qquad \text{(by definition)}$$
$$\Rightarrow \qquad (0 + 0) \otimes \mathbf{a} = 0 \otimes \mathbf{a} \qquad \text{(multiplying both sides by } \mathbf{a}\text{)}$$
$$\Rightarrow \qquad (0 \otimes \mathbf{a}) \oplus (0 \otimes \mathbf{a}) = (0 \otimes \mathbf{a}) \oplus \mathbf{z} \qquad \text{(by D' and N)}$$
$$\Rightarrow \qquad 0 \otimes \mathbf{a} = \mathbf{z} \qquad \text{(cancelling } 0 \otimes \mathbf{a} \text{ on both sides)}$$

(g) If $\lambda \otimes \mathbf{a}$ is the null vector and $\mathbf{a}$ is not the null vector then $\lambda = 0$

Proof

Let $\mathbf{z}$ be the null vector. Suppose $\lambda \neq 0$; then λ has a reciprocal λ^{-1} (i.e. λ^{-1} is a member of the field of scalars, the inverse of λ under multiplication of scalars).

Now
$$\lambda \otimes \mathbf{a} = \mathbf{z}$$
$$\Rightarrow \qquad \lambda^{-1} \otimes (\lambda \otimes \mathbf{a}) = \lambda^{-1} \otimes \mathbf{z} \qquad \text{(multiplying both sides by } \lambda^{-1}\text{)}$$
$$\Rightarrow \qquad (\lambda^{-1}\lambda) \otimes \mathbf{a} = \mathbf{z} \qquad \text{(by A' and (f))}$$
$$\Rightarrow \qquad 1 \otimes \mathbf{a} = \mathbf{z} \qquad \text{(by definition of } \lambda^{-1}\text{)}$$
$$\Rightarrow \qquad \mathbf{a} = \mathbf{z}$$

which contradicts the statement that $\mathbf{a}$ is not the null vector. Therefore $\lambda = 0$.

(h) The inverse of any element a of V is found by multiplying a by -1

Proof

Let **a** be any vector, and **z** the null vector.

Then $\qquad\qquad 1 + (-1) = 0 \qquad$ (by definition)

$\Rightarrow \qquad (1 + (-1)) \otimes \mathbf{a} = 0 \otimes \mathbf{a} \qquad$ (multiplying both sides by **a**)

$\Rightarrow \qquad (1 \otimes \mathbf{a}) \oplus ((-1) \otimes \mathbf{a}) = \mathbf{z} \qquad$ (by D′ and (f))

$\Rightarrow \qquad \mathbf{a} \oplus ((-1) \otimes \mathbf{a}) = \mathbf{z} \qquad$ (by N′)

$\Rightarrow \qquad$ inverse of **a** with respect to $\oplus$ is $(-1) \otimes \mathbf{a}$.

The vector space of 3×3 magic squares

Our introduction to vector spaces was through 3×3 magic squares. To confirm that the 3×3 array $\begin{bmatrix} a & b & c \\ d & e & f \\ g & h & i \end{bmatrix}$ is a magic square you need to check that the sum of the three elements in any row, column or diagonal is always the same 'magic number', m say. If you want to construct a 3×3 magic square by assigning values to $a, b, \ldots, i$ you may choose the value of a freely, and b, and c, but it may be a surprise to find that you then have no further freedom.

To prove this let S be the sum of the following twelve elements:

the three elements in the middle row

the three elements in the middle column

the three elements in one diagonal

the three elements in the other diagonal.

Since S is the sum of one row, one column and two diagonals, $S = 4m$.

But S is the sum of twelve numbers:

the middle element, e, has been used four times

the other eight elements in the array have been used once each.

The sum of all nine elements in the array is $3m$. Therefore $S = 3m + 3e$.

From these two expressions for S, $m = 3e$.

So choosing values for a, b and c determines m, and fixes $e = \frac{1}{3}m$. Because c and e (and m) are known, g is fixed. Once a and g are known, d is fixed. Much the same argument applies to i, f and h. Despite having nine elements, 3×3 magic squares have only three degrees of freedom.

Activity

On page 213 we used:

$\oplus$ to denote adding two magic squares term by term

$\otimes$ to denote multiplying each element of a magic square by a scalar.

We now use $+$ instead of $\oplus$, and suppress $\otimes$, writing the scalar and magic square side by side.

(i) By examining the central elements of the magic squares show that the equation:

$$x\begin{bmatrix} 1 & 1 & 1 \\ 1 & 1 & 1 \\ 1 & 1 & 1 \end{bmatrix} + y\begin{bmatrix} 1 & -1 & 0 \\ -1 & 0 & 1 \\ 0 & 1 & -1 \end{bmatrix} + z\begin{bmatrix} 0 & -1 & 1 \\ 1 & 0 & -1 \\ -1 & 1 & 0 \end{bmatrix} = \begin{bmatrix} 0 & 0 & 0 \\ 0 & 0 & 0 \\ 0 & 0 & 0 \end{bmatrix}$$

can only be satisfied if $x = 0$ and deduce that $x = y = z = 0$ is its only solution.

(ii) Check that $\begin{bmatrix} 35 & 24 & -29 \\ -54 & 10 & 74 \\ 49 & -4 & -15 \end{bmatrix}$ is a magic square and use a technique similar

to that used in (i) to show that if:

$$\begin{bmatrix} 35 & 24 & -29 \\ -54 & 10 & 74 \\ 49 & -4 & -15 \end{bmatrix} = x\begin{bmatrix} 1 & 1 & 1 \\ 1 & 1 & 1 \\ 1 & 1 & 1 \end{bmatrix} + y\begin{bmatrix} 1 & -1 & 0 \\ -1 & 0 & 1 \\ 0 & 1 & -1 \end{bmatrix} + z\begin{bmatrix} 0 & -1 & 1 \\ 1 & 0 & -1 \\ -1 & 1 & 0 \end{bmatrix}$$

then $x = 10$, $y = 25$ and $z = -39$.

(iii) Show that the magic square $\begin{bmatrix} a & b & c \\ d & e & f \\ g & h & i \end{bmatrix}$ is equal to the following expression.

$$e\begin{bmatrix} 1 & 1 & 1 \\ 1 & 1 & 1 \\ 1 & 1 & 1 \end{bmatrix} + (a-e)\begin{bmatrix} 1 & -1 & 0 \\ -1 & 0 & 1 \\ 0 & 1 & -1 \end{bmatrix} + (c-e)\begin{bmatrix} 0 & -1 & 1 \\ 1 & 0 & -1 \\ -1 & 1 & 0 \end{bmatrix}.$$

Representing $\begin{bmatrix} 1 & 1 & 1 \\ 1 & 1 & 1 \\ 1 & 1 & 1 \end{bmatrix}$, $\begin{bmatrix} 1 & -1 & 0 \\ -1 & 0 & 1 \\ 0 & 1 & -1 \end{bmatrix}$, $\begin{bmatrix} 0 & -1 & 1 \\ 1 & 0 & -1 \\ -1 & 1 & 0 \end{bmatrix}$ and $\begin{bmatrix} 0 & 0 & 0 \\ 0 & 0 & 0 \\ 0 & 0 & 0 \end{bmatrix}$

by **L**, **M**, **N**, and **O** respectively, and using V to stand for the set of all 3×3 magic squares, notice the following in the activity above.

- The expression $x\mathbf{L} + y\mathbf{M} + z\mathbf{N}$ is known as a *linear combination* of **L**, **M** and **N**.

- Since $x\mathbf{L} + y\mathbf{M} + z\mathbf{N} = \mathbf{O} \Leftrightarrow x = y = z = 0$ it is not possible to express any of **L**, **M** and **N** in terms of the others. We say that **L**, **M** and **N** are *linearly independent*.

- Since every member of V can be expressed as a linear combination of $\mathbf{L}$, $\mathbf{M}$ and $\mathbf{N}$ we say that V is *spanned* or *generated* by the set $\{\mathbf{L}, \mathbf{M}, \mathbf{N}\}$; V is also spanned by sets with more elements, such as $\{\mathbf{L}, \mathbf{M}, \mathbf{N}, \mathbf{L} + 3\mathbf{M}\}$, for example.

- Since V is spanned by $\{\mathbf{L}, \mathbf{M}, \mathbf{N}\}$ where $\mathbf{L}$, $\mathbf{M}$ and $\mathbf{N}$ are linearly independent we describe $\{\mathbf{L}, \mathbf{M}, \mathbf{N}\}$ as a *basis* for V. (Note the indefinite article.)

The next few pages will show you how this new terminology is applied to other vector spaces.

Linear combinations of vectors

Whenever we can express a vector $\mathbf{v}$ in the form $\mathbf{v} = \lambda_1 \mathbf{a}_1 + \lambda_2 \mathbf{a}_2 + \cdots + \lambda_n \mathbf{a}_n$ we say $\mathbf{v}$ is a *linear combination* of the vectors $\mathbf{a}_1, \mathbf{a}_2, \ldots, \mathbf{a}_n$. The set of all linear combinations of the vectors $\mathbf{a}_1, \mathbf{a}_2, \ldots, \mathbf{a}_n$ is known as the set *spanned* by $\{\mathbf{a}_1, \mathbf{a}_2, \ldots, \mathbf{a}_n\}$.

Activity

The set V_2 is spanned by $\{\mathbf{u}, \mathbf{v}\}$ where $\mathbf{u} = \begin{pmatrix} 1 \\ 0 \end{pmatrix}$ and $\mathbf{v} = \begin{pmatrix} 1 \\ 1 \end{pmatrix}$.

(i) The vectors $\mathbf{u}$ and $\mathbf{v}$ belong to V_2. Write down $\mathbf{w}$, $\mathbf{x}$ and $\mathbf{y}$, three further members of V_2.

(ii) Show that $\mathbf{w} + \mathbf{x}$, $\mathbf{x} + \mathbf{y}$, $5\mathbf{w}$ and $-2\mathbf{x}$ belong to V_2.

(iii) State the values of α and β such that $\alpha\mathbf{u} + \beta\mathbf{v} = \mathbf{0}$. Are there other possible answers?

(iv) Find the inverses of $\mathbf{u}$, $\mathbf{v}$, $\mathbf{w}$, $\mathbf{x}$ and $\mathbf{y}$.

(As you have probably guessed, V_2 is a vector space.)

In general, if $\mathbf{a}_1, \mathbf{a}_2, \ldots, \mathbf{a}_n$ are members of a vector space V then the set V' spanned by $A = \{\mathbf{a}_1, \mathbf{a}_2, \ldots, \mathbf{a}_n\}$ is also a vector space. The vector space V' is sometimes said to be *generated* by the vectors in A. As there are the usual ten axioms to check you may feel that the proof that V' is a vector space will be longwinded. But axioms A, A′, N′, D and D′ apply to all vectors, and vector addition is commutative for all vectors so you only need to check that axioms C, C′, N and I remain true. See Question 8 in Exercise 5J.

Linearly independent vectors

Since the set spanned by $\{\mathbf{a}_1, \mathbf{a}_2, \ldots, \mathbf{a}_n\}$ is a vector space, at least one linear combination of the vectors $\mathbf{a}_1, \mathbf{a}_2, \ldots, \mathbf{a}_n$ is the null vector, $\mathbf{0}$. If the only such linear combination is the one in which all the scalars equal 0 (as in the activity above) we say that the vectors $\mathbf{a}_1, \mathbf{a}_2, \ldots, \mathbf{a}_n$ are *linearly independent*.

That is:

$$\mathbf{a}_1, \mathbf{a}_2, \ldots, \mathbf{a}_n \text{ are linearly independent if and only if}$$

$$\lambda_1 \mathbf{a}_1 + \lambda_2 \mathbf{a}_2 + \cdots + \lambda_n \mathbf{a}_n = \mathbf{0} \quad \Rightarrow \quad \lambda_1 = \lambda_2 = \cdots = \lambda_n = 0.$$

Vectors which are not linearly independent are described as *linearly dependent*.

Activity

(i) By showing that the only solution to $x\mathbf{i} + y\mathbf{j} + z\mathbf{k} = \mathbf{0}$ is $x = y = z = 0$ prove that $\mathbf{i}$, $\mathbf{j}$ and $\mathbf{k}$ are linearly independent vectors.

(ii) Explain why every set of vectors containing the null vector is linearly dependent.

The set V spanned by a set of vectors such as $A = \{\mathbf{a}_1, \mathbf{a}_2, \ldots, \mathbf{a}_n\}$ is a vector space. If A is a linearly independent set of vectors it is known as a *basis* for V. This means that A is a basis of a vector space V if and only if both the following conditions hold:

(a) the vectors $\mathbf{a}_1, \mathbf{a}_2, \ldots, \mathbf{a}_n$ are linearly independent

(b) every vector in V can be expressed as a linear combination of the vectors $\mathbf{a}_1, \mathbf{a}_2, \ldots, \mathbf{a}_n$.

So, for example, since $\{\mathbf{i}, \mathbf{j}, \mathbf{k}\}$ is a set of linearly independent vectors which

span V_3, i.e. the vector space of all vectors of the form $\begin{pmatrix} x \\ y \\ z \end{pmatrix}$, $\{\mathbf{i}, \mathbf{j}, \mathbf{k}\}$ is a basis

for V_3. The basis for a vector space is not unique: for example, since any vector in V_3 can be expressed as a linear combination of the linearly independent vectors $\mathbf{i}, \mathbf{j}$ and $\mathbf{i} + \mathbf{k}$, so the set $\{\mathbf{i}, \mathbf{j}, \mathbf{i} + \mathbf{k}\}$ is also a basis for V_3.

EXAMPLE

Is $\{\mathbf{a}, \mathbf{b}, \mathbf{c}\}$ a basis for a vector space, where $\mathbf{a} = \begin{pmatrix} 2 \\ -2 \\ 1 \end{pmatrix}$, $\mathbf{b} = \begin{pmatrix} 1 \\ 2 \\ 2 \end{pmatrix}$ and $\mathbf{c} = \begin{pmatrix} 1 \\ -3 \\ 1 \end{pmatrix}$?

Solution

We need to show that $x\mathbf{a} + y\mathbf{b} + z\mathbf{c} = \mathbf{0}$ has no solutions other than $x = y = z = 0$.

Rewriting the equation:

$$x\begin{pmatrix} 2 \\ -2 \\ 1 \end{pmatrix} + y\begin{pmatrix} 1 \\ 2 \\ 2 \end{pmatrix} + z\begin{pmatrix} 1 \\ -3 \\ 1 \end{pmatrix} = \begin{pmatrix} 0 \\ 0 \\ 0 \end{pmatrix}$$

$$\Leftrightarrow \left. \begin{array}{r} 2x + y + z = 0 \\ -2x + 2y - 3z = 0 \\ x + 2y + z = 0 \end{array} \right\} \Leftrightarrow \begin{pmatrix} 2 & 1 & 1 \\ -2 & 2 & -3 \\ 1 & 2 & 1 \end{pmatrix} \begin{pmatrix} x \\ y \\ z \end{pmatrix} = \begin{pmatrix} 0 \\ 0 \\ 0 \end{pmatrix}.$$

Evaluating 3×3 determinants is dealt with in Chapter 1; you may, of course, use your calculator if it has that facility.

Since $\begin{vmatrix} 2 & 1 & 1 \\ -2 & 2 & -3 \\ 1 & 2 & 1 \end{vmatrix} = 2(2+6) + 2(1-2) + 1(-3-2) = 9 \neq 0$

the equations have a unique solution, and by inspection it is $x = y = z = 0$, showing that $\mathbf{a}$, $\mathbf{b}$ and $\mathbf{c}$ are linearly independent vectors and so $\{\mathbf{a}, \mathbf{b}, \mathbf{c}\}$ is a basis for a vector space. (The set $\{\mathbf{a}, \mathbf{b}, \mathbf{c}\}$ is another basis for vector space V_3.)

The example above illustrates one way of checking whether a set of n vectors is linearly independent or not, provided the vectors can be expressed in column form. Form the matrix $\mathbf{M}$ using the n vectors as the n columns of the matrix and consider solving the equation $\mathbf{Mx} = \mathbf{0}$. If the only solution is $\mathbf{x} = \mathbf{0}$ (indicated by $\det \mathbf{M} \neq 0$) the vectors are linearly independent, otherwise they are linearly dependent.

Activity

Given $\mathbf{u} = \begin{pmatrix} 1 \\ 0 \end{pmatrix}$ and $\mathbf{v} = \begin{pmatrix} 1 \\ 1 \end{pmatrix}$, $\{\mathbf{u}, \mathbf{v}\}$ is a basis for V_2. Find α and β in terms of p and q such that $\alpha\mathbf{u} + \beta\mathbf{v} = \begin{pmatrix} p \\ q \end{pmatrix}$. Are there other possible answers?

The activity above is an example of the following general result.

If $A = \{\mathbf{a}_1, \mathbf{a}_2, \ldots, \mathbf{a}_n\}$ is a basis of the vector space V, and $\mathbf{v}$ belongs to V then the linear combination of the vectors $\mathbf{a}_1, \mathbf{a}_2, \ldots, \mathbf{a}_n$ that results in $\mathbf{v}$ is unique.

To prove this suppose:

$$\mathbf{v} = \lambda_1 \mathbf{a}_1 + \lambda_2 \mathbf{a}_2 + \cdots + \lambda_n \mathbf{a}_n$$

and

$$\mathbf{v} = \mu_1 \mathbf{a}_1 + \mu_2 \mathbf{a}_2 + \cdots + \mu_n \mathbf{a}_n.$$

then subtraction gives

$$\mathbf{0} = (\lambda_1 - \mu_1)\mathbf{a}_1 + (\lambda_2 - \mu_2)\mathbf{a}_2 + \cdots + (\lambda_n - \mu_n)\mathbf{a}_n.$$

The set $\{\mathbf{a}_1, \mathbf{a}_2, \ldots, \mathbf{a}_n\}$ is a basis for V

$$\Rightarrow \mathbf{a}_1, \mathbf{a}_2, \ldots, \mathbf{a}_n \text{ are linearly independent}$$

$$\Rightarrow (\lambda_1 - \mu_1) = (\lambda_2 - \mu_2) = \cdots = (\lambda_n - \mu_n) = 0$$

$$\Rightarrow \lambda_1 = \mu_1, \lambda_2 = \mu_2, \ldots, \lambda_n = \mu_n$$

showing that there is only one linear combination of the vectors $\mathbf{a}_1, \mathbf{a}_2, \ldots, \mathbf{a}_n$ that results in $\mathbf{v}$, as required.

In mechanics this result means that when you resolve a force into components in three given non-coplanar directions the answer is unique.

Exercise 5J

1. By showing that:

$$ae^x + be^{2x} + ce^{3x} = 0 \Rightarrow a = b = c = 0$$

show that the functions e^x, e^{2x} and e^{3x} are linearly independent.

2. In each case, decide whether the vectors **a**, **b** and **c** are linearly independent or not.

(i) $\mathbf{a} = \begin{pmatrix} 3 \\ 2 \\ -1 \end{pmatrix}$, $\mathbf{b} = \begin{pmatrix} 5 \\ -2 \\ 3 \end{pmatrix}$, $\mathbf{c} = \begin{pmatrix} 1 \\ 2 \\ 0 \end{pmatrix}$

(ii) $\mathbf{a} = \begin{pmatrix} 4 \\ 2 \\ 3 \end{pmatrix}$, $\mathbf{b} = \begin{pmatrix} 1 \\ 1 \\ 1 \end{pmatrix}$, $\mathbf{c} = \begin{pmatrix} 0 \\ 2 \\ 1 \end{pmatrix}$

(iii) $\mathbf{a} = \begin{pmatrix} 2 \\ 1 \\ 1 \end{pmatrix}$, $\mathbf{b} = \begin{pmatrix} 3 \\ -2 \\ 1 \end{pmatrix}$, $\mathbf{c} = \begin{pmatrix} 1 \\ -4 \\ -2 \end{pmatrix}$

(iv) $\mathbf{a} = \begin{pmatrix} 1 \\ -7 \\ -1 \end{pmatrix}$, $\mathbf{b} = \begin{pmatrix} 0 \\ 2 \\ 1 \end{pmatrix}$, $\mathbf{c} = \begin{pmatrix} 1 \\ -1 \\ 2 \end{pmatrix}$

3. Given **a**, **b** and **c** as follows, does $\{\mathbf{a}, \mathbf{b}, \mathbf{c}\}$ form a basis for a vector space?

(i) $\mathbf{a} = \begin{pmatrix} 2 \\ 2 \\ 4 \end{pmatrix}$, $\mathbf{b} = \begin{pmatrix} 1 \\ 2 \\ -1 \end{pmatrix}$, $\mathbf{c} = \begin{pmatrix} -1 \\ -2 \\ 2 \end{pmatrix}$

(ii) $\mathbf{a} = \begin{pmatrix} 3 \\ 2 \\ -2 \end{pmatrix}$, $\mathbf{b} = \begin{pmatrix} 1 \\ 2 \\ -1 \end{pmatrix}$, $\mathbf{c} = \begin{pmatrix} 3 \\ -2 \\ 2 \end{pmatrix}$

(iii) $\mathbf{a} = \begin{pmatrix} 6 \\ 3 \\ 8 \end{pmatrix}$, $\mathbf{b} = \begin{pmatrix} 2 \\ 2 \\ 5 \end{pmatrix}$, $\mathbf{c} = \begin{pmatrix} 1 \\ 4 \\ 7 \end{pmatrix}$

(iv) $\mathbf{a} = \begin{pmatrix} 2 \\ -1 \\ -2 \end{pmatrix}$, $\mathbf{b} = \begin{pmatrix} 0 \\ 1 \\ -1 \end{pmatrix}$, $\mathbf{c} = \begin{pmatrix} 2 \\ 2 \\ -5 \end{pmatrix}$

4. The vector space V is spanned by $\{\mathbf{u}, \mathbf{v}, \mathbf{w}\}$

where $\mathbf{u} = \begin{pmatrix} 1 \\ 1 \\ 0 \end{pmatrix}$, $\mathbf{v} = \begin{pmatrix} 0 \\ 1 \\ 1 \end{pmatrix}$ and $\mathbf{w} = \begin{pmatrix} -1 \\ 0 \\ 1 \end{pmatrix}$.

(i) Show that:
  (a) **u**, **v** and **w** are linearly dependent
  (b) **u** and **v** are linearly independent.

(ii) Show that $\{\mathbf{u}, \mathbf{v}, \mathbf{w}\}$ and $\{\mathbf{u}, \mathbf{v}\}$ span the same vector space.

(iii) Interpret geometrically the vector space spanned by $\{\mathbf{u}, \mathbf{v}\}$.

5. Prove that every non-empty subset of a linearly independent set of vectors is linearly independent.

6. The set V consists of all functions f of the form $f(x) = e^{-x}(a \cos 2x + b \sin 2x)$, where a and b are real numbers. Addition of functions and multiplication of functions by a scalar are defined in the usual way.

(i) Show that V is a vector space over the field of real numbers.

(ii) Given that $g(x) = e^{-x} \cos 2x$ and $h(x) = e^{-x} \sin 2x$, show that the functions g and h are linearly independent and deduce that $\{g, h\}$ is a basis for V.

7. Suppose **M** is an $m \times n$ matrix. Vectors $\mathbf{a}_1, \mathbf{a}_2, \ldots, \mathbf{a}_n$ are the first, second, $\ldots n$th columns of **M**.

(i) Given $m = n$, explain why:
  (a) $\det \mathbf{M} = 0 \Rightarrow \mathbf{a}_1, \mathbf{a}_2, \ldots, \mathbf{a}_n$ are linearly dependent
  (b) $\det \mathbf{M} \neq 0 \Rightarrow \mathbf{a}_1, \mathbf{a}_2, \ldots, \mathbf{a}_n$ are linearly independent.

(ii) What conclusions (if any) can you draw if (a) $m < n$ (b) $m > n$?

8. The members of $A = \{\mathbf{a}_1, \mathbf{a}_2, \ldots, \mathbf{a}_n\}$ are elements of a vector space V. The set V' is spanned by A. Prove that V' is a vector space.

9. The game of Nim is for two players. They set up three piles of matches and then take it in turns to select a pile and remove any number of matches from that pile. The following is a winning strategy for the version of the game in which the aim is to take the last match.

(A) Express the number of matches in each pile in the binary scale, and treat the digits as a row vector ($1 \times n$ matrix) with extra zeros at the left if necessary.

E.g. 13 is represented by $(1\ 1\ 0\ 1)$; 5 is represented by $(0\ 1\ 0\ 1)$.

(B) Choose a pile of matches and draw from it so as to make the three vectors linearly dependent under addition modulo 2. If two vectors are equal choose the third pile and take all the matches from it.

The notation $17, 12, 5 \to 9, 12, 5$ records a move in which a player presented with piles containing $17, 12, 5$ matches chooses to remove 8 matches from the pile of 17.

(i) Decide whether the following moves agree with the winning strategy.

(a) $17, 12, 5 \to 9, 12, 5$

(b) $13, 7, 5 \to 6, 7, 5$

(c) $15, 12, 11 \to 7, 12, 11$

(d) $15, 12, 9 \to 15, 6, 9$

(e) $15, 12, 9 \to 15, 12, 3$

(f) $10, 9, 3 \to 6, 9, 3$

(ii) Explain why you cannot apply the winning strategy if you are presented with piles containing 14, 9, 7 matches. What would you do if you were presented with piles of 9, 9, 7 matches?

(Early computers were programmed to play Nim to demonstrate their ability.)

The dimension of a vector space

In this section you will see that:

- a basis of a vector space is both the largest set of linearly independent vectors belonging to the vector space, and also the smallest set of vectors that span the vector space

- all (finite) bases of V have the same number of elements; it is that number that is defined as the *dimension* of V.

Activity

The vector space V_3 consists of all vectors of the form $\begin{pmatrix} x \\ y \\ z \end{pmatrix}$ where x, y and z are real numbers, with the usual rules for adding vectors or multiplying a vector by a scalar.

(i) Let $\mathbf{a} = \begin{pmatrix} 1 \\ 0 \\ -2 \end{pmatrix}$, $\mathbf{b} = \begin{pmatrix} 2 \\ -3 \\ -7 \end{pmatrix}$ and $\mathbf{c} = \begin{pmatrix} 0 \\ 1 \\ 1 \end{pmatrix}$.

(a) Show that $\mathbf{a}$, $\mathbf{b}$ and $\mathbf{c}$ are linearly dependent.

(b) Show that, in whatever order you take $\mathbf{a}$, $\mathbf{b}$ and $\mathbf{c}$, you can always express the third vector as a linear combination of the other two.

(ii) Select four non-zero vectors, **w**, **x**, **y** and **z**, from V_3.

    (a) Show that **w**, **x**, **y** and **z** are linearly dependent.

    (b) Show that, in whatever order you take **w**, **x**, **y** and **z** you can always express the fourth vector as a linear combination of the other three. (It may also be possible to express the third vector as a linear combination of the first two.)

(iii) Choose a set of n linearly independent vectors from V. What are the possible values of n?

The activity illustrates the following general theorems.

Theorem 1

The sequence of non-zero vectors $(\mathbf{a}_1, \mathbf{a}_2, \ldots, \mathbf{a}_m)$ in vector space V are linearly dependent if and only if at least one of the vectors is a linear combination of the preceding vectors in the sequence.

Proof

Suppose $\mathbf{a}_k$ is a linear combination of the preceding vectors in sequence. Then:

$$\mathbf{a}_k = \lambda_1 \mathbf{a}_1 + \lambda_2 \mathbf{a}_2 + \cdots + \lambda_{k-1} \mathbf{a}_{k-1} \text{ for some } \lambda_1, \lambda_2, \cdots, \lambda_{k-1}$$

$$\Rightarrow \lambda_1 \mathbf{a}_1 + \lambda_2 \mathbf{a}_2 + \ldots + \lambda_{k-1} \mathbf{a}_{k-1} + (-1)\mathbf{a}_k = \mathbf{0}$$

with at least the coefficient of $\mathbf{a}_k$ non-zero

$$\Rightarrow \lambda_1 \mathbf{a}_1 + \lambda_2 \mathbf{a}_2 + \cdots + \lambda_{k-1} \mathbf{a}_{k-1} + (-1)\mathbf{a}_k + 0\mathbf{a}_{k+1} + \cdots + 0\mathbf{a}_m = \mathbf{0}$$

$$\Rightarrow \mathbf{a}_1, \mathbf{a}_2, \ldots, \mathbf{a}_m \text{ are linearly dependent.}$$

Conversely, suppose the vectors are linearly dependent. Then:

$$\lambda_1 \mathbf{a}_1 + \lambda_2 \mathbf{a}_2 + \cdots + \lambda_m \mathbf{a}_m = \mathbf{0} \text{ for some } \lambda_1, \lambda_2, \ldots, \lambda_m, \text{ not all zero.}$$

Suppose the last λ_i which is not zero has subscript k; then $\lambda_k \neq 0$, so that

$$\mathbf{a}_k = \frac{-\lambda_1}{\lambda_k} \mathbf{a}_1 + \frac{-\lambda_2}{\lambda_k} \mathbf{a}_2 + \cdots + \frac{-\lambda_{k-1}}{\lambda_k} \mathbf{a}_{k-1}$$

expressing $\mathbf{a}_k$ as a linear combination of the preceding vectors, unless $k = 1$. If $k = 1$ then $\lambda_1 \mathbf{a}_1 = \mathbf{0}$ with $\lambda_1 \neq 0 \Rightarrow \mathbf{a}_1 = \mathbf{0}$ which contradicts the hypothesis of the theorem that all $\mathbf{a}_1, \mathbf{a}_2, \ldots, \mathbf{a}_m$ are non-zero vectors.

Theorem 2

If r linearly independent vectors can be chosen from a vector space V, spanned by n vectors, then $r \leqslant n$.

Proof

Let $A_0 = (\mathbf{a}_1, \mathbf{a}_2, \ldots, \mathbf{a}_n)$ be a sequence of n vectors spanning V, and let $X = \{\mathbf{x}_1, \mathbf{x}_2, \ldots, \mathbf{x}_r\}$ be a set of r linearly independent vectors in V.

A_0 spans $V \Rightarrow \mathbf{x}_1$ is a linear combination of the vectors in A_0

$$\Rightarrow B_1 = (\mathbf{x}_1, \mathbf{a}_1, \mathbf{a}_2, \ldots, \mathbf{a}_n) \text{ spans } V \text{ with } (n+1)$$
linearly dependent vectors

$$\Rightarrow \text{ at least one of } \mathbf{x}_1, \mathbf{a}_1, \mathbf{a}_2, \ldots, \mathbf{a}_n \text{ is linearly dependent}$$
on its predecessors (by Theorem 1).

This cannot be $\mathbf{x}_1$ as it has no predecessors. Therefore, some vector, $\mathbf{a}_k$ say, is dependent on its predecessors $\mathbf{x}_1, \mathbf{a}_1, \mathbf{a}_2, \ldots, \mathbf{a}_{k-1}$.

Delete this vector from B_1 to form $A_1 = (\mathbf{x}_1, \mathbf{a}_1, \mathbf{a}_2, \ldots, \mathbf{a}_{k-1}, \mathbf{a}_{k+1}, \ldots, \mathbf{a}_n)$, another sequence of n vectors spanning V.

Repeat the argument:

construct $B_2 = (\mathbf{x}_2, \mathbf{x}_1, \mathbf{a}_1, \mathbf{a}_2, \ldots, \mathbf{a}_{k-1}, \mathbf{a}_{k+1}, \ldots, \mathbf{a}_n)$, another sequence of $n+1$ linearly dependent vectors spanning V;

form A_2 by deleting the first vector in B_2 which is dependent on its predecessors.

This cannot be $\mathbf{x}_2$, which has no predecessors, or $\mathbf{x}_1$, because its only predecessor is $\mathbf{x}_2$ and both $\mathbf{x}_1$ and $\mathbf{x}_2$ belong to X, a set of linearly independent vectors.

Repeat the argument until all the members of X have been used. Each time an element of A_0 is thrown out. Thus A_0 must have originally contained at least r elements, proving that $r \leqslant n$.

An immediate and important consequence of Theorem 2 is that all (finite) bases of V have the same number of elements. For if both $A = \{\mathbf{a}_1, \mathbf{a}_2, \ldots, \mathbf{a}_n\}$ and $B = \{\mathbf{b}_1, \mathbf{b}_2, \ldots, \mathbf{b}_m\}$ are bases of V, then the vectors $\mathbf{a}_1, \mathbf{a}_2, \ldots, \mathbf{a}_n$ span V and the vectors $\mathbf{b}_1, \mathbf{b}_2, \ldots, \mathbf{b}_m$ are linearly independent, so, by Theorem 2, $m \leqslant n$. Similarly, $n \leqslant m$, so that $m = n$.

A basis of a vector space is the largest set of linearly independent vectors belonging to the vector space. It is also the smallest set of vectors that span the vector space.

There are many possible bases for a vector space V, infinitely many if we are working with an infinite field such as the real numbers. We have just proved that if any one of these bases is finite, then all the possible bases have the same number of elements: that number is known as the *dimension* of V, denoted by $\dim V$.

Since a basis for the vector space of 3×3 magic squares has three elements, the dimension of that vector space is 3.

Be careful with your use of the word 'dimension'. If V is the vector space of all arithmetic sequences, it contains the two sequences $(1, 1, 1, 1, \ldots)$ and $(0, 1, 2, 3, \ldots)$. These elements of V may (wrongly) suggest that V is an infinite vector space. But every arithmetic sequence can be expressed in the form $(a, a+d, a+2d, a+3d, \ldots)$, which is just $a(1, 1, 1, 1, \ldots) + d(0, 1, 2, 3, \ldots)$ showing that V is spanned by the linearly independent sequences $(1, 1, 1, 1, \ldots)$ and $(0, 1, 2, 3, \ldots)$, so $\dim V = 2$.

Although a sequence in V is infinitely long, we can define that sequence by just two numbers, the first term and the common difference. The dimension of a vector space is, in effect, the number of degrees of freedom of that vector space.

If V is the vector space spanned by $\mathbf{i} = \begin{pmatrix} 1 \\ 0 \\ 0 \end{pmatrix}$, $\mathbf{j} = \begin{pmatrix} 0 \\ 1 \\ 0 \end{pmatrix}$ and $\mathbf{k} = \begin{pmatrix} 0 \\ 0 \\ 1 \end{pmatrix}$ then

$\dim V = 3$ as $\mathbf{i}, \mathbf{j}$ and $\mathbf{k}$ are linearly independent, so that $\{\mathbf{i}, \mathbf{j}, \mathbf{k}\}$ is a basis

of V. However, if S is the vector space spanned by $\mathbf{a}_1 = \begin{pmatrix} 1 \\ 2 \\ 3 \end{pmatrix}$, $\mathbf{a}_2 = \begin{pmatrix} 1 \\ 3 \\ 4 \end{pmatrix}$

and $\mathbf{a}_3 = \begin{pmatrix} 1 \\ 4 \\ 5 \end{pmatrix}$ its dimension is not quite so obvious, though clearly not

more than 3. Vectors $\mathbf{a}_1$ and $\mathbf{a}_2$ are linearly independent, but $\mathbf{a}_3 = 2\mathbf{a}_2 - \mathbf{a}_1$ so that $\mathbf{a}_1, \mathbf{a}_2$ and $\mathbf{a}_3$ are not linearly independent. The set $\{\mathbf{a}_1, \mathbf{a}_2\}$ is a basis of S so $\dim S = 2$. Don't be deceived by the fact that S consists of 'three-dimensional' vectors! We can (of course) represent S by a plane (two dimensions) in 'three-dimensional space'.

The vector spaces S and V discussed in the last paragraph illustrate another concept. Set S is a non-empty subset of the set V, and the same (unmentioned) rules apply for adding vectors, and for multiplying vectors by scalars. If S is vector space, it is called a *subspace* of vector space V. It is important to note that both space and subspace are over the same field of scalars.

Notice that if V consists of only the null vector, V has no basis, and its dimension is 0.

HISTORICAL NOTE

Abstract vector spaces are a twentieth century invention, despite the fact that many of our examples of vector spaces have been around much longer. In a letter dated 1739 Leonhard Euler (1707–83) described to John Bernoulli (1667–1748) how to find all the solutions of homogeneous linear differential equations such as

$$\frac{d^2 y}{dx^2} - 5\frac{dy}{dx} + 6y = 0;$$ *his method involved expressing the solutions as linear*

combinations of n solutions if the equation was of order n. Bernoulli replied that he had obtained similar results in 1700! The idea of a vector was introduced (independently) by William Rowan Hamilton (1805–65) and Hermann Günther Grassmann (1809–77) around the middle of the nineteenth century, and Grassmann's book of 1862 contains the concepts of basis and dimension. Arthur Cayley (1821–95) introduced n-dimensional vectors. But the first formal definition of a vector space was given by Hermann Weyl (1885–1955) in his book on the theory of relativity, Space, Time and Matter, *published in 1918.*

Exercise 5K

1. Find the dimension of the vector space of 3×3 magic squares with 0 as the central element.

2. In a *semi-magic square* the sum of the entries in any row or column is identical. Find the dimension of the vector space of 3×3 semi-magic squares (with addition and multiplication by a scalar defined as for magic squares or matrices).

3. Show that the set of polynomials:

$$P = a_0 + a_1 t + a_2 t^2 + a_3 t^3$$

   with real coefficients forms a vector space V of dimension 4.

4. Let V be a vector space over the real numbers and S be a subset of V.

   (i) (a) Explain how you know that the commutative axiom and axioms A, A′, N′, D and D′ are valid for S without having to check them individually.

   (b) By considering $0\mathbf{a}$ show that:

   C' is valid in $S \Rightarrow$ N is valid in S.

   (c) Show that:

   C' is valid in $S \Rightarrow$ I is valid in S.

   (ii) Prove that:

   S is a subspace of V if and only if both axioms C and C′ are valid for S.

5. A vector space V has dimension n; X is the set $\{\mathbf{x}_1, \mathbf{x}_2, \dots, \mathbf{x}_n\}$ of vectors belonging to V. Are the following true or false?

   (i) Vectors $\mathbf{x}_1, \mathbf{x}_2, \dots, \mathbf{x}_n$ are linearly independent $\Rightarrow X$ is a basis of V.

   (ii) Vectors $\mathbf{x}_1, \mathbf{x}_2, \dots, \mathbf{x}_n$ span V $\Rightarrow X$ is a basis of V.

   (iii) If vectors $\mathbf{x}_1, \mathbf{x}_2, \dots, \mathbf{x}_n$ are linearly dependent, then each member of X can be expressed as a linear combination of the other members of X.

6. A vector space V has dimension n and X is the set $\{\mathbf{x}_1, \mathbf{x}_2, \dots, \mathbf{x}_r\}$ of linearly independent vectors in V, where $r < n$. Prove that there is a basis of V that has X as a subset. (We say X can be *extended* to form a basis of V.)

7. (i) Find conditions on the components of the vector $\mathbf{u}$ such that it belongs to the vector space spanned by the vectors $\mathbf{x}$, $\mathbf{y}$ and $\mathbf{z}$, where:

$$\mathbf{u} = \begin{pmatrix} a \\ b \\ c \end{pmatrix}, \quad \mathbf{x} = \begin{pmatrix} 2 \\ 1 \\ 0 \end{pmatrix}, \quad \mathbf{y} = \begin{pmatrix} 1 \\ -1 \\ 2 \end{pmatrix}, \quad \mathbf{z} = \begin{pmatrix} 0 \\ 3 \\ -4 \end{pmatrix}.$$

   (ii) Determine the dimension of this vector space.

   (iii) Show that $\mathbf{v} = \begin{pmatrix} 5 \\ 1 \\ 2 \end{pmatrix}$ lies in the vector space and find a basis, including $\mathbf{v}$, of the vector space spanned by $\mathbf{x}$, $\mathbf{y}$ and $\mathbf{z}$. [MEI]

8. The four polynomials:

$$P_1 = 1 + 4t - 2t^2 + t^3$$
$$P_2 = -1 + 9t - 3t^2 + 2t^3$$
$$P_3 = -5 + 6t + t^3$$
$$P_4 = 5 + 7t - 5t^2 + 2t^3$$

   generate a vector space V.

   (i) Show that P_1, P_2, P_3 and P_4 are not linearly independent.

   (ii) Find a basis for V and the dimension of V. [MEI, adapted]

Linear transformations and their matrices

You will be familiar with transformations of the plane, such as reflection in a line, or rotation about a point. These transformations map $\mathbb{R}^2$ into $\mathbb{R}^2$. The activity below deals with two transformations mapping $\mathbb{R}^3$ into $\mathbb{R}^3$.

Activity

The diagram shows the plane π: $x - y - z = 0$ and the point $P(X, Y, Z)$. P' is the foot of the perpendicular from P to π. P'' is the image of P by reflection in π. Show that:

(i) P' is at $\left(\frac{2}{3}X + \frac{1}{3}Y + \frac{1}{3}Z, \frac{1}{3}X + \frac{2}{3}Y - \frac{1}{3}Z, \frac{1}{3}X - \frac{1}{3}Y + \frac{2}{3}Z\right)$

(ii) P'' is at $\left(\frac{1}{3}X + \frac{2}{3}Y + \frac{2}{3}Z, \frac{2}{3}X + \frac{1}{3}Y - \frac{2}{3}Z, \frac{2}{3}X - \frac{2}{3}Y + \frac{1}{3}Z\right)$.

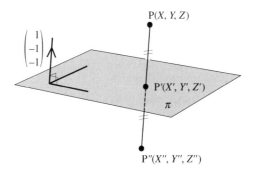

A matrix is often used to represent a transformation. In part (i) of the activity above you showed that:

$$\begin{cases} X' = \frac{2}{3}X + \frac{1}{3}Y + \frac{1}{3}Z \\ Y' = \frac{1}{3}X + \frac{2}{3}Y - \frac{1}{3}Z \\ Z' = \frac{1}{3}X - \frac{1}{3}Y + \frac{2}{3}Z \end{cases}$$

which you can write in the form:

$$\begin{pmatrix} X' \\ Y' \\ Z' \end{pmatrix} = \begin{pmatrix} \frac{2}{3}X + \frac{1}{3}Y + \frac{1}{3}Z \\ \frac{1}{3}X + \frac{2}{3}Y - \frac{1}{3}Z \\ \frac{1}{3}X - \frac{1}{3}Y + \frac{2}{3}Z \end{pmatrix} = \begin{pmatrix} \frac{2}{3} & \frac{1}{3} & \frac{1}{3} \\ \frac{1}{3} & \frac{2}{3} & -\frac{1}{3} \\ \frac{1}{3} & -\frac{1}{3} & \frac{2}{3} \end{pmatrix} \begin{pmatrix} X \\ Y \\ Z \end{pmatrix}.$$

Thus the 3×3 matrix which represents the orthogonal (i.e. perpendicular) projection of $\mathbb{R}^3$ onto the plane $x - y - z = 0$ is $\begin{pmatrix} \frac{2}{3} & \frac{1}{3} & \frac{1}{3} \\ \frac{1}{3} & \frac{2}{3} & -\frac{1}{3} \\ \frac{1}{3} & -\frac{1}{3} & \frac{2}{3} \end{pmatrix}$.

Similarly, since $\begin{pmatrix} X'' \\ Y'' \\ Z'' \end{pmatrix} = \begin{pmatrix} \frac{1}{3} & \frac{2}{3} & \frac{2}{3} \\ \frac{2}{3} & \frac{1}{3} & -\frac{2}{3} \\ \frac{2}{3} & -\frac{2}{3} & \frac{1}{3} \end{pmatrix} \begin{pmatrix} X \\ Y \\ Z \end{pmatrix}$ the matrix $\begin{pmatrix} \frac{1}{3} & \frac{2}{3} & \frac{2}{3} \\ \frac{2}{3} & \frac{1}{3} & -\frac{2}{3} \\ \frac{2}{3} & -\frac{2}{3} & \frac{1}{3} \end{pmatrix}$

represents reflection of $\mathbb{R}^3$ in the plane $x - y - z = 0$.

In both these examples the transformation maps domain $\mathbb{R}^3$ into codomain $\mathbb{R}^3$. But transformations exist where the domain and codomain are not the same set. For example: when we draw a plan of a building with scale $1 : 100$, we start with an object in $\mathbb{R}^3$ and finish with a plane drawing, in $\mathbb{R}^2$. If the plane $z = 0$ in $\mathbb{R}^3$ is horizontal, then this transformation maps (X, Y, Z) to (X', Y') where $\begin{cases} X' = \frac{1}{100} X \\ Y' = \frac{1}{100} Y \end{cases}$

The matrix for this transformation is $\begin{pmatrix} \frac{1}{100} & 0 & 0 \\ 0 & \frac{1}{100} & 0 \end{pmatrix}$.

Notice how the shape of the matrix depends on the dimensions of the domain and codomain. When mapping from $\mathbb{R}^3$ to $\mathbb{R}^2$, we use a 2×3 matrix, with two rows and three columns. Similarly a $m \times n$ matrix defines a transformation from $\mathbb{R}^n$ to $\mathbb{R}^m$.

EXAMPLE

The set $\mathbb{C}$ of complex numbers $z = x + yj$ is a vector space over $\mathbb{R}$, where $z_1 + z_2$ and kz ($k \in \mathbb{R}$) are defined in the usual way. The transformation T maps z to $(1 - 3j)z$. Find the matrix $\mathbf{M}$ which corresponds to T when $\{1, j\}$ is the basis used for the domain $\mathbb{C}$ and the range, also $\mathbb{C}$.

Solution

Using the basis $\{1, j\}$ the complex number $z = x + yj$ may be written as $z = \begin{pmatrix} x \\ y \end{pmatrix}$.

Its image under T is $(1 - 3j)z = (1 - 3j)(x + yj) = (x + 3y) + (-3x + y)j$.

This may be written as $\begin{pmatrix} x + 3y \\ -3x + y \end{pmatrix} = \begin{pmatrix} 1 & 3 \\ -3 & 1 \end{pmatrix} \begin{pmatrix} x \\ y \end{pmatrix}$

so T may be represented by the matrix $\begin{pmatrix} 1 & 3 \\ -3 & 1 \end{pmatrix}$.

For the general theory, let $(V, \oplus, \otimes)$ and $(W, \boxplus, \boxtimes)$ be two vector spaces over the same field: a function T which maps V to W is called a *linear transformation* if and only if:

$$T((\lambda \otimes \mathbf{u}) \oplus (\mu \otimes \mathbf{v})) = (\lambda \boxtimes T(\mathbf{u})) \boxplus (\mu \boxtimes T(\mathbf{v}))$$

for all $\mathbf{u}$ and $\mathbf{v}$ in V and all scalars λ and μ.

Here different symbols are used for addition in the two vector spaces, and different symbols for multiplying by a scalar, since these operations may be different in the two vector spaces (for example the two spaces may have different dimensions). It is unnecessary and tedious to keep showing these distinctions, since the meaning is always clear from the context, so from now on + will denote the addition of vectors in either space, and multiplication by a scalar will be shown by juxtaposition, e.g. $\lambda \mathbf{u}$. Then the condition for a transformation to be linear, which has just been stated, can be written as:

$$T(\lambda \mathbf{u} + \mu \mathbf{v}) = \lambda T(\mathbf{u}) + \mu T(\mathbf{v})$$

for all $\mathbf{u}$ and $\mathbf{v}$ in V and all scalars λ and μ.

Activity

Prove by induction that if T is a linear transformation then

$$T(\lambda_1 \mathbf{a}_1 + \lambda_2 \mathbf{a}_2 + \cdots + \lambda_n \mathbf{a}_n) = \lambda_1 T(\mathbf{a}_1) + \lambda_2 T(\mathbf{a}_2) + \cdots + \lambda_n T(\mathbf{a}_n).$$

There is a very important connection between linear transformations and matrices, which is dealt with in Theorem 3.

Theorem 3

Transformation T is linear if and only if we can represent T by means of a matrix $\mathbf{M}$.

Proof

Suppose T has domain V and codomain W

V has basis $A = \{\mathbf{a}_1, \mathbf{a}_2, \ldots, \mathbf{a}_n\}$

W has basis $C = \{\mathbf{c}_1, \mathbf{c}_2, \ldots, \mathbf{c}_m\}$

T maps $\mathbf{v} = \lambda_1 \mathbf{a}_1 + \lambda_2 \mathbf{a}_2 + \cdots + \lambda_n \mathbf{a}_n$ to $\mathbf{v}' = T(\mathbf{v})$.

Part 1: Proof that if T is a linear transformation then $T(\mathbf{v}) = \mathbf{Mv}$ where $\mathbf{M}$ is a matrix

$$
\begin{aligned}
T(\mathbf{v}) &= T(\lambda_1 \mathbf{a}_1 + \lambda_2 \mathbf{a}_2 + \cdots + \lambda_n \mathbf{a}_n) \\
&= \lambda_1 T(\mathbf{a}_1) + \lambda_2 T(\mathbf{a}_2) + \cdots + \lambda_n T(\mathbf{a}_n) \\
&= \lambda_1(\alpha_1 \mathbf{c}_1 + \alpha_2 \mathbf{c}_2 + \cdots + \alpha_m \mathbf{c}_m) \\
&\quad + \lambda_2(\beta_1 \mathbf{c}_1 + \beta_2 \mathbf{c}_2 + \cdots + \beta_m \mathbf{c}_m) \\
&\quad + \cdots + \lambda_n(\nu_1 \mathbf{c}_1 + \nu_2 \mathbf{c}_2 + \cdots + \nu_m \mathbf{c}_m) \\
&= \lambda_1(\alpha_1 \mathbf{c}_1) + \lambda_2(\beta_1 \mathbf{c}_1) + \cdots + \lambda_n(\nu_1 \mathbf{c}_1) \\
&\quad + \lambda_1(\alpha_2 \mathbf{c}_2) + \lambda_2(\beta_2 \mathbf{c}_2) + \cdots + \lambda_n(\nu_2 \mathbf{c}_2) \\
&\quad + \cdots + \lambda_1(\alpha_m \mathbf{c}_m) + \lambda_2(\beta_m \mathbf{c}_m) + \cdots + \lambda_n(\nu_m \mathbf{c}_m) \\
&= (\lambda_1 \alpha_1 + \lambda_2 \beta_1 + \cdots + \lambda_n \nu_1)\mathbf{c}_1 \\
&\quad + (\lambda_1 \alpha_2 + \lambda_2 \beta_2 + \cdots + \lambda_n \nu_2)\mathbf{c}_2 \\
&\quad + \cdots + (\lambda_1 \alpha_m + \lambda_2 \beta_m + \cdots + \lambda_n \nu_m)\mathbf{c}_m.
\end{aligned}
$$

(1)

> Expressing $T(\mathbf{a}_1)$, $T(\mathbf{a}_2)$, $\ldots$ in terms of the basis C.

You should now distinguish between a vector $\mathbf{v}$ and its representation as a column of numbers, which will vary according to the basis in use. We use $\mathbf{v}_A$ to denote the column vector (i.e. column of numbers) giving $\mathbf{v}$ in terms of the basis A, and similarly $\mathbf{v}'_C$ for the column vector giving $\mathbf{v}'$ in terms of basis C.

Thus $\mathbf{v}_A = \begin{pmatrix} \lambda_1 \\ \lambda_2 \\ \vdots \\ \lambda_n \end{pmatrix}$ and the result obtained in line ① means that

$$\mathbf{v}'_C = \begin{pmatrix} \lambda_1\alpha_1 + \lambda_2\beta_1 + \cdots + \lambda_n\nu_1 \\ \lambda_1\alpha_2 + \lambda_2\beta_2 + \cdots + \lambda_n\nu_2 \\ \vdots \\ \lambda_1\alpha_m + \lambda_2\beta_m + \cdots + \lambda_n\nu_m \end{pmatrix} = \begin{pmatrix} \alpha_1 & \beta_1 & \cdots & \nu_1 \\ \alpha_2 & \beta_2 & \cdots & \nu_2 \\ \vdots & \vdots & & \vdots \\ \alpha_m & \beta_m & \cdots & \nu_m \end{pmatrix} \begin{pmatrix} \lambda_1 \\ \lambda_2 \\ \vdots \\ \lambda_n \end{pmatrix}$$

showing that $\mathbf{v}'_C = \mathbf{M}\mathbf{v}_A$ and proving that T can be represented by the matrix

$$\mathbf{M} = \begin{pmatrix} \alpha_1 & \beta_1 & \cdots & \nu_1 \\ \alpha_2 & \beta_2 & \cdots & \nu_2 \\ \vdots & \vdots & & \vdots \\ \alpha_m & \beta_m & \cdots & \nu_m \end{pmatrix}.$$

Note that $\mathbf{M}$ has m rows and n columns, and that the columns are (respectively) the images of the vectors in the basis of V; you can expect a different matrix if you use a different basis for V or W.

Part 2: Proof that if $T(\mathbf{v}) = \mathbf{M}\mathbf{v}$ where $\mathbf{M}$ is a matrix then T is a linear transformation

The vector $\lambda\mathbf{u} + \mu\mathbf{v}$ is represented by $\lambda\mathbf{u}_A + \mu\mathbf{v}_A$ and $T(\lambda\mathbf{u} + \mu\mathbf{v})$ is represented by the product $\mathbf{M}(\lambda\mathbf{u}_A + \mu\mathbf{v}_A) = \mathbf{M}(\lambda\mathbf{u}_A) + \mathbf{M}(\mu\mathbf{v}_A)$

$$= \lambda(\mathbf{M}\mathbf{u}_A) + \mu(\mathbf{M}\mathbf{v}_A)$$

$$= \lambda\mathbf{u}'_C + \mu\mathbf{v}'_C.$$

But $\lambda\mathbf{u}'_C + \mu\mathbf{v}'_C$ represents $\lambda T(\mathbf{u}) + \mu T(\mathbf{v})$, proving that T is a linear transformation.

Kernel and range

Figure 5.6 shows the plane $\pi: x - y - z = 0$. A transformation $\boldsymbol{T}$ maps any point P to P′, the foot of the perpendicular from P to π. On page 232 you saw that this transformation is represented by a matrix, so $\boldsymbol{T}$ is a linear transformation.

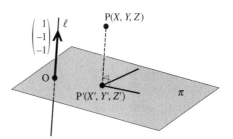

Figure 5.6

The domain of $\boldsymbol{T}$ is the vector space $\mathbb{R}^3$, with dimension 3.

The range of $\boldsymbol{T}$ is π, points with position vectors of the form

$$\beta \begin{pmatrix} 1 \\ 1 \\ 0 \end{pmatrix} + \gamma \begin{pmatrix} 1 \\ 0 \\ 1 \end{pmatrix}, \text{ a vector space with dimension 2.}$$

> You can easily check that $\begin{pmatrix} 1 \\ 1 \\ 0 \end{pmatrix}$ and $\begin{pmatrix} 1 \\ 0 \\ 1 \end{pmatrix}$ are in the plane $x - y - z = 0$.

All points on the line $\ell: \mathbf{r} = t(\mathbf{i} - \mathbf{j} - \mathbf{k})$ are mapped to the origin O, and the set of points which are mapped to O are known as the *null space* or *kernel* of $\boldsymbol{T}$. In this example they have position vectors of the form $\alpha \begin{pmatrix} 1 \\ -1 \\ -1 \end{pmatrix}$, a vector space with dimension 1. It is no accident that the sum of the dimensions of the range and the kernel is the dimension of the domain.

Whenever $\boldsymbol{T}$ is a linear transformation which maps domain V into codomain W, where V and W are vector spaces, then K, the set of vectors which map to $\mathbf{0}$ in W, is called the *null space* or *kernel* of $\boldsymbol{T}$. The set, R, of images of V is the *range* of $\boldsymbol{T}$. The situation is illustrated in figure 5.7. The kernel is found by solving $\boldsymbol{T}(\mathbf{v}) = \mathbf{0}$. If you know the matrix $\mathbf{M}$ which represents $\boldsymbol{T}$ you may solve $\mathbf{M}\mathbf{v}_A = \mathbf{0}$, where, as before, $\mathbf{v}_A$ is the column vector representing $\mathbf{v}$ in the basis being used for the domain.

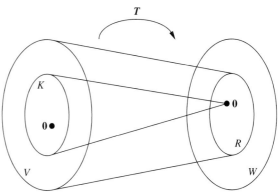

Figure 5.7

Activity

Find the kernel and range of the other transformations examined on pages 232 and 233:

(i) reflecting in $x - y - z = 0$ (ii) drawing a $1:100$ plan.

Activity

Prove that K and R are vector spaces.

Theorem 4

If dim V is finite, $\dim R + \dim K = \dim V$.

Proof

Let $\dim K = k$ and $\dim V = n$, and let $E = \{\mathbf{e}_1, \mathbf{e}_2, \ldots, \mathbf{e}_k\}$ be a basis for K.

Using the result of Question 6, Exercise 5K, we can extend E to form $B = \{\mathbf{e}_1, \mathbf{e}_2, \ldots, \mathbf{e}_k, \mathbf{f}_1, \mathbf{f}_2, \ldots, \mathbf{f}_{n-k}\}$, a basis of V.

Let $\mathbf{g}_i = \boldsymbol{T}(\mathbf{f}_i), i = 1, 2, \ldots, n - k$. We will now prove that $\{\mathbf{g}_i\}$ is a basis of R, so that $\dim R = n - k$ as required.

(i) Suppose $\mathbf{b}$ belongs to R. Then $\mathbf{b} = \boldsymbol{T}(\mathbf{a})$ for some $\mathbf{a}$ belonging to V. Since B is a basis of V:

$$\mathbf{a} = \lambda_1 \mathbf{e}_1 + \lambda_2 \mathbf{e}_2 + \cdots + \lambda_k \mathbf{e}_k + \mu_1 \mathbf{f}_1 + \mu_2 \mathbf{f}_2 + \cdots + \mu_{n-k} \mathbf{f}_{n-k}$$

$$\Rightarrow \quad \mathbf{b} = \boldsymbol{T}(\lambda_1 \mathbf{e}_1 + \lambda_2 \mathbf{e}_2 + \cdots + \lambda_k \mathbf{e}_k + \mu_1 \mathbf{f}_1 + \mu_2 \mathbf{f}_2 + \cdots + \mu_{n-k} \mathbf{f}_{n-k})$$

$$= \lambda_1 \boldsymbol{T}(\mathbf{e}_1) + \lambda_2 \boldsymbol{T}(\mathbf{e}_2) + \cdots + \lambda_k \boldsymbol{T}(\mathbf{e}_k)$$

$$+ \mu_1 \boldsymbol{T}(\mathbf{f}_1) + \mu_2 \boldsymbol{T}(\mathbf{f}_2) + \cdots + \mu_{n-k} \boldsymbol{T}(\mathbf{f}_{n-k})$$

$$= \mu_1 \boldsymbol{T}(\mathbf{f}_1) + \mu_2 \boldsymbol{T}(\mathbf{f}_2) + \cdots + \mu_{n-k} \boldsymbol{T}(\mathbf{f}_{n-k}), \quad \text{since each } \boldsymbol{T}(\mathbf{e}_i) = \mathbf{0}$$

$$= \mu_1 \mathbf{g}_1 + \mu_2 \mathbf{g}_2 + \cdots + \mu_{n-k} \mathbf{g}_{n-k}.$$

Therefore $\{\mathbf{g}_i\}$ spans R.

(ii) $\sum \lambda_i \mathbf{g}_i = \mathbf{0} \Rightarrow \sum \lambda_i \boldsymbol{T}(\mathbf{f}_i) = \mathbf{0}$

$\Rightarrow \boldsymbol{T}(\sum \lambda_i \mathbf{f}_i) = \mathbf{0}$

$\Rightarrow \sum \lambda_i \mathbf{f}_i$ belongs to K

$\Rightarrow \sum \lambda_i \mathbf{f}_i = \sum \mu_j \mathbf{e}_j$

$\Rightarrow \sum \mu_j \mathbf{e}_j - \sum \lambda_i \mathbf{f}_i = \mathbf{0}$

$\Rightarrow \lambda_i = 0$ for each i (and $\mu_j = 0$ for each j) since

> In the summations, i takes values from 1 to $n - k$ and j takes values from 1 to k.

$\{\mathbf{e}_1, \mathbf{e}_2, \ldots, \mathbf{e}_k, \mathbf{f}_1, \mathbf{f}_2, \ldots, \mathbf{f}_{n-k}\}$ is a set of linearly independent vectors, a basis for V.

Therefore $\{\mathbf{g}_i\}$ is a set of linearly independent vectors.

From (i) and (ii) the set of $(n - k)$ vectors $\{\mathbf{g}_i\}$ is a basis of R so that $\dim R = n - k = \dim V - \dim K$, completing the proof.

Exercise 5L

1. Find the matrices which represent the following transformations.

   (i) T maps the quadratic $ax^2 + bx + c$ to the quadratic $cx^2 + bx + a$, given that the basis in use for the domain and range is $\{1, x, x^2\}$.

   (ii) T maps the function $y = a\,e^x + b\,e^{-x} + cx^2$ to $\dfrac{dy}{dx}$, given that the basis in use for the domain is $\{e^x, e^{-x}, x^2\}$ and the basis in use for the range is $\{e^x, e^{-x}, x\}$.

   (iii) T maps $\begin{pmatrix} a & c \\ b & d \end{pmatrix}$ to its *trace*, i.e. $a + d$, given that the basis in use for the domain is
   $$\left\{ \begin{pmatrix} 1 & 0 \\ 0 & 0 \end{pmatrix}, \begin{pmatrix} 0 & 0 \\ 1 & 0 \end{pmatrix}, \begin{pmatrix} 0 & 1 \\ 0 & 0 \end{pmatrix}, \begin{pmatrix} 0 & 0 \\ 0 & 1 \end{pmatrix} \right\}$$
   and for the range is $\{1\}$.

2. D represents differentiation with respect to x.

   (i) The domain of D is the set of differentiable functions.

   (a) By considering $D(a f(x) + b g(x))$ show that D is a linear transformation.

   (b) Identify the kernel of D.

   (ii) The domain of D is now restricted to polynomials of degree $\leqslant 3$, together with the zero polynomial.

   (a) What is the range of D?

   (b) What are the dimensions of the domain and range?

   (c) Show that $\{1, x, x^2, x^3\}$ is a basis for the domain, and, using $\{1, x, x^2\}$ as the basis for the range, find the matrix which represents D.

3. Show that T which maps the point with coordinates (x, y) to the point with coordinates $(x + 2, y - 3)$ is not a linear transformation.

4. The vector space V consists of all functions f such that $\displaystyle\int_0^1 f(x)\,dx$ exists. Transformation T maps f to $\displaystyle\int_0^1 f(x)\,dx$.

   Is T a linear transformation?

5. A transformation is represented by $\begin{pmatrix} 2 & 1 \\ 6 & 3 \end{pmatrix}$.

   Show that its kernel is $\left\{ \begin{pmatrix} x \\ y \end{pmatrix} : 2x + y = 0 \right\}$.

6. Find the null space of the transformations given by the following matrices.

   (i) $\begin{pmatrix} 4 & 2 \\ -5 & 3 \end{pmatrix}$

   (ii) $\begin{pmatrix} 2 & -3 & 1 \\ 1 & -1 & 2 \\ 1 & -2 & -1 \end{pmatrix}$

   (iii) $\begin{pmatrix} 2 & 2 & -1 \\ 2 & 2 & -1 \\ 2 & 2 & -1 \end{pmatrix}$

   (iv) $\begin{pmatrix} 2 & 3 & -1 \\ 5 & 2 & 0 \end{pmatrix}$

   (v) $\begin{pmatrix} 1 & 3 \\ 2 & -1 \\ 1 & 1 \end{pmatrix}$

   (vi) $\begin{pmatrix} 4 & 3 & 2 & -4 \\ 3 & 2 & 1 & -1 \\ 1 & 1 & 1 & -3 \\ 2 & 1 & 0 & 2 \end{pmatrix}$

7. The vector space V consists of all functions f of the form $f(x) = e^{-2x}(a\cos 3x + b\sin 3x)$, where a and b are real numbers. Addition of functions and multiplication of functions by a scalar are defined in the usual way.

   (i) Given that $u_1(x) = e^{-2x}\cos 3x$ and $u_2(x) = e^{-2x}\sin 3x$, show that $\{u_1, u_2\}$ is a basis for V.

   A mapping $T: V \to V$ is defined by $T(f) = f'$, the derivative of f.

   (ii) Show that T is a linear mapping, and find the matrix $\mathbf{M}$ associated with T and the basis $\{u_1, u_2\}$.

   (iii) Write down the inverse matrix $\mathbf{M}^{-1}$, and use it to find $\displaystyle\int e^{-2x}(7\cos 3x + 2\sin 3x)\,dx$.

   [MEI]

8. The linear transformation T maps V to V and T^2 denotes the transformation that maps $\mathbf{v}$ to $T(T(\mathbf{v}))$. Prove that:

   (i) the kernel of T is a subset of the kernel of T^2

   (ii) the range of T^2 is a subset of the range of T.

9. The linear transformation T maps $\mathbb{R}^3$ into $\mathbb{R}^3$. The kernel and range of T have dimensions k and r respectively. Explain why $k + r = 3$ and interpret the situation geometrically for each of the possible values of k.

10. Suppose T is a one-to-one linear transformation, mapping V onto W.

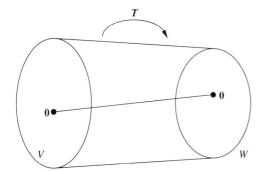

(i) Show that the kernel K consists of just $\mathbf{0}$, $\dim K = 0$, and $\dim V = \dim W$.

(ii) Show that $\mathbf{a} + \mathbf{b} = \mathbf{c} \Leftrightarrow \mathbf{a}' + \mathbf{b}' = \mathbf{c}'$ and $\lambda\mathbf{a} = \mathbf{d} \Leftrightarrow \lambda\mathbf{a}' = \mathbf{d}'$, where $\mathbf{a}$, $\mathbf{b}$, $\mathbf{c}$ and $\mathbf{d}$ belong to V and $\mathbf{a}' = T(\mathbf{a})$, $\mathbf{b}' = T(\mathbf{b})$, etc. Hence show that vector spaces V and W have the same form. Such vector spaces are described as *isomorphic*.

(iii) Deduce that any vector space of dimension n over $\mathbb{R}$ is isomorphic to $\mathbb{R}^n$ over $\mathbb{R}$.

11. Data are transmitted between computers in binary code, blocks of noughts and ones. External 'noise' can corrupt the signal, replacing a nought with a one, or vice versa. Provided there are comparatively few errors there are various ways they may be corrected automatically. One such is known as the *Hamming code*. The binary digits are grouped in fours; there are 16 such blocks, which you should treat as vectors:

$$\begin{pmatrix} 0 \\ 0 \\ 0 \\ 0 \end{pmatrix}, \begin{pmatrix} 1 \\ 0 \\ 0 \\ 0 \end{pmatrix}, \begin{pmatrix} 0 \\ 1 \\ 0 \\ 0 \end{pmatrix}, \begin{pmatrix} 1 \\ 1 \\ 0 \\ 0 \end{pmatrix}, \dots, \begin{pmatrix} 1 \\ 1 \\ 1 \\ 1 \end{pmatrix}.$$

Expand these 4-vectors to 7-vectors as follows.

The 4-vector $\begin{pmatrix} a \\ b \\ c \\ d \end{pmatrix}$ becomes the 7-vector

$$\begin{pmatrix} x_1 \\ x_2 \\ x_3 \\ x_4 \\ x_5 \\ x_6 \\ x_7 \end{pmatrix}, \text{ where } x_3 = a, x_5 = b, x_6 = c, x_7 = d,$$

and x_1, x_2, x_4 are chosen such that:

$$\begin{cases} x_1 + x_3 + x_5 + x_7 = 0 \\ x_2 + x_3 + x_6 + x_7 = 0 \quad (\text{mod } 2). \\ x_4 + x_5 + x_6 + x_7 = 0 \end{cases} \quad \text{①}$$

Data would be transmitted (and received) as 7-vectors.

(i) Find a matrix $\mathbf{C}$ such that pre-multiplying a 4-vector by $\mathbf{C}$ gives the 7-vector.

(ii) Find $\mathbf{D}$ so that pre-multiplying an error-free 7-vector by $\mathbf{D}$ restores the original 4-vector.

(iii) To check whether the digits of a received 7-vector have been corrupted they are substituted into the equations labelled ①. Find a matrix $\mathbf{E}$ so that this can be performed by pre-multiplying the 7-vector by $\mathbf{E}$.

(iv) Pre-multiplying a 7-vector by $\mathbf{E}$ gives a 3-vector. If a single digit in the 7-vector has been corrupted, the error and its position are indicated by the 3-vector. Explain how.

(v) The set V_4 consists of all 4-vectors with components 0 or 1, and V_7 and V_3 are similarly defined. Check that V_4, V_7 and V_3 are vector spaces over the field $\{0,1\}$, where addition and multiplication of scalars are performed $\text{mod } 2$.

(vi) State the dimensions of V_4, V_7 and V_3, and the number of vectors in these three vector spaces.

(vii) What is the significance of the kernel of the transformation represented by $\mathbf{E}$?

Changing bases

Let V be a vector space of dimension 2 and let $A = \{\mathbf{a}_1, \mathbf{a}_2\}$ and $B = \{\mathbf{b}_1, \mathbf{b}_2\}$ be two bases for V, where $\mathbf{b}_1 = 2\mathbf{a}_1 + \mathbf{a}_2$ and $\mathbf{b}_2 = 2\mathbf{a}_1 + 3\mathbf{a}_2$. Then a vector $\mathbf{v}$ belonging to V can be expressed as both $\alpha_1\mathbf{a}_1 + \alpha_2\mathbf{a}_2$ and $\beta_1\mathbf{b}_1 + \beta_2\mathbf{b}_2$ so that:

$$\mathbf{v} = \alpha_1\mathbf{a}_1 + \alpha_2\mathbf{a}_2 = \beta_1\mathbf{b}_1 + \beta_2\mathbf{b}_2$$
$$= \beta_1(2\mathbf{a}_1 + \mathbf{a}_2) + \beta_2(2\mathbf{a}_1 + 3\mathbf{a}_2)$$
$$= (2\beta_1 + 2\beta_2)\mathbf{a}_1 + (\beta_1 + 3\beta_2)\mathbf{a}_2.$$

Since the expression of a vector as a linear combination of the elements of a basis is unique, $\quad \alpha_1 = 2\beta_1 + 2\beta_2$

and $\qquad\qquad\qquad \alpha_2 = \beta_1 + 3\beta_2.$

Thus the vector $\mathbf{v}$ may be written as $\begin{pmatrix} \beta_1 \\ \beta_2 \end{pmatrix}$ when referred to basis B or as

$\begin{pmatrix} 2\beta_1 + 2\beta_2 \\ \beta_1 + 3\beta_2 \end{pmatrix}$ when referred to basis A. This result $\begin{pmatrix} 2\beta_1 + 2\beta_2 \\ \beta_1 + 3\beta_2 \end{pmatrix}$ may be

written as the matrix product $\begin{pmatrix} 2 & 2 \\ 1 & 3 \end{pmatrix}\begin{pmatrix} \beta_1 \\ \beta_2 \end{pmatrix}$, where the columns of the 2×2

matrix are the vectors $\mathbf{b}_1$, $\mathbf{b}_2$ respectively expressed in components by reference to basis A; we shall denote such a matrix by $_A\mathbf{P}_B$, the matrix which changes from basis B to basis A. As previously, attaching a subscript to a

vector can make it clear which basis is in use. For instance, $\mathbf{v}_A = \begin{pmatrix} \alpha_1 \\ \alpha_2 \end{pmatrix}$, the

column vector which contains the components of $\mathbf{v}$ when expressed as a

linear combination of basis A. Similarly, $\mathbf{v}_B = \begin{pmatrix} \beta_1 \\ \beta_2 \end{pmatrix}$, and you have just seen

that $\mathbf{v}_A = {_A\mathbf{P}_B}\mathbf{v}_B$. It is perhaps more useful to note that $\mathbf{v}_B = (_A\mathbf{P}_B)^{-1}\mathbf{v}_A$. In short:

> to change the basis from A to B you pre-multiply the column vector expressing $\mathbf{v}$ relative to A by the inverse of the matrix $_A\mathbf{P}_B$; the columns of $_A\mathbf{P}_B$ are the vectors forming basis B, in the right order, expressed by reference to basis A.

Activity

Given that $\dim V = 3$ and that $A = \{\mathbf{a}_1, \mathbf{a}_2, \mathbf{a}_3\}$ and $B = \{\mathbf{p}, \mathbf{q}, \mathbf{r}\}$ are two bases for

V, where $\mathbf{p}_A = \begin{pmatrix} 1 \\ 1 \\ 2 \end{pmatrix}$, $\mathbf{q}_A = \begin{pmatrix} 2 \\ -1 \\ 3 \end{pmatrix}$, $\mathbf{r}_A = \begin{pmatrix} 0 \\ 2 \\ 1 \end{pmatrix}$, show that $_A\mathbf{P}_B = \begin{pmatrix} 1 & 2 & 0 \\ 1 & -1 & 2 \\ 2 & 3 & 1 \end{pmatrix}$.

The argument is readily generalised: if $A = \{a_1, a_2, \ldots, a_n\}$ and $B = \{b_1, b_2, \ldots, b_n\}$ are two bases for V, then a vector $\mathbf{v}$ may be written as

$$\begin{pmatrix} \alpha_1 \\ \alpha_2 \\ \vdots \\ \alpha_n \end{pmatrix}_A \quad \text{or as} \quad \begin{pmatrix} \beta_1 \\ \beta_2 \\ \vdots \\ \beta_n \end{pmatrix}_B \quad \text{where} \quad \begin{pmatrix} \alpha_1 \\ \alpha_2 \\ \vdots \\ \alpha_n \end{pmatrix} = {}_A\mathbf{P}_B \begin{pmatrix} \beta_1 \\ \beta_2 \\ \vdots \\ \beta_n \end{pmatrix},$$

> Attaching a subscript to the column vector clarifies which basis is being used.

the columns of the matrix being, respectively, the vectors $b_1, b_2, \ldots, b_n$ referred to basis A. That is, $\mathbf{v}_A = {}_A\mathbf{P}_B\mathbf{v}_B$.

Similarly $\mathbf{v}_B = {}_B\mathbf{P}_A\mathbf{v}_A$ and it follows that $({}_A\mathbf{P}_B)^{-1} = {}_B\mathbf{P}_A$.

Activity

(i) The vector $\mathbf{v}$ belongs to vector space $(V, \oplus, \otimes)$. Show that if $\mathbf{v}_A = \begin{pmatrix} \alpha_1 \\ \alpha_2 \\ \vdots \\ \alpha_n \end{pmatrix}_A$

then $\lambda \otimes \mathbf{v}$ may be written as $\lambda\mathbf{v}_A$ however $\oplus$ and $\otimes$ are defined.

(ii) Explain why:
   (a) ${}_A\mathbf{P}_B$ is always square   (b) $\det({}_A\mathbf{P}_B)$ is never 0.

(iii) Given that A, B and C are three bases for V, show that ${}_B\mathbf{P}_C = ({}_B\mathbf{P}_A)({}_A\mathbf{P}_C)$.

Similar matrices

Activity

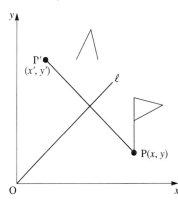

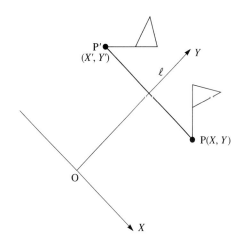

In both diagrams the line ℓ passes through O and the point P$'$ is the image of P under reflection in ℓ. The first diagram shows the usual x- and y-axes; ℓ is inclined at $45°$ to both axes. In the second diagram the axes are X and Y, where the X-axis is perpendicular to ℓ and the Y-axis is along ℓ.
Show that the matrix representing reflection in ℓ is:

(i) $\begin{pmatrix} 0 & 1 \\ 1 & 0 \end{pmatrix}$ when we are referring to the first diagram

(ii) $\begin{pmatrix} -1 & 0 \\ 0 & 1 \end{pmatrix}$ when we are referring to the second diagram.

The above activity demonstrates that the matrix representing a transformation depends on which axes are being used. In the language of vector spaces, the matrix representing a transformation depends on the bases being used. If we change the basis of either the domain or the range of a transformation we must expect to change the matrix associated with that transformation.

The details are simplest in the case where the domain and codomain of the transformation are the same vector space, as in the activity above, and as illustrated in the example below. (The general case, with domain and codomain different vector spaces, is left to Question 10, Exercise 5M.)

EXAMPLE

Matrix $\mathbf{M} = \begin{pmatrix} 3 & 7 \\ 2 & 5 \end{pmatrix}$ represents a transformation relative to the usual basis $A = \{\mathbf{i}, \mathbf{j}\}$.

Find the matrix which represents the same transformation relative to

$$B = \left\{ \begin{pmatrix} 2 \\ 1 \end{pmatrix}, \begin{pmatrix} 2 \\ 3 \end{pmatrix} \right\}.$$

Solution

Pre-multiplying by $_A\mathbf{P}_B = \begin{pmatrix} 2 & 2 \\ 1 & 3 \end{pmatrix}$ changes a position vector from basis B to basis A, so:

$$\mathbf{v}_A = {}_A\mathbf{P}_B\mathbf{v}_B \text{ and } \mathbf{v}'_A = {}_A\mathbf{P}_B\mathbf{v}'_B.$$

But $\mathbf{v}'_A = \mathbf{M}\mathbf{v}_A$ so:

$$_A\mathbf{P}_B\mathbf{v}'_B = \mathbf{M}({}_A\mathbf{P}_B\mathbf{v}_B)$$

$$\Rightarrow \mathbf{v}'_B = ({}_A\mathbf{P}_B)^{-1}\mathbf{M}({}_A\mathbf{P}_B)\mathbf{v}_B$$

and the matrix representing the transformation when using basis B is:

$$({}_A\mathbf{P}_B)^{-1}\mathbf{M}({}_A\mathbf{P}_B) = \begin{pmatrix} 2 & 2 \\ 1 & 3 \end{pmatrix}^{-1} \begin{pmatrix} 3 & 7 \\ 2 & 5 \end{pmatrix} \begin{pmatrix} 2 & 2 \\ 1 & 3 \end{pmatrix}$$

$$= \frac{1}{4}\begin{pmatrix} 3 & -2 \\ -1 & 2 \end{pmatrix}\begin{pmatrix} 13 & 27 \\ 9 & 19 \end{pmatrix} = \frac{1}{4}\begin{pmatrix} 21 & 43 \\ 5 & 11 \end{pmatrix} = \begin{pmatrix} \frac{21}{4} & \frac{43}{4} \\ \frac{5}{4} & \frac{11}{4} \end{pmatrix}.$$

Activity

The results of the previous activity showed that reflection in ℓ is represented by

$\mathbf{M} = \begin{pmatrix} 0 & 1 \\ 1 & 0 \end{pmatrix}$ when we use x- and y-axes, i.e. basis $A = \{\mathbf{i}, \mathbf{j}\}$ and by

$\mathbf{N} = \begin{pmatrix} -1 & 0 \\ 0 & 1 \end{pmatrix}$ when we use X- and Y-axes, i.e. basis $B = \left\{ \begin{pmatrix} \frac{1}{\sqrt{2}} \\ -\frac{1}{\sqrt{2}} \end{pmatrix}, \begin{pmatrix} \frac{1}{\sqrt{2}} \\ \frac{1}{\sqrt{2}} \end{pmatrix} \right\}$.

Write down $_A\mathbf{P}_B$, the matrix which changes $\mathbf{v}_B$ to $\mathbf{v}_A$, and check that $(_A\mathbf{P}_B)^{-1}\mathbf{M}(_A\mathbf{P}_B) = \mathbf{N}$.

Matrices which represent the same transformation referred to different bases are called *similar*, so by a suitable choice of basis, a given transformation can be represented by any one of a set of similar matrices. One obvious question which arises is how to choose the basis so that the transformation matrix is as simple as possible. A partial answer to this has already been given in the work on eigenvectors (see page 37) which showed you how to 'reduce' a square matrix to 'diagonal form', i.e. convert it so that its only non-zero elements are on the leading diagonal.

The process of diagonalising a matrix is essentially the same as representing the transformation by reference to the basis of eigenvectors. This is not always possible, since linearly independent eigenvectors may not exist: for example the shear matrix $\begin{pmatrix} 1 & 1 \\ 0 & 1 \end{pmatrix}$ cannot be reduced to diagonal form.

HISTORICAL NOTE

In 1870 Camille Jordan (1838–1922) proved that every square matrix is similar to a matrix in 'canonical form' in which the only non-zero elements are eigenvalues along the leading diagonal and 1's in particular places immediately above the leading diagonal.

Exercise 5M

1. A point Q has position vector $\begin{pmatrix} 3 \\ 4 \end{pmatrix}$ relative to basis A. Find the position vector of Q relative to basis B when:

   (i) $A = \{\mathbf{i}, \mathbf{j}\}$ and $B = \left\{ \begin{pmatrix} 4 \\ 2 \end{pmatrix}, \begin{pmatrix} 3 \\ 1 \end{pmatrix} \right\}$

   (ii) $A = \left\{ \begin{pmatrix} 3 \\ 2 \end{pmatrix}, \begin{pmatrix} 5 \\ 3 \end{pmatrix} \right\}$ and $B = \left\{ \begin{pmatrix} 1 \\ 1 \end{pmatrix}, \begin{pmatrix} -1 \\ 1 \end{pmatrix} \right\}$.

2. Matrix $\begin{pmatrix} 3 & 5 \\ 2 & 4 \end{pmatrix}$ represents a transformation relative to the usual basis $\{\mathbf{i}, \mathbf{j}\}$. Find the matrix which represents the same transformation relative to $A = \left\{ \begin{pmatrix} 2 \\ 1 \end{pmatrix}, \begin{pmatrix} 1 \\ 2 \end{pmatrix} \right\}$.

3. Matrix $\begin{pmatrix} 1 & 1 & 2 \\ 3 & 2 & 3 \\ -1 & 1 & 2 \end{pmatrix}$ represents a transformation relative to the usual basis $\{\mathbf{i}, \mathbf{j}, \mathbf{k}\}$. Find the matrix which represents the same transformation relative to basis $A = \left\{ \begin{pmatrix} 1 \\ 1 \\ 1 \end{pmatrix}, \begin{pmatrix} 0 \\ 1 \\ 1 \end{pmatrix}, \begin{pmatrix} 0 \\ 0 \\ 1 \end{pmatrix} \right\}$.

Exercise 5M continued

4. The point Q has position vector $\begin{pmatrix} 3 \\ 2 \end{pmatrix}$ relative to basis $\left\{ \begin{pmatrix} 3 \\ 1 \end{pmatrix}, \begin{pmatrix} 1 \\ 2 \end{pmatrix} \right\}$. If Q has position vector $\begin{pmatrix} 1 \\ 1 \end{pmatrix}$ relative to $\left\{ \begin{pmatrix} a \\ 5 \end{pmatrix}, \begin{pmatrix} -2 \\ b \end{pmatrix} \right\}$, find a and b.

5. Show that square matrices **A** and **B** are similar if and only if there exists a matrix **P** such that $\mathbf{A} = \mathbf{P}^{-1}\mathbf{BP}$. See Ex. 1F, Question 13.

6. The vectors **u** and **v** are non-zero. Does the property

   u and **v** are perpendicular $\Leftrightarrow \mathbf{u} \cdot \mathbf{v} = 0$

   depend on the basis in use?

7. The vector space V consists of vectors
   $\begin{pmatrix} a \\ b \\ c \end{pmatrix}$ where a, b and c are real numbers.

   Addition of vectors, and multiplication of a vector by a scalar are defined in the usual way.

   (i) Show that the vectors $\begin{pmatrix} 4 \\ 1 \\ 1 \end{pmatrix}$, $\begin{pmatrix} 3 \\ 2 \\ -1 \end{pmatrix}$ and $\begin{pmatrix} 1 \\ 4 \\ -5 \end{pmatrix}$ are linearly dependent.

   (ii) Given that $\mathbf{e}_1 = \begin{pmatrix} 0 \\ -1 \\ 0 \end{pmatrix}$, $\mathbf{e}_2 = \begin{pmatrix} -1 \\ 0 \\ 1 \end{pmatrix}$ and $\mathbf{e}_3 = \begin{pmatrix} 0 \\ -1 \\ 1 \end{pmatrix}$, show that $\{\mathbf{e}_1, \mathbf{e}_2, \mathbf{e}_3\}$ is a basis for V.

   The mapping T: $V \to V$ is defined by $\mathrm{T}\begin{pmatrix} a \\ b \\ c \end{pmatrix} = \begin{pmatrix} -b \\ a \\ 2c \end{pmatrix}$.

   (iii) Show that T is a linear mapping.

   (iv) Find the matrix associated with T and the basis $\{\mathbf{e}_1, \mathbf{e}_2, \mathbf{e}_3\}$. [MEI]

8. The set $\mathbb{C}$ of complex numbers can be regarded as a real vector space, where addition of complex numbers, and multiplication of a complex number by a real scalar, are defined in the usual way.

   (i) Show that $\{1, j\}$ is a basis for this vector space.

   Let $u = a + bj$ and $v = c + dj$ (where a, b, c and d are real) be fixed numbers. A mapping T: $\mathbb{C} \to \mathbb{C}$ is defined by $\mathrm{T}(z) = uz + vz^*$ (where z^* is the complex conjugate of z).

   (ii) Show that T is a linear mapping.

   (iii) Find the matrix **M** associated with T and the basis $\{1, j\}$.

   (iv) Given that $\mathbf{M} = \begin{pmatrix} 6 & 1 \\ 11 & 2 \end{pmatrix}$, find the complex number z for which $uz + vz^* = 1 + 4j$. [MEI]

9. In this question the position vector of a point is $\begin{pmatrix} x \\ y \end{pmatrix}$ relative to basis $A = \{\mathbf{i}, \mathbf{j}\}$ and $\begin{pmatrix} X \\ Y \end{pmatrix}$ relative to basis $B = \{\mathbf{u}, \mathbf{v}\}$ where **u** and **v** are linear combinations of **i** and **j**.

   (i) Show that $\begin{pmatrix} x \\ y \end{pmatrix} = \mathbf{P} \begin{pmatrix} X \\ Y \end{pmatrix}$, where **P** is the matrix with first column **u** and second column **v**, and that $(x \ \ y) = (X \ \ Y)\mathbf{P}^\mathrm{T}$.

   Hint: See activity on page 25.

   (ii) Now take $\mathbf{u} = \begin{pmatrix} 2 \\ -1 \end{pmatrix}$ and $\mathbf{v} = \begin{pmatrix} 1 \\ 2 \end{pmatrix}$.

   (a) Show that **u** and **v** are perpendicular and of equal length and that $\mathbf{P}^\mathrm{T} = 5\mathbf{P}^{-1}$.

   (b) A transformation is represented by matrix $\mathbf{M} = \begin{pmatrix} 11 & 2 \\ 2 & 14 \end{pmatrix}$ when its domain and range are specified relative to basis A, and by **N** when the domain and range are specified relative to basis B. Find **N**.

(c) A curve C has equation
$11x^2 + 4xy + 14y^2 = 100$. Show that
this equation may be written as
$$(x \ \ y)\begin{pmatrix} 11 & 2 \\ 2 & 14 \end{pmatrix}\begin{pmatrix} x \\ y \end{pmatrix} = (100) \text{ and}$$
deduce that it may also be written as
$2X^2 + 3Y^2 = 4$.

(d) Draw a sketch to show $\mathbf{i}, \mathbf{j}, \mathbf{u}, \mathbf{v}$ and the
curve C. Calculate the length of the
longest chord that can be drawn
between two points on C.

10. Suppose $\mathbf{M}$ represents linear transformation T
from $\mathbb{R}^n$ to $\mathbb{R}^m$, where vectors in $\mathbb{R}^n$ are referred
to basis A, and vectors in $\mathbb{R}^m$ are referred to basis
C, so that $\mathbf{v}_C = \mathbf{M}\mathbf{v}_A$. Suppose B and D are
alternative bases for $\mathbb{R}^n$, $\mathbb{R}^m$, respectively, with
$\mathbf{v}_A = {}_A\mathbf{P}_B\mathbf{v}_B$ and $\mathbf{v}_C = {}_C\mathbf{P}_D\mathbf{v}_D$. Prove that when
vectors in $\mathbb{R}^n$ and $\mathbb{R}^m$ are referred to bases C and
D respectively the matrix representing
transformation T is $({}_C\mathbf{P}_D)^{-1}\mathbf{M}({}_A\mathbf{P}_B)$.

Investigation

The symmetries of the five platonic solids

You will find it a great help to have models of the five platonic solids
(tetrahedron, cube, octahedron, dodecahedron and icosahedron) available to
handle as you work through this investigation. Here we are concerned only with
regular solids.

(i) **The tetrahedron**
Any particular vertex can be moved to any one of the four vertices of the
tetrahedron, and then the other three vertices can be permuted in any of six
ways. Therefore the tetrahedron has $4 \times 6 = 24$ symmetries, corresponding
to the 24 permutations of the four vertices, and so the group of symmetries
of the tetrahedron is isomorphic to S_4. Some of these are *opposite*
symmetries, which involve a change of handedness. It is not possible to
perform these operations on a three-dimensional solid in the space of three
dimensions, though it could be done if the solid were embedded in four-
dimensional space (just as reflecting a rigid two-dimensional shape in a
line cannot be done without using a half-turn about this line, which moves
the shape into the third dimension and then back into its plane). How
many of these opposite symmetries are reflections?

The others are *direct* symmetries, which are rotations, and can be done
physically with a model. There are two distinct types of axis of rotation:

(a) a line joining a vertex to the centre of the opposite face,

(b) a line joining the midpoints of opposite edges.

Find how many rotations of type (a) there are, giving the angles of
rotation, and do the same with type (b). Hence show that the direct
symmetries (including the identity) form a subgroup of order 12, which
is often denoted by A_4. Explain why the opposite symmetries form a coset
of A_4.

(ii) **The cube**
Any particular vertex A can be moved to any one of the eight vertices of
the cube, and then the three vertices nearest to A can be permuted in any
of six ways. The position of the whole cube is determined by the positions

of these four vertices, so there are $8 \times 6 = 48$ symmetries. Here we concentrate on the direct symmetries (rotations). Describe the possible axes of rotation (there are three types) and give the possible angles of rotation. Hence show that the direct symmetries form a group of order 24; this is called the *octahedral* group (see (iv) below).

(iii) Show that the octahedral group is isomorphic to S_4, the group of permutations of four objects. **Hint:** Don't make this difficult – consider the effects of the rotations on the space diagonals of the cube.

(iv) **The octahedron**
Show that the midpoints of the faces of a cube are the vertices of an octahedron, and that the midpoints of the faces of an octahedron are the vertices of a cube. Hence these two solids have the same symmetry properties – if one is transformed into itself then so is the other. Thus the symmetry group of the octahedron is also of order 48, and the rotations form a subgroup (the octahedral group) of order 24, isomorphic to S_4. Check that you can find these 24 rotations of an octahedron.

(v) **The dodecahedron**
The dodecahedron has 12 pentagonal faces, with three faces meeting at each of the 20 vertices. By adapting the argument for the cube given in (ii), show that the dodecahedron has $20 \times 6 = 120$ symmetries. Describe the possible axes of rotation (there are again three types) and give the possible angles of rotation. Hence show that the direct symmetries form a group of order 60; this is called the *dodecahedral* group.

(vi) The dodecahedron has the property that it is possible to choose one diagonal of each pentagonal face so that these 12 lines are the edges of a cube, as in the diagram below. By using each of the five diagonals of one particular face in turn we obtain five cubes inscribed in the dodecahedron.

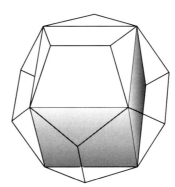

By considering the effect of the symmetry transformations on these cubes, show that the symmetry group of the dodecahedron is isomorphic to S_5, with the dodecahedral group of rotations forming a subgroup A_5, and the set of opposite symmetries forming a coset of A_5.

(vii) **The icosahedron**
Show that the centres of the 12 faces of a dodecahedron may be joined to form an icosahedron, having 12 vertices and 20 triangular faces, and that the midpoints of the faces of an icosahedron are the vertices of a dodecahedron. Deduce that an icosahedron has the same symmetry group as a dodecahedron.

K E Y   P O I N T S

- A *group* $(S, \bullet)$ is a non-empty set S with a binary operation $\bullet$ which is closed and associative, which has an identity element, and in which each element has an inverse element. The group is *Abelian* if and only if $\bullet$ is commutative.

- Two groups $(S, \bullet)$ and $(T, \diamond)$ are *isomorphic* if there is a one-to-one mapping between them which preserves their structure, i.e.

$$a \leftrightarrow x \text{ and } b \leftrightarrow y \Rightarrow a \bullet b \leftrightarrow x \diamond y.$$

- A *subgroup* of a group $(S, \bullet)$ is a non-empty subset of S which forms a group under the same binary operation $\bullet$.

- Lagrange's theorem: the order (i.e. number of elements) of any subgroup of a finite group is a factor of the order of the group.

- The *order* of an element x of a group is the smallest positive power k such that $x^k = e$ (the identity element). The set of powers of x forms a subgroup of order k, called the *cyclic* group *generated* by x.

- A *field* $(S, +, \times)$ is a non-empty set S with two binary operations such that $(S, +)$ and $(S \backslash \{0\}, \times)$ are Abelian groups and $\times$ is distributive over $+$.

- A *vector space* consists of
  (i)   an Abelian group $(V, +)$ whose elements are called vectors;

  (ii) a field F whose elements are called scalars;

  (iii) a binary operation combining a scalar λ with a vector **a** to give a vector λ**a**, with the properties listed on page 215.

- A *basis* of a vector space is a set of linearly independent vectors which spans the space. If a vector space V has a finite basis then every basis of V has the same number of elements; this number is called the *dimension* of V.

- A transformation of a finite-dimensional vector space is *linear* if and only if it can be represented by a matrix. Matrices which represent the same transformation using different bases are *similar*.

Appendix to Chapter 2, Limiting Processes

In Chapter 2 we gave formal definitions of the limiting process:

- We say that $f(x) \to \ell$ as $x \to +\infty$ if and only if, given a positive number ε, however small, there exists X such that $|f(x) - \ell| < \varepsilon$ for all $x > X$; the limit ℓ is denoted by $\lim\limits_{x \to \infty} f(x)$.

- We say that $f(x) \to +\infty$ as $x \to +\infty$ if and only if, given K, however large, there exists X such that $f(x) > K$ for all $x > X$.

- We say that $f(x) \to \ell$ as $x \to a^+$ if and only if, given a positive number ε, however small, there exists δ such that $|f(x) - \ell| < \varepsilon$ whenever $0 < x - a < \delta$; the limit ℓ is denoted by $\lim\limits_{x \to a^+} f(x)$; $\lim\limits_{x \to a^-} f(x)$ is defined similarly.

- If $\lim\limits_{x \to a^-} f(x) = \lim\limits_{x \to a^+} f(x)$ then the common limit is denoted by $\lim\limits_{x \to a} f(x)$.

From these formal definitions you can prove the basic addition, subtraction, multiplication, and division properties of limits that are so important. This appendix indicates the kind of work involved.

Properties of limits

These properties and proofs apply whether the argument of the function is continuous or discrete. In the following: $f_1(x) \to \ell_1$ as $x \to \infty$ and $f_2(x) \to \ell_2$ as $x \to \infty$.

1. Multiplying a function by a constant multiplies the limit by that constant

$$kf_1(x) \to k\ell_1 \text{ as } x \to \infty \text{ where } k \text{ is a constant}$$

This is a special case of Property 3, with $f_2(x) \equiv k = \ell_2$ so a separate proof is unnecessary.

2. The limit of a sum (or difference) is the sum (or difference) of the limits

$$f_1(x) + f_2(x) \to \ell_1 + \ell_2 \text{ and } f_1(x) - f_2(x) \to \ell_1 - \ell_2 \text{ as } x \to \infty$$

The proof for differences follows from Property 1 (with $k = -1$) and the result for sums.

To prove the property for sums we wish to show that:

given $\varepsilon > 0$, $|(f_1(x) + f_2(x)) - (\ell_1 + \ell_2)| < \varepsilon$ for all x greater than some X.

Now $|(f_1(x) + f_2(x)) - (\ell_1 + \ell_2)| = |(f_1(x) - \ell_1) + (f_2(x) - \ell_2)|$

$$\leqslant |(f_1(x) - \ell_1)| + |(f_2(x) - \ell_2)|. \qquad \text{①}$$

Since $f_1(x) \to \ell_1$ and $f_2(x) \to \ell_2$ as $x \to \infty$ we know that, given $\varepsilon' > 0$,

$$\begin{cases} \text{there exists } X_1 \text{ such that } |f_1(x) - \ell_1| < \varepsilon' \text{ for all } x > X_1 & ② \\ \text{there exists } X_2 \text{ such that } |f_2(x) - \ell_2| < \varepsilon' \text{ for all } x > X_2 & ③ \end{cases}$$

If we take X as the larger of X_1, X_2, and ε' as $\tfrac{1}{2}\varepsilon$ (which satisfies $\varepsilon' > 0$) then

$$①, \; ② \text{ and } ③ \; \Rightarrow |(f_1(x) + f_2(x)) - (\ell_1 + \ell_2)| < \tfrac{1}{2}\varepsilon + \tfrac{1}{2}\varepsilon = \varepsilon \text{ for all } x > X$$

which proves that $f_1(x) + f_2(x) \to \ell_1 + \ell_2$ as $x \to \infty$.

3. The limit of a product is the product of the limits

$$f_1(x)f_2(x) \to \ell_1\ell_2 \text{ as } x \to \infty$$

To prove this we wish to show that:

given $\varepsilon > 0$, $|f_1(x)f_2(x) - \ell_1\ell_2| < \varepsilon$ for all x greater than some X. Now

$$\begin{aligned} |f_1(x)f_2(x) - l_1 l_2| &= |(f_1(x) - \ell_1)(f_2(x) - \ell_2) + \ell_1 f_2(x) + \ell_2 f_1(x) - 2\ell_1 \ell_2| \\ &= |(f_1(x) - \ell_1)(f_2(x) - \ell_2) + (f_2(x) - \ell_2)\ell_1 + (f_1(x) - \ell_1)\ell_2| \\ &\leqslant |(f_1(x) - \ell_1)(f_2(x) - \ell_2)| + |(f_2(x) - \ell_2)\ell_1| + |(f_1(x) - \ell_1)\ell_2| \\ &= |(f_1(x) - \ell_1)||(f_2(x) - \ell_2)| + |(f_2(x) - \ell_2)||\ell_1| + |(f_1(x) - \ell_1)||\ell_2| \end{aligned}$$

$$④$$

As before: since $f_1(x) \to \ell_1$ and $f_2(x) \to \ell_2$ as $x \to \infty$ we know that, given $\varepsilon' > 0$,

$$\begin{cases} \text{there exists } X_1 \text{ such that } |f_1(x) - \ell_1| < \varepsilon' \text{ for all } x > X_1 \\ \text{there exists } X_2 \text{ such that } |f_2(x) - \ell_2| < \varepsilon' \text{ for all } x > X_2 \end{cases}$$

so that for all $x > X$, the larger of X_1, X_2:

$$\begin{aligned} ④ \Rightarrow |f_1(x)f_2(x) - \ell_1\ell_2| &< (\varepsilon')^2 + \varepsilon'|\ell_1| + \varepsilon'|\ell_2| \\ &< (\varepsilon')^2 + 2\varepsilon'L, \text{ where } L = \text{larger of } |\ell_1| \text{ and } |\ell_2| \\ &= \varepsilon \text{ if } (\varepsilon')^2 + 2\varepsilon'L - \varepsilon = 0 \text{ i.e. if } \varepsilon' = -L \pm \sqrt{L^2 + \varepsilon} \end{aligned}$$

If we set $\varepsilon' = \sqrt{L^2 + \varepsilon} - L$, where $L = $ larger of $|\ell_1|$ and $|\ell_2|$ we satisfy the requirement that $\varepsilon' > 0$, and have then shown that $|f_1(x)f_2(x) - \ell_1\ell_2| < \varepsilon$ for all $x > X$, which proves that $f_1(x)f_2(x) \to \ell_1\ell_2$ as $x \to \infty$.

4. The limit of a quotient is the quotient of the limits

$$\frac{f_1(x)}{f_2(x)} \to \frac{\ell_1}{\ell_2} \text{ as } x \to \infty \text{ provided } \ell_2 \neq 0.$$

We start by proving that if $f(x) \to \ell$ as $x \to \infty$ then $\dfrac{1}{f(x)} \to \dfrac{1}{\ell}$ as $x \to \infty$ provided $\ell \neq 0$.

As before, $f(x) \to \ell$ as $x \to \infty$ implies that:

given $\varepsilon' > 0$, there exists X such that $|f(x) - \ell| < \varepsilon'$ for all $x > X$

Then
$$\left| \frac{1}{f(x)} - \frac{1}{\ell} \right| = \left| \frac{\ell - f(x)}{\ell f(x)} \right|$$

$$= \frac{|\ell - f(x)|}{|\ell||f(x)|}$$

$$\leqslant \frac{\varepsilon'}{|\ell|(|\ell| - \varepsilon')} \qquad \text{provided } |\ell| - \varepsilon' > 0$$

$$\leqslant \frac{\varepsilon'}{|\ell| \times \frac{1}{2}|\ell|} \qquad \text{provided } \varepsilon' < \frac{1}{2}|\ell|$$

$$= \frac{2\varepsilon'}{\ell^2} < \varepsilon \qquad \text{provided } \varepsilon' < \frac{1}{2}\varepsilon\ell^2$$

So if we are given (small) positive ε we select (small) positive ε' less than the smaller of $\frac{1}{2}|\ell|$ and $\frac{1}{2}\varepsilon\ell^2$. The argument above shows that there exists X such that $|f(x) - \ell| < \varepsilon'$ and $\left| \dfrac{1}{f(x)} - \dfrac{1}{\ell} \right| < \varepsilon$ for all $x > X$, completing the proof that $\dfrac{1}{f(x)} \rightarrow \dfrac{1}{\ell}$ as $x \rightarrow \infty$ provided $\ell \neq 0$.

Combining this result with Property 3 shows that the limit of a quotient is the quotient of the limits.

Similar methods can be used to prove the other properties of limits (when x tends to a or a^+ or a^-).

Answers

Activity (p. 6) The points are on opposite sides of the plane.

Exercise 1A

1. (i) $\sqrt{29}$; (ii) 7; (iii) $\sqrt{104}$
2. (i) 13; (ii) $\sqrt{45}$; (iii) 1
3. (i) 5; (ii) 5; (iii) $5\sqrt{2}$
4. (i) No; (ii) yes; (iii) no (Q is on π)
5. (i) 4; (ii) 0.4; (iii) 0 (ℓ intersects ℓ')
6. (i) (10, 8, 14); (ii) (7, 9, −5); (iii) (−3, 2, 6)
7. (i) A(4, 8, 3), B(9, −2, 8); (ii) Lines meet at A = B = (1, 4, 11);
   (iii) A(1, 8, 20), B(7, −14, 10)
8. $\dfrac{|(\mathbf{a} - \mathbf{c}) \cdot ((\mathbf{a} - \mathbf{b}) \times (\mathbf{c} - \mathbf{d}))|}{|(\mathbf{a} - \mathbf{b}) \times (\mathbf{c} - \mathbf{d})|}$; $89/\sqrt{521}$
11. (ii) $x - 2 = \dfrac{y - 5}{-2} = \dfrac{z - 4}{-3}$
12. (ii) $7 + \tfrac{2}{3}k$; (iii) $-10\tfrac{1}{2}$; (iv) $\mathbf{r} = \begin{pmatrix} 2 \\ 2 \\ -6 \end{pmatrix} + t \begin{pmatrix} 2 \\ 10 \\ 11 \end{pmatrix}$
14. $-\dfrac{(\mathbf{v}_1 - \mathbf{v}_2) \cdot (\mathbf{a}_1 - \mathbf{a}_2)}{|\mathbf{v}_1 - \mathbf{v}_2|^2}$

Exercise 1B

1. (i) 30; (ii) −77; (iii) 26
2. (i) Yes; (ii) no; (iii) no
3. (i) Yes; (ii) yes; (iii) no
4. (i) Left; (ii) neither (coplanar vectors); right
5. (i) 35; (ii) 5; (iii) 0; (iv) (−)36
6. (i) 1; (ii) 2; (iii) $(-)24\tfrac{1}{3}$
9. (ii) $\mathbf{b} + \dfrac{\mathbf{a} \cdot (\mathbf{b} \times \mathbf{d})}{\mathbf{b} \cdot (\mathbf{d} \times \mathbf{e})}\mathbf{e}$
10. (ii) $\mathbf{b} = \sqrt{3}\mathbf{i} + \mathbf{j}$; $\mathbf{c} = \dfrac{2}{\sqrt{3}}\mathbf{i} + \dfrac{2\sqrt{2}}{\sqrt{3}}\mathbf{k}$; (iii) $4\sqrt{2}$

Exercise 1C

1. (i) 30; (ii) −33; (iii) −15; (iv) −2
2. (i) $5(h - 2)(4 - k)$; (ii) 0; (iii) $abc + 2fgh - af^2 - bg^2 - ch^2$
4. (ii) Stretching by scale factor k in one direction only multiplies volume by k.
   (iii) Multiplying any one column by constant k multiplies determinant by k.
5. (i) 430; (ii) −6020; (iii) $86x^3 y$
6. (ii) A shear does not change volume.
7. (i) 43; (ii) 7; (iii) −1
8. $|\mathbf{a} \quad \mathbf{b} \quad \mathbf{a} + \mathbf{b}| = |\mathbf{a} \quad \mathbf{b} \quad \mathbf{a}| = 0$
9. (iii) $a_1 B_1 + a_2 B_2 + a_3 B_3$; $b_1 C_1 + b_2 C_2 + b_3 C_3$; $c_1 A_1 + c_2 A_2 + c_3 A_3$; $c_1 B_1 + c_2 B_2 + c_3 B_3$

10. (i) $\begin{pmatrix} \Delta & 0 & 0 \\ 0 & \Delta & 0 \\ 0 & 0 & \Delta \end{pmatrix}$, $\dfrac{1}{\Delta}\begin{pmatrix} A_1 & A_2 & A_3 \\ B_1 & B_2 & B_3 \\ C_1 & C_2 & C_3 \end{pmatrix}$

(ii) No. (Yes if we restrict ourselves to matrices whose determinant is not 0.)

11. Equation of the straight line joining (x_1, y_1), (x_2, y_2).

12. 5, −8

14. (i) $(a-b)(b-c)(c-a)$; (ii) $(y-z)(z-x)(x-y)$; (iii) $(y-z)(z-x)(x-y)(x+y+z)$;
(iv) $(y-z)(z-x)(x-y)(xy+yz+zx)$

15. $x(x+1)(x-1)^3$

Exercise 1D

1. (i) $\dfrac{1}{3}\begin{pmatrix} 3 & 0 & -6 \\ -4 & 2 & 3 \\ 2 & -1 & 0 \end{pmatrix}$; (ii) none; (iii) $\begin{pmatrix} -0.06 & -0.1 & -0.1 \\ 0.92 & 0.2 & 0.7 \\ 0.66 & 0.1 & 0.6 \end{pmatrix}$; (iv) $\dfrac{1}{21}\begin{pmatrix} 34 & 11 & 32 \\ 9 & 6 & 6 \\ -38 & -16 & -37 \end{pmatrix}$

2. (i) $\begin{pmatrix} -1 & 0 & 2 \\ 9 & -7 & 3 \\ -5 & 7 & -4 \end{pmatrix}$; (ii) $7\mathbf{I}$; (iii) $7\mathbf{I}$; (iv) 7; (v) 49; (vi) $7\mathbf{M}$; $\text{adj}(\text{adj }\mathbf{M}) = (\det \mathbf{M})\mathbf{M}$

3. $\dfrac{1}{7}\begin{pmatrix} 2 & 18 & -11 \\ 2 & 39 & -25 \\ 3 & 41 & -27 \end{pmatrix}$; $x = 8$, $y = 4$, $z = -3$

4. $49\mathbf{I}$; (i) $\frac{1}{49}\mathbf{A}^{\mathrm{T}}$, (ii) 343, (iii) $49\mathbf{I}$

5. (i) $\begin{pmatrix} -5 & 4 & 6 \\ 14 & 16 & 13 \\ 4 & 9 & 8 \end{pmatrix}$, 77; (ii) $\begin{pmatrix} -9 & -4 & -9 \\ 10 & 7 & 14 \\ 8 & 14 & 21 \end{pmatrix}$, 77; (iii) 11; (iv) 7

6. (i) −1.5; (ii) −10; (iii) −2, 4; (iv) −1, 1, 2

9. Yes

Exercise 1E

1. (i) 7, $k(3\mathbf{i} + 2\mathbf{j})$; 2, $k(\mathbf{i} - \mathbf{j})$ where $k \neq 0$
(ii) 4, $k(2\mathbf{i} - 3\mathbf{j})$; −1, $k(\mathbf{i} - 4\mathbf{j})$ where $k \neq 0$
(iii) $1 + \sqrt{2}$, $k(\sqrt{2}\mathbf{i} + \mathbf{j})$; $1 - \sqrt{2}$, $k(\sqrt{2}\mathbf{i} - \mathbf{j})$ where $k \neq 0$
(iv) 2 (repeated), $k(\mathbf{i} - \mathbf{j})$ where $k \neq 0$
(v) 1, $k(4\mathbf{i} + \mathbf{j})$; 0.3, $k(\mathbf{i} + 2\mathbf{j})$ where $k \neq 0$
(vi) p, $k\mathbf{i}$; q, $k\mathbf{j}$ where $k \neq 0$

2. (i) 3, $c\mathbf{i}$; 2, $c\mathbf{j}$; −1, $c(\mathbf{j} - 3\mathbf{k})$ where $c \neq 0$
(ii) 4, $c(\mathbf{i} - \mathbf{j} + 2\mathbf{k})$; 3, $c(\mathbf{i} - 2\mathbf{j} + 2\mathbf{k})$; −1, $c(9\mathbf{i} - 14\mathbf{j} - 2\mathbf{k})$ where $c \neq 0$
(iii) 4, $c(\mathbf{i} + \mathbf{j} + \mathbf{k})$; −3, $c(\mathbf{i} - 6\mathbf{j} + \mathbf{k})$; 0, $c(5\mathbf{i} + 9\mathbf{j} - 7\mathbf{k})$ where $c \neq 0$
(iv) 1, $c(\mathbf{i} + 5\mathbf{j} + \mathbf{k})$; 2 (repeated), $c(p\mathbf{i} + (2p + 4q)\mathbf{j} + q\mathbf{k})$ where $c \neq 0$ and p and q are not both 0
(v) 3, $c\mathbf{j}$; 2, $c(\mathbf{i} + \mathbf{k})$; −2, $c(\mathbf{i} - \mathbf{k})$ where $c \neq 0$
(vi) 9, $c(\mathbf{i} - \mathbf{k})$; 4, $c(\mathbf{i} + \mathbf{j} - 2\mathbf{k})$; 1, $c(\mathbf{i} + \mathbf{j} - \mathbf{k})$ where $c \neq 0$

3. (i) 1, $k(\mathbf{i} + \tan\theta\,\mathbf{j})$; −1, $k(\tan\theta\,\mathbf{i} - \mathbf{j})$
(ii) $\theta \neq n\pi \Rightarrow$ no real eigenvalues;
$\theta = n\pi \Rightarrow$ eigenvalues are $(-1)^n$, and all non-zero vectors are eigenvectors.

4. (i) $\alpha + \beta$; (ii) $\alpha\beta$

6. (i) 9; (ii) $c(2\mathbf{i} + 5\mathbf{j} + 4\mathbf{k})$, 9; $c(\mathbf{i} - 2\mathbf{j} + 2\mathbf{k})$, −9 where $c \neq 0$

8. (i) (a) 2, 3; (b) 4, 9; (c) 32, 243; (d) $\frac{1}{2}, \frac{1}{3}$

(ii) (a) 1, 2, 3; (b) 1, 4, 9; (c) 1, 32, 243; (d) 1, $\frac{1}{2}, \frac{1}{3}$

9. (i) $\mathbf{M}^n\mathbf{v}$ converges to $\mathbf{0}$;

(ii) If $\lambda_1 = 1$, $\lambda\mathbf{M}^n\mathbf{v}$ converges to $\mathbf{s}_1$; if $\lambda_1 = -1$, $\mathbf{M}^n\mathbf{v}$ eventually alternates between $\pm\mathbf{s}_1$;

(iii) The magnitude of $\mathbf{M}^n\mathbf{v}$ increases without limit;

the direction of $\mathbf{M}^n\mathbf{v}$ becomes parallel to $\mathbf{s}_1$.

11. (i) $\mathbf{M} = \begin{pmatrix} 0.5 & 0.5 \\ 0.3 & 0.7 \end{pmatrix}$

(ii) $(100 \quad 100)\mathbf{M}^2 = (76 \quad 124)$ so 76 at Calgary, 124 at Vancouver;

(iii) $\mathbf{x} = (75 \quad 125)$; $\mathbf{xM} = \lambda\mathbf{x}$ with $\lambda = 1$; (iv) 75 at Calgary, 125 at Vancouver.

12. (iv) 4, $c(3\mathbf{i} + \mathbf{j})$; 2, $c(\mathbf{i} + \mathbf{j})$; $\begin{pmatrix} r \\ w \end{pmatrix} = 475e^{4t}\begin{pmatrix} 3 \\ 1 \end{pmatrix} - 425e^{2t}\begin{pmatrix} 1 \\ 1 \end{pmatrix}$

13. (i) Under $\mathbf{M}$ the image of a vector is attracted towards the eigenvector with the numerically largest eigenvalue; each multiplication by $\mathbf{M}$ maps the image closer to a multiple of that eigenvector. This does not happen if $\mathbf{v}_0 = \beta\mathbf{s}_2 + \gamma\mathbf{s}_3$, where $\mathbf{s}_2$, $\mathbf{s}_3$ are eigenvectors with other (numerically smaller) eigenvalues; rounding errors may also cause failures.

(ii) $\dfrac{x_n}{x_{n-1}}$, $\dfrac{y_n}{y_{n-1}}$ and $\dfrac{z_n}{z_{n-1}} \to$ a limit as $n \to \infty$. This limit is an eigenvalue, generally k, the largest eigenvalue (numerically).

Exercise 1 F

1. Note: the columns of $\mathbf{S}$ may be reversed provided the eigenvalues are also reversed.
Each column of $\mathbf{S}$ may (independently) be multiplied by a non-zero constant.

(i) $\mathbf{S} = \begin{pmatrix} 4 & 1 \\ -3 & 1 \end{pmatrix}$, $\boldsymbol{\Lambda} = \begin{pmatrix} 2 & 0 \\ 0 & 9 \end{pmatrix}$

(ii) $\mathbf{S} = \begin{pmatrix} 4 & 5 \\ 2 & 3 \end{pmatrix}$, $\boldsymbol{\Lambda} = \begin{pmatrix} 2 & 0 \\ 0 & 1 \end{pmatrix}$

(iii) $\mathbf{S} = \begin{pmatrix} 1 & 5 \\ 1 & -3 \end{pmatrix}$, $\boldsymbol{\Lambda} = \begin{pmatrix} 1 & 0 \\ 0 & 0.2 \end{pmatrix}$

2. $\begin{pmatrix} 5 & 3 \\ 3 & 2 \end{pmatrix}\begin{pmatrix} 1 & 0 \\ 0 & 0.9 \end{pmatrix}\begin{pmatrix} 2 & -3 \\ -3 & 5 \end{pmatrix}$, $\begin{pmatrix} 4.0951 & -5.1585 \\ 2.0634 & -2.4390 \end{pmatrix}$; approximates to $\begin{pmatrix} 10 & -15 \\ 6 & -9 \end{pmatrix}$

3. (i) $\begin{pmatrix} 876 & -1266 \\ 422 & -601 \end{pmatrix}$; (ii) $\begin{pmatrix} 524\,800 & -523\,776 \\ -523\,776 & 524\,800 \end{pmatrix}$; (iii) $\begin{pmatrix} 0.6667 & 0.3333 \\ 0.6666 & 0.3334 \end{pmatrix}$

4. e.g. (i) $\begin{pmatrix} 3 & 1 \\ 0 & 3 \end{pmatrix}$; (ii) $\begin{pmatrix} 1 & 0 \\ 0 & 1 \end{pmatrix}$; (iii) $\begin{pmatrix} 0 & 1 \\ 0 & 0 \end{pmatrix}$;

(iv) $\begin{pmatrix} 1 & 3 \\ 0 & 0 \end{pmatrix} = \begin{pmatrix} 3 & 1 \\ -1 & 0 \end{pmatrix}\begin{pmatrix} 0 & 0 \\ 0 & 1 \end{pmatrix}\begin{pmatrix} 0 & -1 \\ 1 & 3 \end{pmatrix}$ or $\begin{pmatrix} 6 & 4 \\ 3 & 2 \end{pmatrix} = \begin{pmatrix} 2 & 2 \\ -3 & 1 \end{pmatrix}\begin{pmatrix} 0 & 0 \\ 0 & 8 \end{pmatrix}\begin{pmatrix} \frac{1}{8} & -\frac{1}{4} \\ \frac{3}{8} & \frac{1}{4} \end{pmatrix}$

7. (i) 4; (ii) $k\begin{pmatrix} 1 \\ 0 \\ -1 \end{pmatrix}$, $k \neq 0$; (iii) (b) $p = \frac{1}{8}$, $q = -\frac{1}{8}$, $r = \frac{5}{4}$; (iv) $\frac{1}{2}\begin{pmatrix} -1 \\ -1 \\ 1 \end{pmatrix}$

8. 5; $\begin{pmatrix} 1 & 1 & 1 \\ 1 & 1 & 0 \\ 0 & 1 & -1 \end{pmatrix}\begin{pmatrix} 5 & 0 & 0 \\ 0 & 4 & 0 \\ 0 & 0 & 2 \end{pmatrix}\begin{pmatrix} 1 & 1 & 1 \\ 1 & 1 & 0 \\ 0 & 1 & -1 \end{pmatrix}^{-1}$

10. (i) There are four such products, generally distinct, but **AB** has at most two distinct eigenvalues which may or may not be the product of an eigenvalue of **A** and an eigenvalue of **B**.

(ii) 'Proof' assumes that eigenvector of **A** is eigenvector of **B**.

Exercise 2A

1. (i) (a) Increases without limit; (b) converges to 0; (c) converges to 1;
(d) converges to 1; (e) converges to 1; (f) oscillates infinitely;
(g) converges to 0; (h) converges to $\frac{1}{2}$.

(ii) The least integral N are: (a) $1\,000\,001$; (b) 7; (c) 70; (d) 69; (e) 652; (g) 95; (h) 99

2. (i) Converges to e^3; (ii) converges to 1; (iii) converges to $e^{-1} \approx 0.368$;
(iv) increases without limit; (v) converges to $\ln 20 \approx 3$; (vi) converges to 0.

3. The larger of p or q.

4. $k < -1$: oscillates infinitely; $k = -1$: oscillates finitely; $-1 < k < 1$: converges to 0;
$k = 1$: converges to 1; $k > 1$: increases without limit.

6. If r_n = radius of C_n then $r_{n+1} = r_n \cos\left(\dfrac{\pi}{n}\right)$.

7. (i) (a) 4, 4.5, 4.472 222, 4.472 136;

(b)

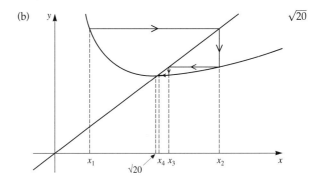

(ii) (b) converges to 0.

8. (i) 0.6429; (ii) alternates between 0.5130 and 0.7995;
(iii) (a) cycles around 0.3828, 0.8269, 0.5009, 0.8750; (b) cycles around 0.1561, 0.5047, 0.9574;
(c) appears chaotic, but remains between 0 and 1.

Exercise 2B

1. (i) $\frac{2}{3}$; (ii) no limit; (iii) 0; (iv) 0; (v) e^3

2. (ii) $\frac{1}{2}$

3. (i) 0; (ii) 0; (iii) no limit; (iv) no limit; (v) -3

4. (i) 0; (ii) $\dfrac{1}{b_n} \to \pm\infty$ as $n \to \infty$

5. e.g.: (i) $a_n = n,\, b_n = 2n$; (ii) $a_n = n + n^{-1},\, b_n = n$; (iii) $a_n = n + 3,\, b_n = n$;
(iv) $a_n = 2n,\, b_n = n$; (v) $a_n = n^2 + (-1)^n n,\, b_n = n^2$; (vi) $a_n = n,\, b_n = n^2$;
(vii) $a_n = 5n + 1,\, b_n = n$; (viii) $a_n = n^2,\, b_n = n$

8. (i) $1 + 2\sqrt{n} + 2n - 2$;

(ii) $1 + 1 + \dfrac{n(n-1)}{2!} \times \dfrac{1}{n^2} + \ldots + \dfrac{1}{n^n} < 1 + 1 + \dfrac{1}{2!} + \ldots + \dfrac{1}{n!}$

$$< 1 + 1 + \tfrac{1}{2}(n-1) = \tfrac{1}{2}n + \tfrac{3}{2} < n \text{ if } n > 3$$

12. (i) (a) 1.82

r	0	1	2	3	4	5	6	7	8
x_r	2.000000	1.732051	1.861210	1.795469	1.828044	1.811683	1.819845	1.815760	1.817801
e_r	0.182879	−0.085070	0.044089	−0.221651	0.010923	−0.005437	0.002725	−0.001361	0.000681
e_{r+1}/e_r	−0.465169	−0.518270	−0.491081	−0.504513	−0.497757	−0.501125	−0.499438	−0.500281	

(ii) (b) 4.472

r	0	1	2	3	4	5	6
x_r	1.000000	10.500000	6.202381	4.713475	4.478314	4.472140	4.472136
e_r	−3.472136	6.027864	1.730245	0.241339	0.006178	0.000004	0.000000

Exercise 2C

1. (i) $2 - \dfrac{1}{n + \frac{1}{2}} \to 2$; (ii) $\dfrac{1}{2} - \dfrac{1}{(n+1)(n+2)} \to \dfrac{1}{2}$;

(iii) $\ln(n+1)$, diverges; (iv) $1 \pm \dfrac{1}{n+1} \to 1$

2. (i) Inconclusive; (ii) converges; (iii) diverges; (iv) converges.

3. $\displaystyle\sum_{r=1}^{\infty} \dfrac{1}{\sqrt{r}}$ diverges.

4. $x \leqslant 0 \Rightarrow$ terms do not tend to 0; $x > 0$ gives geometric progression with common ratio between 0 and 1.

6. (i) Diverges; (ii) still converges; (iii) still converges.

9. Series diverges; calculation assumes it converges.

10. (ii) First series: satisfies conditions for alternating series to converge; second series: terms do not tend to 0.

13. (ii) No: e.g. $\displaystyle\sum_{r=1}^{\infty} \dfrac{(-1)^r}{\sqrt{r}}$ converges, but $\displaystyle\sum_{r=1}^{\infty} \dfrac{1}{r}$ diverges.

For discussion (p. 71)

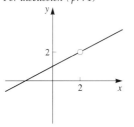

$y = \mathrm{f}(x) = \frac{1}{2}x + 1,\ x \neq 2$
(f(2) is undefined)

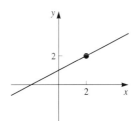

$y = \mathrm{f}(x) = \frac{1}{2}x + 1$

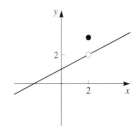

$y = \mathrm{f}(x) = \begin{cases} \frac{1}{2}x + 1,\ x \neq 2 \\ 3,\ x = 2 \end{cases}$

Exercise 2E

1.

	(a) $\lim\limits_{x \to a^-} \mathrm{f}(x)$	(b) $\lim\limits_{x \to a^+} \mathrm{f}(x)$	(c) $\lim\limits_{x \to a} \mathrm{f}(x)$
(i)	no limit	no limit	no limit
(ii)	1	1	1
(iii)	no limit	no limit	no limit
(iv)	0	0	0
(v)	0	0	0
(vi)	3	3	3
(vii)	0	2	no limit

2.

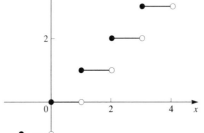

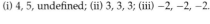
(i) 4, 5, undefined; (ii) 3, 3, 3; (iii) −2, −2, −2.

3.

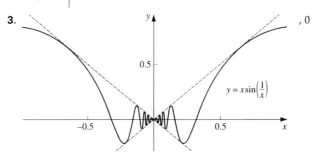

, 0

$y = x \sin\left(\frac{1}{x}\right)$

4.

	ℓ	δ
(i)	e	0.003 67
(ii)	4	0.002 50
(iii)	0	0.009 90
(iv)	2	0.086 66
(v)	0	0.001 54

6. No; e.g. $f(x) = x \Rightarrow \lim_{x \to a} f(x) = 0$, but $\sqrt{f(x)}$ does not exist when $x < 0$.

7. (i) e.g.: $f(x) = \begin{cases} x + 1, \ x > 0 \\ x - 1, \ x < 0 \end{cases}$ gets closer to 0 (and any y, $-1 < y < 1$) as x approaches 0 (from left or from right) but $\lim_{x \to 0} f(x)$ is undefined.

    (ii) e.g.: if $f(x) = [x]$, the largest integer that does not exceed x, then given $\delta = 0.5 > 0$, there exists $\varepsilon = 0.6 > 0$ such that $|f(x) - 4.5| < \varepsilon$ whenever $|x - 5| < \delta$. But $\lim_{x \to 5} f(x)$ is undefined.

8.

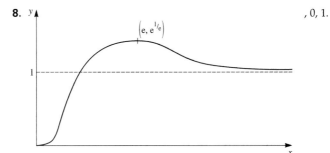

, 0, 1.

10. (i) $y \to e$ as $x \to \infty$; (iii) $\dfrac{dy}{dx} \to 0$ as $x \to \infty$.

(iv)
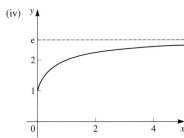

Exercise 2F

1. e.g.: $y = |(x-1)(x-2)|$

2. (iii)
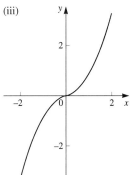
(iv) $2|x|$; (v) no.

3. (i) 2; (ii) 1; (iii) 6; (iv) $\dfrac{1}{3k^{2/3}}$; (v) $\dfrac{\pi}{180}$; (vi) $-\dfrac{1}{2\pi}$

5. (i) Numerator and denominator of $\dfrac{2x-2}{2x-\cos x}$ are not 0 when $x = 0$ so you cannot apply L'Hôpital a second time here. (ii) 2

6. $2\sqrt{a}\cos a$

7. (ii) False, e.g.: $f(x) \equiv |x|$ is continuous but not differentiable at $x = 0$.

8. (i) One application of the rule for differentiating a product.

10.
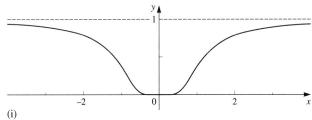
(i)

Exercise 2G

1. 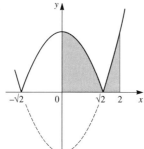 $4(2\sqrt{2}-1)/3$

2. 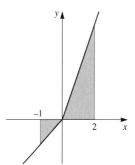 7

3. 0, no area; $(15 \pm \sqrt{33})/4$, area above x-axis $=$ area below x-axis.

4. (i) $4(\sqrt{3}-1) - \dfrac{\pi}{3} \approx 1.88$; (ii) 0; (iii) 0; (iv) $(1+e^2)/2 \approx 4.19$;

(v) $\dfrac{4\pi}{3} \approx 4.19$; (vi) $9\frac{1}{3}$

8. (i) (a) e^{-x^2}; (b) $\dfrac{e^{-x}}{2\sqrt{x}}$

9. (iv) (a) $\dfrac{\pi^2}{4}$; (b) $\dfrac{\pi \ln 2}{8}$

Exercise 2H

1. (i) $m < -1$; (ii) $m > -1$; (iii) none.

2. (i) $\frac{1}{8}$; (ii) $\dfrac{\pi}{3\sqrt{3}}$; (iii) divergent; (iv) 1; (v) $\dfrac{\ln 2}{3}$; (vi) divergent; (vii) $\frac{1}{2}$; (viii) 2; (ix) $\dfrac{\pi}{2}$

3. $\dfrac{1}{x^2} > 0$ for all $x \Rightarrow$ area under curve is positive. Integral should be treated as

$$\lim_{p \to 0^-} \left(\int_{-1}^{p} \frac{1}{x^2}\, dx \right) + \lim_{q \to 0^+} \left(\int_{q}^{1} \frac{1}{x^2}\, dx \right),$$ but neither limit exists.

6. (i) $X > 3 \Rightarrow \dfrac{\ln x}{x} > \dfrac{1}{x} \Rightarrow \displaystyle\int_{3}^{X} \frac{\ln x}{x}\, dx > \cdot \int_{3}^{X} \frac{1}{x}\, dx = [\ln x]_{3}^{X} \to \infty$ as $X \to \infty \Rightarrow \displaystyle\int_{3}^{\infty} \frac{\ln x}{x}\, dx$
diverges.

(ii)

	(a)	(b)	(c)	(d)
Conclusion:	divergent	convergent	divergent	convergent
By comparison with	$\displaystyle\int_{3}^{16} \frac{1}{x-3}\, dx$	$\displaystyle\int_{2}^{\infty} \frac{1}{x^6}\, dx$	$\displaystyle\int_{0}^{\pi} \frac{x - \frac{1}{6}x^3}{x^2}\, dx$	$\displaystyle\int_{1}^{\infty} x\, e^{-x^2}\, dx$

7. (i) $A = \dfrac{\lambda}{2}$; (ii) $B = \dfrac{1}{\pi}$, infinitely large.

8. (i)   (ii) $-\dfrac{\pi \ln 2}{2}$

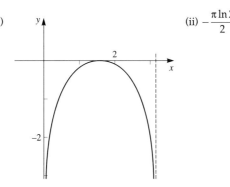

Activity (p. 93)

(i)   (a) Area contained in n rectangles, from rectangle with left edge at $x = 1$ to rectangle with right edge at $x = n + 1$. (b) Area under curve from $x = 1$ to $x = n + 1$;

(iii)

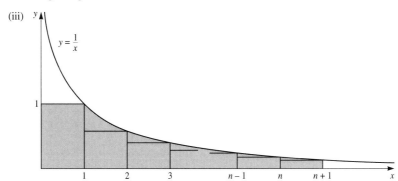

Exercise 2I

1. (i)

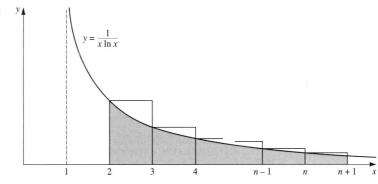

(iii)

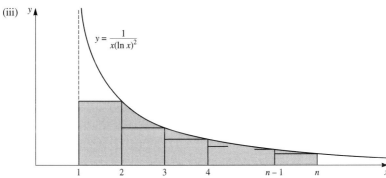

2. Converges to limit between $\dfrac{\pi}{4}$ and $\dfrac{\pi}{2}$.

4. 2

6. (ii) $\dfrac{14}{3}$; (iii) 126 000

(iv) area of rectangles > area under $y = \sqrt{1 + x} \Rightarrow$ exact value of sum > estimate.

7. (i) (b) 101

Exercise 3A

1. (i)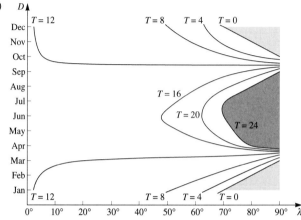

(ii)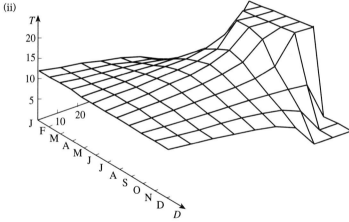

The surface has flat regions since the length of day cannot be negative or more than 24.

2. (i) B; (ii) A; (iii) B.

3.

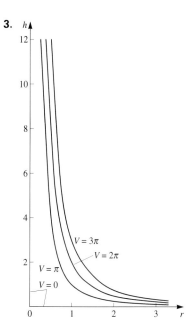

4. If $x^2 + y^2 = a^2$ then $z = f(a^2)$, which is constant. So circles centre O are contours.

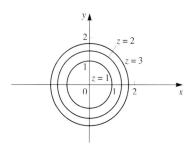

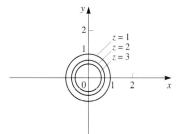

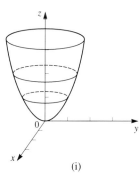

(i)

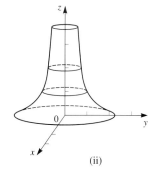

(ii)

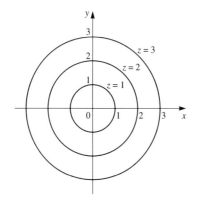

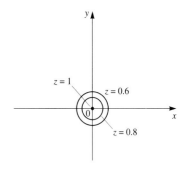

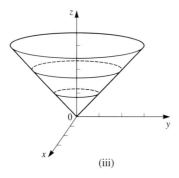

(iii)

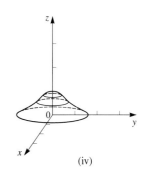

(iv)

5. Contour $z = k\,(\neq 0)$ is the circle $x^2 + y^2 - (x + 2y)/k = 0$, through O with centre $(1/2k, 1/k)$; contour $z = 0$ is the line $x + 2y = 0$.

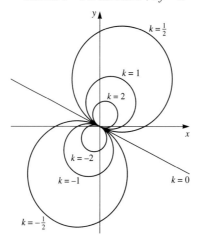

6. (i)

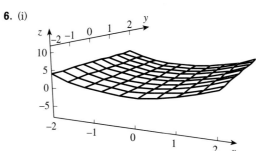

(ii)

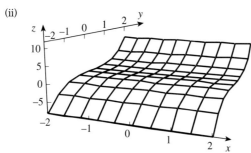

(iii)

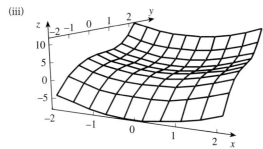

7.

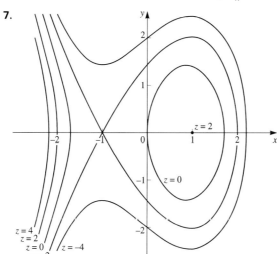

A curved cliff opposite an island.

8. (i)

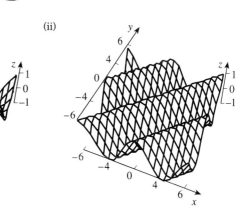

(ii)

(iii)

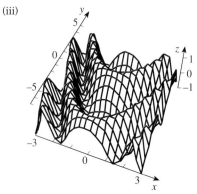

9. (i) $(ct, ct, 0)$; (ii) NE (if the x and y axes are E and N, respectively); (iii) ellipses; (iv) pressure falls exponentially with height.

10.

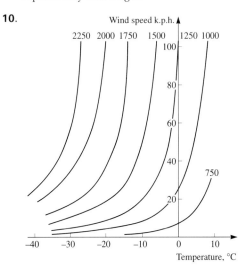

Exercise 3B

1.

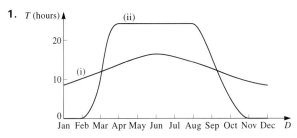

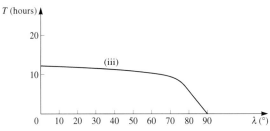

2.

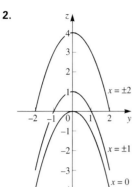

(i)

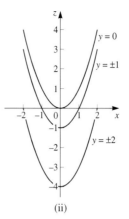

(ii)

3. (i)

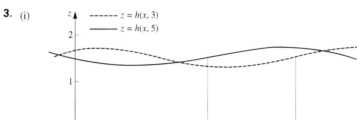

Shows shape of water surface at times $t = 3$ and $t = 5$.

(ii)

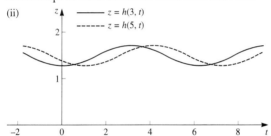

Shows variation of depth at positions $x - 3$ and $x - 5$.

(iii) $2\,\text{m s}^{-1}$

4. (i) $z = 27 - 11y^2$, $z = x^3 - 4x + 1$;

(ii)

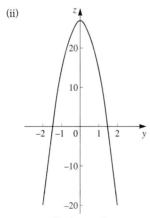

Section $x = 3$

(iii) -22, 23

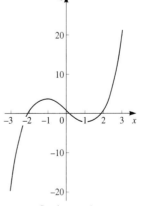

Section $y = 1$

5. Corrugations parallel to $0.4x - 0.3y = 0$, i.e. $y = \dfrac{4x}{3}$. To F: $7.7°$, to N: $-5.8°$.

6. (i) 12.5; (ii) $z = \dfrac{125}{y^2 + 1}$, $z = \dfrac{x^3}{10}$; (iii) -7.5, 7.5

7. $z = (a - 2y)\ln(a^2 + 3y)$, $z = (x - 2b)\ln(x^2 + 3b)$;

$$\frac{3(a - 2b)}{a^2 + 3b} - 2\ln(a^2 + 3b), \quad \frac{2a(a - 2b)}{a^2 + 3b} + \ln(a^2 + 3b).$$

8. (ii) $(10, \mu, \mu/5)$, $(-10, \mu, -\mu/5)$

(iii) The point P dividing EG in the ratio $\alpha : 1 - \alpha$ is $(\lambda, 10 - 20\alpha, (1 - 2\alpha)\lambda/5)$.
If $10 - 20\alpha = \mu$ then P is $(\lambda, \mu, \lambda\mu/50)$. By a similar argument P also lies on FH.

(v) The contour $z = c$ is the hyperbola $xy = 50c$. If $c = 0$ the 'hyperbola' degenerates to the x- and y-axes.

(vi) If $y = x$ then $z = x^2/50$, i.e. $z = u^2/100$, where $u (= \sqrt{2}x)$ is the distance from O to $(x, x, 0)$.

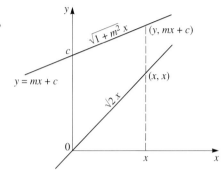

Similarly, on the section cut by the plane $y = mx + c$,

$$z = \frac{mx^2 + cx}{50} = \frac{mu^2 + \sqrt{1 + m^2}cu}{50\sqrt{1 + m^2}}, \text{ where } u (= \sqrt{1 + m^2}x) \text{ is the distance from}$$

$(0, c, 0)$ to $(x, mx + c, 0)$. So every vertical section not parallel to the x- or y-axis is a parabola.

Exercise 3C

1. (i) $\dfrac{2x}{y^3}, -\dfrac{3x^2}{y^4}$;   (ii) $-\dfrac{y}{(x^2 + y^2)}, \dfrac{x}{(x^2 + y^2)}$;

(iii) $\cos y - y\sin x$, $-x\sin y + \cos x$;   (iv) $(1 + \tfrac{1}{2}\sqrt{xy})\,e^{\sqrt{xy}}, \dfrac{x\sqrt{x}}{2\sqrt{y}}\,e^{\sqrt{xy}}$

2. 111, 190

3. At a point $3\,\mathrm{m}$ from the heater 10 minutes after switching the heater on
(i) the temperature decreases by $0.3°\mathrm{C/m}$ as you move away from the heater;
(ii) the temperature increases by $0.1°\mathrm{C/minute}$.

4. $d_x(0.4, 2) = -0.0029$, the gradient of the string at $x = 0.4$ after 2 seconds;
$d_t(0.4, 2) = 1.1$, the speed in $\mathrm{m\,s}^{-1}$ of the point of the string at $x = 0.4$ after 2 seconds.

6. $\dfrac{\partial T}{\partial p} = \dfrac{V - b}{R}, \dfrac{\partial p}{\partial V} = \dfrac{-RT}{(V - b)^2} + \dfrac{2a}{V^3} = \dfrac{aV - 2ab - pV^3}{V^3(V - b)}, \dfrac{\partial V}{\partial T} = \dfrac{RV^3}{pV^3 - aV + 2ab}$

Activity (p. 112) (i) (a) 1200, (b) -24, (c) 0.12; $dy = 12dx$ (ii) $dy = 3x^2\,dx$
Activity (p. 114) (i) (A′) $\Leftarrow$ (B′) (ii) (B′) $\Leftrightarrow$ (C′) (iii) (C′) $\Rightarrow$ (A′)

Activity (p. 115)

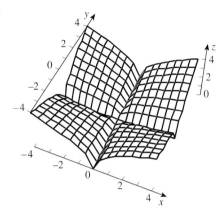

Exercise 3D

1. (i) $\begin{pmatrix} x \\ y \\ z \end{pmatrix} = \begin{pmatrix} 3 \\ 1 \\ 13 \end{pmatrix} + \lambda \begin{pmatrix} 9 \\ -1 \\ -1 \end{pmatrix}$, $z = 9x - y - 13$; (ii) $\begin{pmatrix} x \\ y \\ z \end{pmatrix} = \begin{pmatrix} 4 \\ -8 \\ 1 \end{pmatrix} + \lambda \begin{pmatrix} 3 \\ 1 \\ -4 \end{pmatrix}$, $z = \frac{3}{4}x + \frac{1}{4}y$;

(iii) $\begin{pmatrix} x \\ y \\ z \end{pmatrix} = \begin{pmatrix} 2 \\ 4 \\ 1 \end{pmatrix} + \lambda \begin{pmatrix} 1 \\ 2 \\ 5 \end{pmatrix}$, $z = -\frac{1}{5}x - \frac{2}{5}y + 3$; (iv) $\begin{pmatrix} x \\ y \\ z \end{pmatrix} = \begin{pmatrix} 1 \\ \pi \\ -e \end{pmatrix} + \lambda \begin{pmatrix} e \\ 0 \\ 1 \end{pmatrix}$, $z = -ex$

2. (i) Not a linear equation. (ii) Forgot to evaluate the partial derivatives at (3, 1, 9).
  (iii) $z = 6(x - 3) + 36(y - 1) + 9 = 6x + 36y - 45$

3. -1878

4. Yes. Such dx, dy give the direction of the horizontal line through the particular point and lying in the tangent plane at that point. If this tangent plane is itself horizontal then any dx, dy will do.

5. $(5, -3, -32)$

7. $\begin{pmatrix} x \\ y \\ z \end{pmatrix} = \begin{pmatrix} 3 \\ 4 \\ 75 \end{pmatrix} + \lambda \begin{pmatrix} 6 \\ 8 \\ -1 \end{pmatrix}$, $z = 6x + 8y + 25$

8. (i) $2acx + 6bcy - (a^2 + 3b^2)z = \dfrac{c^2}{k}$

Exercise 3E

1. (i) $\begin{pmatrix} y \\ x \end{pmatrix}$ 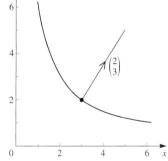 ; (ii) $\begin{pmatrix} 1/y \\ -x/y^2 \end{pmatrix}$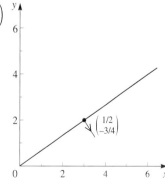

(iii) $\begin{pmatrix} 2x \\ 6y \end{pmatrix}$

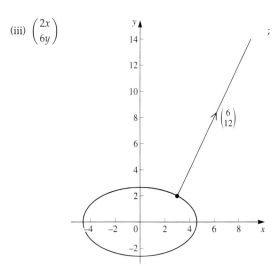

(iv) $\begin{pmatrix} \dfrac{y^2 - x^2}{(x^2 + y^2)^2} \\ \dfrac{-2xy}{(x^2 + y^2)^2} \end{pmatrix}$

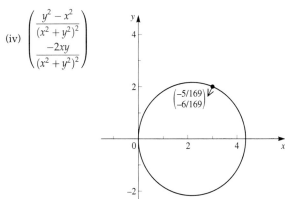

2. N: -6, NE: $\sqrt{2}$, E: 8, SE: $7\sqrt{2}$, S: 6, SW: $-\sqrt{2}$, W: -8, NW: $-7\sqrt{2}$.

3. (i) $-1/\sqrt{2}$, $-35.3°$;   (ii) $9/5\sqrt{5} \approx 0.805$, $38.8°$

4. (i) $(0, 0)$, $(20, 10)$;   (ii) $\begin{pmatrix} 8/\sqrt{65} \\ 1/\sqrt{65} \end{pmatrix} \approx \begin{pmatrix} 0.992 \\ 0.124 \end{pmatrix}$, 0.089;

(iii) $\pm \begin{pmatrix} 7/\sqrt{449} \\ -20/\sqrt{449} \end{pmatrix} \approx \pm \begin{pmatrix} 0.330 \\ -0.944 \end{pmatrix}$

5. $p \sin \theta + q \cos \theta$

Activity (p. 121) $\dfrac{18x + 6y - 20x^3 - 4y^2}{8xy - 6x - 2y}$

Exercise 3F

1. (i) $\dfrac{18x - 7y}{7x + 10y}$;   (ii) $-\dfrac{\cos x}{\cosh y}$;   (iii) $\dfrac{2x^4 + 2xy^3 - x^2}{y^2 - 2x^3y^2 - 2y^5}$;   (iv) $\dfrac{y^2 - xy \ln y}{x^2 - xy \ln x}$

2. $x - 2y = 11$, $2x + y = 7$

3. (i) $\dfrac{xx_1}{a^2} + \dfrac{yy_1}{b^2} = 1$; (ii) $yy_1 = 2a(x + x_1)$; (iii) $xy_1 + x_1 y = 2c^2$

4. (iii) $\left(\dfrac{fh - gb}{ab - h^2}, \dfrac{gh - af}{ab - h^2} \right)$; (iv) $ab - h^2 = 0$ with $fh - gb$ and $gh - af$ not both zero.

Activity (p. 123) (i) Minimum; (ii) saddle; (iii) maximum; (iv) saddle.

Exercise 3G

1. Minimum   2. $(4, 2, 5)$, saddle
3. $(0, 0, 0)$, saddle; $(144, 24, -6912)$, minimum
4. (i) $(1, -1, 0)$, saddle; $(-1, 1, 0)$, saddle; (ii) $(1, 2, 7)$, minimum; $(1, -2, -1)$, saddle;
   (iii) $(2, -\frac{1}{2}, \ln 2 - \frac{1}{2})$, saddle; $(-2, \frac{1}{2}, \ln 2 - \frac{1}{2})$, saddle
5. $(0, \pm 2)$, $(\pm\sqrt{2}, 0)$; maximum
6. (i) $B^2 < 4AC$ and $A < 0$; (ii) $B^2 < 4AC$ and $A > 0$; (iii) $B^2 > 4AC$
7. (ii) Edges in ratio $2 : 2 : 3$, with a square base.
8. Maxima at $(2m\pi, 0, 2)$, minima at $((2m + 1)\pi, 0, -2)$ for all integer m.
9. Each of the lines $y = 2m\pi, z = 0$ (where m is an integer) is a horizontal line of stationary points.
10. $(\pm 2^{1/4}, 2^{-1/4}, 2^{-1/4})$ or $(\pm 2^{1/4}, -2^{-1/4}, -2^{-1/4})$
11. $\theta = \pi/3$, $d = \sqrt{20/\sqrt{3}} \approx 3.40$, $w = 3.92$
12. (iv) $m = 0.7$, $c = 2.15$; (v) Q is a quadratic expression in m, c which is large when the point (m, c) is far from $(0, 0)$ in any direction, so the surface is a 'bowl', and its single stationary point is a minimum point.

Activity (p. 129) (i) $58.7\,\text{m}^2$; (ii) $43.0\,\text{m}^2$

Exercise 3H

1. $(p - q - r)\%$
2. 100; $\delta T \approx 2.64$ (actually $\delta T = 2.69$)
3. $\delta V \approx S\varepsilon$
4. 1%, 2%
5. (i) (A) $V = \frac{1}{3}\pi r^3 \cot\alpha$;   (B) $V = \frac{1}{3}\pi r^2\sqrt{\ell^2 - r^2}$

   (ii) (A) $\dfrac{\delta V}{V} \approx 3\dfrac{\delta r}{r} - \dfrac{\text{cosec}^2\alpha}{\cot\alpha}\delta\alpha = 3\dfrac{\delta r}{r} - 2\,\text{cosec}\,2\alpha \cdot \delta\alpha$;

   (B) $\dfrac{\delta V}{V} \approx \dfrac{2\ell^2 - 3r^2}{r(\ell^2 - r^2)}\delta r + \dfrac{\ell}{\ell^2 - r^2}\delta\ell$

   (iii) (A) $\text{cosec}\,2\alpha$ is least when $\alpha = \pi/4$;
   (B) the coefficient of δr is zero when $2\ell^2 = 3r^2$, i.e. when $\alpha = \arcsin\sqrt{\frac{2}{3}}$
6. (i) $2b$ $2c\cos A$, $2c - 2b\cos A$, $2bc\sin A$,
   (ii) $\delta(a^2) \approx 2(b - c\cos A)\delta b + 2(c - b\cos A)\delta c + 2bc\sin A\delta A$; (iii) $11.2\,\text{cm}$;
   (iv) 11.0%

7. $\delta b \approx b\left(\dfrac{h}{a} + \cot B\,\delta B - \cot(B + C)(\delta B + \delta C)\right)$.

   If $B + C < \pi/2$ the greatest error is $b\left(\dfrac{h}{a} + \alpha\cot B\right)$, when $\delta B = \alpha$ and $\delta C = -\alpha$.

   If $B + C > \pi/2$ the greatest error is $b\left(\dfrac{h}{a} + \alpha(\cot B - 2\cot(B + C))\right)$,

   when $\delta B = \delta C = \alpha$, since $\cot(B + C) < 0$.

Exercise 3I

1. (i) $-14/\sqrt{3} \approx -8.08$;   (ii) $-1/3$;   (iii) $-9/5$
2. (i) $yz + zx + xy + c$;   (ii) $\frac{1}{2}\ln(x^2 + y^2 + z^2) + c$
4. Along the generators of a cone with vertex P, axis in the direction of **grad** g, and semi-vertical angle $\pi/3$.

5. (i) $\begin{pmatrix} 2/\sqrt{38} \\ 5/\sqrt{38} \\ 3/\sqrt{38} \end{pmatrix}$; (ii) 18.5 unit s^{-1}

6. (i) $\dfrac{1}{\sqrt{2}}(-\sin t\, \mathbf{i} + \cos t\, \mathbf{j} + \mathbf{k})$; (ii) $\dfrac{1000\sqrt{2}(6\sin t - t)}{(12\cos t + t^2 + 37)^2}$; (iv) 29.9

Exercise 3J

1. $3x - 10y + 4z = 39$

2. (i) $\begin{pmatrix} x \\ y \\ z \end{pmatrix} = \begin{pmatrix} 3 \\ -1 \\ 1 \end{pmatrix} + \lambda \begin{pmatrix} 6 \\ -1 \\ 2 \end{pmatrix}$, $6x - y - 2z = 17$;

(ii) $\begin{pmatrix} x \\ y \\ z \end{pmatrix} = \begin{pmatrix} -4 \\ 24 \\ -2 \end{pmatrix} + \lambda \begin{pmatrix} -8 \\ -\frac{1}{4} \\ -6 \end{pmatrix}$, $32x + y + 24z = -152$

3. $(12, 14, 6)$ or $(-12, -14, -6)$ **5.** (i) (a) Point $(-a, -b, -c)$; (b) no real points.

7. $66.8°$, parallel to $\begin{pmatrix} 4 \\ -5 \\ 0 \end{pmatrix}$

8. (i) $6x - 2y, \ -2x + 4y, \ 2z + 4$; (ii) $\begin{pmatrix} x \\ y \\ z \end{pmatrix} = \begin{pmatrix} 2 \\ 1 \\ 3 \end{pmatrix} + \lambda \begin{pmatrix} 1 \\ 0 \\ 1 \end{pmatrix}$;

(iii) $(-3, 1, -2)$; (iv) $5, -9$.

9. (iv) $\begin{pmatrix} -2v/bc \\ (v^2 - 1)/ca \\ (v^2 + 1)/ab \end{pmatrix}$; (v) $\begin{pmatrix} 2(1 - u^2v^2)/a^2bc \\ -2(u + v + u^2v + uv^2)/ab^2c \\ 2(uv^2 - u + v - u^2v)/abc^2 \end{pmatrix}$,

which has $-2(1 + uv)/abc$ as a common factor. The tangent plane contains both generators, and $\mathbf{d} \times \mathbf{e}$ is normal to this.

Investigation (p. 137)
(i) (a) $4u^3, 12u^2$; (b) $1 + u\ln u, 1 + \ln u$;

(c) $\arctan\left(\dfrac{2u}{1 - u^2}\right) = 2\arctan u, \ \dfrac{2}{1 + u^2}$

(iv) $\delta W \approx \dfrac{\mathrm{d}W}{\mathrm{d}t}\delta t = \left(\dfrac{\partial W}{\partial T}\dfrac{\mathrm{d}T}{\mathrm{d}t} + \dfrac{\partial W}{\partial R}\dfrac{\mathrm{d}R}{\mathrm{d}t}\right)\delta t$, where t is the time in years

$= (-4.8 \times 0.07 + 1.2 \times (-0.16)) \times 1 = -0.528$ unit/year.

Activity (p. 141) (i) $x\sec\theta + y\,\mathrm{cosec}\,\theta - 15 = 0$; (ii) $\dfrac{x}{L} + \dfrac{y}{\sqrt{225 - L^2}} - 1 = 0$

Exercise 4A

1.

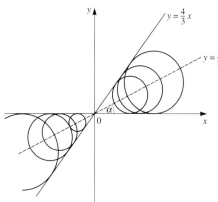

$y = \frac{4}{3}x$

$y = \frac{1}{2}x$

The circles touch $y = 0$ and $y = 4x/3$ (the reflection of $y = 0$ in $y = x/2$).

2. $y^2 = 4ax$.

4. $4(x - 2a)^3 = 27ay^2$

5. (ii) $y = \dfrac{kx}{p} + hp$; (iii) Further steps give lines touching the parabola beyond the points where it touches AX and BX.

6. (ii) $A^2 + B^2 = C^2$

7. (ii) $y = \dfrac{V^2}{2g} - \dfrac{gx^2}{2V^2}$, vertex $(0, V^2/2g)$, focus $(0, 0)$

8. Ellipse $\dfrac{x^2}{2} + y^2 = 1$

9. Hint: Show that the centre of one of the family of circles can be expressed as $(1 + \cos\alpha, \sin\alpha)$. Find the equation of this circle and then use Question 6(ii).

11. (i) The given diameter; (ii) $x^2 + y^2 - 4ax\cos\theta - 4ay\sin\theta + 4a^2\cos^2\theta = 0$, $x = 2a\cos\theta(2\sin^2\theta + 1)$, $y = 4a\sin^3\theta$; or $y = 0$.

[**Hint:** Start by eliminating x between $f = 0$ and $\dfrac{\partial f}{\partial \theta} = 0$.]

Activity (p. 147)

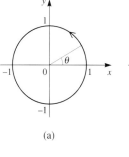

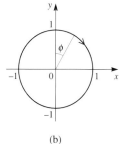

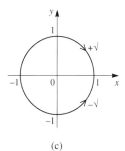

(a)   (b)   (c)

Exercise 4B

1. $8(10^{3/2} - 1)/27 \approx 9.07$. **2.** 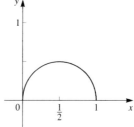 $\pi/4$. **3.** $c\sinh(X/c)$. **6.** $24a$.

7. $\sqrt{1 + k^2}/k$. **8.** $8a$. **11.** (i)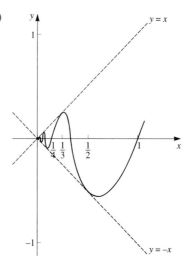

Exercise 4C

1. (i) 45π; (ii) $4\pi a^2$; (iii) $2\pi ac + \pi c^2 \sinh(2a/c)$; (iv) $8\pi a^2(5^{3/2} - 2^{3/2})/3 \approx 70a^2$; (v) $64\pi a^2/3$;
  (vi) $12\pi a^2/5$; (vii) $(2 - \sqrt{2})\pi a^2$; (viii) $32\pi a^2/5$

2. (i) $2\pi a^2(1 - \cos\alpha)$; (ii) $4\pi a^2(\sin\alpha - \alpha\cos\alpha)$

4. (i) $4\pi a^2$; (ii) $2.03 \times 10^8 \text{ km}^2$

6. Hint: The surface area generated by rotating the part of the curve from $x = \dfrac{1}{2m + 1}$ to $x = \dfrac{1}{2m - 1}$

is less than that of the frustum formed by rotating the region bounded by the lines $x = \dfrac{1}{2m + 1}$,

$x = \dfrac{1}{2m - 1}$, $y = \pm x$. Show that the series formed by the areas of such frusta converges, and hence

that the surface area generated by the curve is finite.

Investigation (p. 157)

(vi) $2\pi^2 a^2 b^2$, $4\pi^2 ab$

(vii) On axis of symmetry (a) $\dfrac{4r}{3\pi}$ (b) $\dfrac{2r}{\pi}$ from the diameter

(viii) $2\pi chk$, $4\pi c(h + k)$

Activity (p. 158)

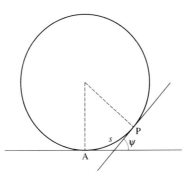

Exercise 4D

1. $s = \ln|\sec\psi + \tan\psi|$ with $s = 0$ when $\psi = 0$.

2. $s = 2a(\sec^3\psi - 1)$ with $s = 0$ when $\psi = 0$.

4. (vi) $x = a(\phi + \sin\phi)$, $y = a(1 - \cos\phi)$, where $\phi = \theta - \pi$.

6. $x = \ln\sin y$

8. Choose the origin so that $z = \dfrac{ak(k-j)}{k^2+1}$ when $\psi = 0$.

Then $z = \dfrac{ak\,e^{(k+j)\psi}}{k+j} = \dfrac{ak\,e^{k\psi}}{\sqrt{k^2+1}}\,e^{j(\psi - \alpha)}$, where $\alpha = \text{arccot}\,k$.

Therefore $r = |z| = \dfrac{ak\,e^{k\psi}}{\sqrt{k^2+1}}$ and $\theta = \arg z = \psi - \alpha$, so $r = \dfrac{ak}{\sqrt{k^2+1}}\,e^{k(\theta+\alpha)}$.

Activity (p. 163) ψ decreases as s increases. *Activity* (p. 164) $\dfrac{-\dfrac{d^2x}{dy^2}}{\left\{\left(\dfrac{dx}{dy}\right)^2 + 1\right\}^{3/2}}$

Exercise 4E

1. $6/10^{3/2} \approx 0.190$, $-12/145^{3/2} \approx -0.006\,87$

2. $-2^{-3/2} \approx -0.354$

5. $\arctan\sqrt{\frac{2}{33}} \approx 13.8°$

6. (i) $2\,\text{cosec}\,2\psi/3a$; (ii) $-2\,\text{cosec}\,2p/3a$;

$\psi = \pi - p$, with positive sense clockwise in (i) and anticlockwise in (ii).

9. $ab(a^2\sin^2\theta + b^2\cos^2\theta)^{-3/2}$ $(= 1/a$ when $a = b)$

11. $2/3\sqrt{3}$, at $(-\frac{1}{2}\ln 2, 1/\sqrt{2})$

15. (iii) $PK = r\,\delta\theta$, $KQ = \delta r$, and $\angle PQK \approx \phi$; (v) **Hint:** Show that $\tan\phi = -\cot(\theta/2)$, and deduce that $\psi = 3\theta/2 + \pi/2$. Then proceed as in (iv).

Exercise 4F

1. $(8, 8)$

2. $\begin{pmatrix} 1/\sqrt{5} \\ 2/\sqrt{5} \end{pmatrix}$, $\begin{pmatrix} -2/\sqrt{5} \\ 1/\sqrt{5} \end{pmatrix}$, $\begin{pmatrix} \pi/4 - 5/2 \\ 9/4 \end{pmatrix}$

4. $(-k\,e^\theta\sin\theta, \, k\,e^\theta\cos\theta)$

6. $a = 3p^2 + 6p^4$, $b = -6p - 8p^3$, $c = -3p^4 - 8p^6$

7. (i) $e < 1/\sqrt{2}$; (ii) $e = 1/\sqrt{2}$; (iii) $e > 1/\sqrt{2}$

For discussion (p. 171)

(i) If the evolute is continuous it passes through the cusp (e.g. $x = p^2$, $y = p^3$), but the evolute may have a discontinuity corresponding to the cusp point (e.g. $x = p^2$, $y = p^5$).

(ii) The evolute is discontinuous; the normal to the curve at the point of inflection is an asymptote of the evolute (e.g. $x = p$, $y = p^3$ at $p = 0$).

(iii) The evolute has a cusp (e.g. $x = p^2$, $y = p$ at $p = 0$).

Exercise 4G

1. $x = \dfrac{c}{2}\left(3t + \dfrac{1}{t^3}\right)$, $y = \dfrac{c}{2}\left(\dfrac{3}{t} + t^3\right)$

2.

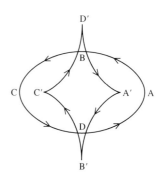

3.

4. $(a(\cos\theta + \sin\theta),\, a(\cos\theta + \sin\theta)),\ (a(\cos\theta - \sin\theta),\, a(\sin\theta - \cos\theta))$

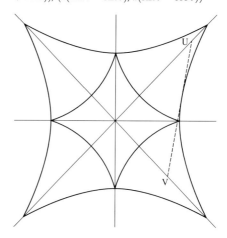

5. (i) $\angle ONP = \alpha$, so $ON = OP\cot\alpha$, so the locus of N is the original spiral enlarged $\times\cot\alpha$ and turned through $\pi/2$ about O.

   (ii) $\angle ONP = \alpha$, which is the angle between ON and the tangent at N, so PN is the tangent at N.

   (iv) Yes, e.g. if α is increased until the normal touches the *original* spiral (when $\tan\alpha = k^{3\pi/2}$).

6.

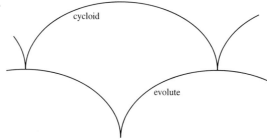

cycloid

evolute

Exercise 4H

1. $4(a^3 - b^3)/ab$

2. (i) $12a$; (ii) $6a$

4. $x = a(\cos\theta + \theta\sin\theta)$, $y = a(\theta\cos\theta - \sin\theta)$.

Activity (p. 180) (i) Finding the modulus of a vector, differentiating a function.
(ii) E.g. finding the mean of 3 numbers, finding the centroid of the triangle formed by 3 points.

Activity (p. 181) (i) Vectors, scalar product; points, finding the distance between 2 points. (ii) Functions (polynomial, rational, or with f(0) = f(1) = 0), composition of functions. (iii) Points, finding the midpoint. (iv) 3-dimensional vectors, vector product.

Activity (p. 181) (i) E.g. 89, 65, −7; (ii) Any 3 factors of 48 (except 1 and 12).

Exercise 5A

1. (i)

$+_5$	0	1	2	3	4
0	0	1	2	3	4
1	1	2	3	4	0
2	2	3	4	0	1
3	3	4	0	1	2
4	4	0	1	2	3

(ii)

$\times_7$	0	1	2	3	4	5	6
0	0	0	0	0	0	0	0
1	0	1	2	3	4	5	6
2	0	2	4	6	1	3	5
3	0	3	6	2	5	1	4
4	0	4	1	5	2	6	3
5	0	5	3	1	6	4	2
6	0	6	5	4	3	2	1

2. (i) Wednesday; (ii) 03.17 next day;
(iii) No (no perfect square has 3 as its final digit).

3. (ii) False, e.g. $x = 0$, $y = 3$.

4. (ii) Neither converse is true.

5. x is divisible by 3 $\Leftrightarrow$ the digit sum of x is divisible by 3.

Examples (p. 183)
A

$+_6$	0	1	2	3	4	5
0	0	1	2	3	4	5
1	1	2	3	4	5	0
2	2	3	4	5	0	1
3	3	4	5	0	1	2
4	4	5	0	1	2	3
5	5	0	1	2	3	4

B $PU = W$

C

followed by	first transformation					
	i	p	q	u	v	w
second transformation i	i	p	q	u	v	w
p	p	q	i	w	u	v
q	q	i	p	v	w	u
u	u	v	w	i	p	q
v	v	w	u	q	i	p
w	w	u	v	p	q	i

D (i) $1, \omega, \omega^2, -1, -\omega, -\omega^2$; (ii)

$\times$	1	ω	ω^2	-1	$-\omega$	$-\omega^2$
1	1	ω	ω^2	-1	$-\omega$	$-\omega^2$
ω	ω	ω^2	1	$-\omega$	$-\omega^2$	-1
ω^2	ω^2	1	ω	$-\omega^2$	-1	$-\omega$
-1	-1	$-\omega$	$-\omega^2$	1	ω	ω^2
$-\omega$	$-\omega$	$-\omega^2$	-1	ω	ω^2	1
$-\omega^2$	$-\omega^2$	-1	$-\omega$	ω^2	1	ω

Activity (p. 187) Only $+$.

Exercise 5B

1. (i)

$\sim$	0	1	2	3	4
0	0	1	2	3	4
1	1	0	1	2	3
2	2	1	0	1	2
3	3	2	1	0	1
4	4	3	2	1	0

(ii) Yes; (iii) no; (iv) yes, 0; (v) yes, each element is self-inverse.

2. (i) Yes; (ii) yes, e.g. $S = \{x : x \geqslant 0\}$, identity 0, no element except 0 has an inverse in S.

3. Both operations are associative.

4. (i) Closed, associative, no identity (0 is not an identity, since $-j \bullet 0 = (-1)^{1/2} = +j$), no inverses.

(ii) Closed, associative, identity 0, no inverses (except that 0 is self-inverse).

(iii) $m = 2k + 1$ (k a positive integer), $q^2 = 2pr$.

5. If $y = \pm k$ then every x except $\mp k$ is a solution.

6. (i) $xy + 3x + 3y + 6$; (ii) $\dfrac{3xy - 2x - 2y + 1}{2xy - x - y}$; (iii) $\dfrac{x + y}{1 - xy}$

7. Not associative; $P = (0, 0)$, $Q = (a_1 - a_2, a_2 - a_1)$, $I = (0, 0)$, no J.

The argument is invalid since it uses associativity.

Activity (p. 189) (i) 0 has no inverse under $\times$

(ii) there is no identity under $+$ (0 is not positive).

Activity (p. 192) E.g., in group B, $PU = UQ$ ($= W$), but $P \neq Q$.

Activity (p. 192) (i) $x = ba^{-1}$ (ii) $x = a^{-1}cb^{-1}$

Exercise 5C

1.

					first transformation				
followed by		I	R	H	S	X	Y	A	B
second transformation	I	I	R	H	S	X	Y	A	B
	R	R	H	S	I	A	B	Y	X
	H	H	S	I	R	Y	X	B	A
	S	S	I	R	H	B	A	X	Y
	X	X	B	Y	A	I	H	S	R
	Y	Y	A	X	B	H	I	R	S
	A	A	X	B	Y	R	S	I	H
	B	B	Y	A	X	S	R	H	I

2. E.g., with no identity,

	a	b	c	d
a	d	c	b	a
b	a	d	c	b
c	b	a	d	c
d	c	b	a	d

4. $x = a^{-1}b, y = ab$

5. (ii), (iii), (iv) are groups, (iii) is Abelian.

6. (ii) $X = \{1, 3, 4, 6, 8\}$

7. E.g. the group of Question 6(i).

8. (ii)

$\times_8$	1	2	3	4	5	6	7
1	1	2	3	4	5	6	7
2	2	4	6	0	2	4	6
3	3	6	1	4	7	2	5
4	4	0	4	0	4	0	4
5	5	2	7	4	1	6	3
6	6	4	2	0	6	4	2
7	7	6	5	4	3	2	1

Not closed. E.g. $\{1, 3\}, \{1, 3, 5, 7\}$

9. All elements of G are self-inverse.

Activity (p. 196) (ii) No.
Activity (p. 197) 6 isomorphisms.

Exercise 5D

1. E.g. $n \leftrightarrow 5n, n \leftrightarrow -5n$

5. $e\ a\ b$, e.g. ($\{1, \omega, \omega^2\}, \times$)
  $a\ b\ e$
  $b\ e\ a$

9. (i), (ii), (iii), (vi) are isomorphic; (iv), (v) are isomorphic.

10. (iii) (a) gives

	e	a	b	c
e	e	a	b	c
a	a	c	e	b
b	b	e	c	a
c	c	b	a	e

(d) gives

	e	a	b	c
e	e	a	b	c
a	a	b	c	e
b	b	c	e	a
c	c	e	a	b

or

	e	a	b	c
e	e	a	b	c
a	a	e	c	b
b	b	c	e	a
c	c	b	a	e

or

	e	a	b	c
e	e	a	b	c
a	a	e	c	b
b	b	c	a	e
c	c	b	e	a

Activity (p. 199) No, not closed.

Exercise 5E

1. $\{I\}$, $\{I, H\}$, $\{I, X\}$, $\{I, Y\}$, $\{I, A\}$, $\{I, B\}$, $\{I, R, H, S\}$, $\{I, H, X, Y\}$, $\{I, H, A, B\}$

2. (i) (1 2 3 4), (2 1 4 3), (3 4 1 2), (4 3 2 1);
    (ii) (1 2 3 4 5 6 7 8), (2 3 4 1 7 8 6 5), (3 4 1 2 6 5 8 7), (4 1 2 3 8 7 5 6),
         (5 8 6 7 1 3 4 2), (6 7 5 8 3 1 2 4), (7 5 8 6 2 4 1 3), (8 6 7 5 4 2 3 1)

3. E.g. G = (integers, +), H = {multiples of 2}, K = {multiples of 3};
    $H \cup K$ is not closed under $+$, since $2 + 3 \notin H \cup K$.

5. (iii) E.g. {all rotations about any fixed point},
    {rotations through multiples of $2\pi/n$ about any fixed point}.

6. Matrices of the form $\begin{pmatrix} 1 & 0 & q \\ 0 & 1 & 0 \\ 0 & 0 & 1 \end{pmatrix}$.

7. E.g. $ba \notin \{e, a, b, c, d\}$.

Activity (p. 202) (i) $IQ = QH = H$, $VH = WH = \{U, V, W\}$. The union is S.

Exercise 5F

1. $XK = RK = \{X, Y, R, S\}$, $AK = \{I, A, H, B\}$

4. (iv) E.g. the subgroup $\{I, X\}$ of the symmetry group of the square is not normal, since
    $A^{-1}XA = Y \notin \{I, X\}$.

Investigation (p. 204)

(i)

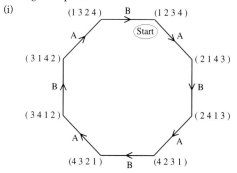

(iv)

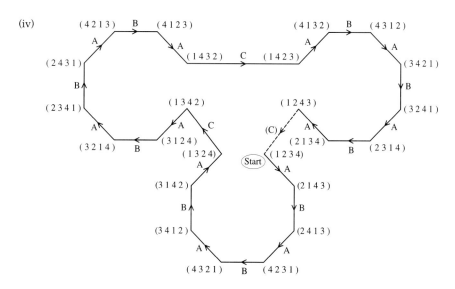

(v)

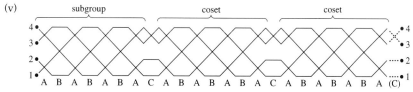

Activity (p. 207) (i) 6; (ii) 2; (iii) 3; (iv) 6

Activity (p. 207) Yes, *A* and *D* are cyclic.

Exercise 5G

1.

Element	I	R	H	S	X	Y	A	B
Order	1	4	2	4	2	2	2	2

$\{I\}$, $\{I, H\}$, $\{I, X\}$, $\{I, Y\}$, $\{I, A\}$, $\{I, B\}$, $\{I, R, H, S\}$

4. (ii) True, false; (iii) k, n have no common factor.

5. 9 and 20 have no common factor, so rotation through $9\pi/10$ generates the whole set of rotations. Other possible values of n (< 20) are 1, 3, 7, 11, 13, 17, 19.

7. (i) $H = R^2$, $S = R^3$, $Y = R^2 X$, $A = RX$, $B = R^3 X$

10. Numbers in G are the products of distinct primes with no repetitions. Show that, if A is the set of prime factors of a, etc., then the correspondence $a \leftrightarrow A$ is an isomorphism between G with operation • and the set of subsets of prime numbers with operation Δ, the symmetric difference (as in Exercise 5C, Question 6).

Activity (p. 210) Irrationals not closed under $+$ or $\times$.

Activity (p. 211) If $a \neq 0$ then $ab = 0 \Rightarrow a^{-1}(ab) = a^{-1} \times 0 = 0$ (by the Example)
$\Rightarrow (a^{-1}a)b = 0 \Rightarrow 1b = 0 \Rightarrow b = 0$.

Activity (p. 211) Rotation about O through θ, where $\cos\theta : \sin\theta : 1 = x : y : \sqrt{x^2 + y^2}$, followed by enlargement with scale factor $\sqrt{x^2 + y^2}$.

Activity (p. 211) Rotation about O through $\pi/2$.

Exercise 5H

4. If and only if m is prime.

5. Identities: 1 for $\oplus$, 0 for $\otimes$; inverses: $2 - m$ for $\oplus$, $m/(m-1)$ for $\otimes$ ($m \neq 1$).

6. $(2 + j)z + (4 - 5j) = (1 + 3j)z$, $\begin{pmatrix} -2.8 & 0.6 \\ -0.6 & -2.8 \end{pmatrix}$

7. $\begin{pmatrix} 5 & 3 \\ -3 & 5 \end{pmatrix}$ or $\begin{pmatrix} 4 & -1 \\ 1 & 4 \end{pmatrix}$

8. $\begin{pmatrix} -1 & 0 \\ 0 & -1 \end{pmatrix}$

9. (i) Taking the conjugate; (iii) $\det(\mathbf{Z}_1\mathbf{Z}_2) = \det\mathbf{Z}_1 \det\mathbf{Z}_2$

Exercise 5I

2. 1   **3.** D and D'.   **5.** (i) Yes;   (ii) no: fails N;   (iii) yes;   (iv) yes;   (v) no: fails N.

7. No: fails D and D'.

Exercise 5J

2. Linearly independent: (i), (iii); not linearly independent: (ii), (iv).

3. (i) Yes; (ii) yes; (iii) yes; (iv) no.

4. (iii) The set of all vectors contained in the plane $x - y + z = 0$.

7. (ii) (a) $m < n \Rightarrow \mathbf{a}_1, \mathbf{a}_2, \ldots, \mathbf{a}_n$ are not linearly independent vectors.

       (b) $m > n$ tells you nothing: $\mathbf{a}_1, \mathbf{a}_2, \ldots, \mathbf{a}_n$ may or may not be linearly independent.

9. (i) (a) Yes; (b) no; (c) yes; (d) yes; (e) yes; (f) no.

    (ii) The vectors are already linearly dependent; decreasing one pile will make them linearly independent. Take the whole of the pile of 7 matches.

Exercise 5K

1. 2

2. 5

5. (i) True; (ii) true; (iii) false.

7. (i) $2a - 4b - 3c = 0$; (ii) 2; (iii) $\mathbf{v} = 2\mathbf{x} + \mathbf{y}$; e.g. $\{\mathbf{v}, \mathbf{x}\}$ or $\{\mathbf{v}, \mathbf{y}\}$.

8. (i) e.g. $P_4 = P_1 + P_2 - P_3$ or $P_3 = 2P_2 - 3P_1$; (ii) e.g. $\{P_1, P_2\}$, 2.

Exercise 5L

1. (i) $\begin{pmatrix} 0 & 0 & 1 \\ 0 & 1 & 0 \\ 1 & 0 & 0 \end{pmatrix}$; (ii) $\begin{pmatrix} 1 & 0 & 0 \\ 0 & -1 & 0 \\ 0 & 0 & 2 \end{pmatrix}$; (iii) $(1 \quad 0 \quad 0 \quad 1)$

2. (i) (b) {f: f(x) = constant}

(ii) (a) {Polynomials of degree $\leqslant 2$ together with the zero polynomial}; (b) 4, 3;

(c) $\begin{pmatrix} 0 & 1 & 0 & 0 \\ 0 & 0 & 2 & 0 \\ 0 & 0 & 0 & 3 \end{pmatrix}$

4. Yes

6. (i) $\left\{ \begin{pmatrix} 0 \\ 0 \end{pmatrix} \right\}$; (ii) $\left\{ \mathbf{r}: \mathbf{r} = t \begin{pmatrix} 5 \\ 3 \\ -1 \end{pmatrix} \right\}$; (iii) $\left\{ \mathbf{r}: \mathbf{r} = t \begin{pmatrix} 1 \\ 0 \\ 2 \end{pmatrix} + s \begin{pmatrix} 0 \\ 1 \\ 2 \end{pmatrix} \right\}$;

(iv) $\left\{ \mathbf{r}: \mathbf{r} = t \begin{pmatrix} 2 \\ -5 \\ -11 \end{pmatrix} \right\}$; (v) $\left\{ \begin{pmatrix} 0 \\ 0 \end{pmatrix} \right\}$; (vi) $\left\{ \mathbf{r}: \mathbf{r} = t \begin{pmatrix} 1 \\ -2 \\ 1 \\ 0 \end{pmatrix} + s \begin{pmatrix} 0 \\ -2 \\ 5 \\ 1 \end{pmatrix} \right\}$

7. (ii) $\begin{pmatrix} -2 & 3 \\ -3 & -2 \end{pmatrix}$; (iii) $\frac{1}{13}\begin{pmatrix} -2 & -3 \\ 3 & -2 \end{pmatrix}$, $\frac{1}{13}e^{-2x}(-20\cos 3x + 17\sin 3x) + c$

9.

r	3	2	1	0
Range	The whole of $\mathbb{R}^3$	A plane containing O	A line through O	The origin O only
Kernel	The origin O only	A line through O	A plane containing O	The whole of $\mathbb{R}^3$

11. (i) $\begin{pmatrix} 1 & 1 & 0 & 1 \\ 1 & 0 & 1 & 1 \\ 1 & 0 & 0 & 0 \\ 0 & 1 & 1 & 1 \\ 0 & 1 & 0 & 0 \\ 0 & 0 & 1 & 0 \\ 0 & 0 & 0 & 1 \end{pmatrix}$; (ii) $\begin{pmatrix} 0 & 0 & 1 & 0 & 0 & 0 & 0 \\ 0 & 0 & 0 & 0 & 1 & 0 & 0 \\ 0 & 0 & 0 & 0 & 0 & 1 & 0 \\ 0 & 0 & 0 & 0 & 0 & 0 & 1 \end{pmatrix}$; (iii) $\begin{pmatrix} 1 & 0 & 1 & 0 & 1 & 0 & 1 \\ 0 & 1 & 1 & 0 & 0 & 1 & 1 \\ 0 & 0 & 0 & 1 & 1 & 1 & 1 \end{pmatrix}$;

(iv) Error indicated by non-zero product; the product, read from bottom to top and interpreted as a number in binary, gives position of error. (v) Yes. (vi) 4, 7, 3; 16, 128, 8.

(vii) The kernel consists of the 16 error-free 7-vectors which represent the original 16 4-vectors.

Exercise 5M

1. (i) $\begin{pmatrix} \frac{9}{2} \\ -5 \end{pmatrix}$; (ii) $\begin{pmatrix} \frac{47}{2} \\ -\frac{11}{2} \end{pmatrix}$

2. $\begin{pmatrix} \frac{14}{3} & \frac{16}{3} \\ \frac{5}{3} & \frac{7}{3} \end{pmatrix}$

3. $\begin{pmatrix} 4 & 3 & 2 \\ 4 & 2 & 1 \\ -6 & -2 & -1 \end{pmatrix}$

4. $a = 13, b = 2$

6. Yes

7. (iv) $\begin{pmatrix} -1 & -1 & -3 \\ -1 & 0 & -1 \\ 1 & 2 & 3 \end{pmatrix}$

8. (iii) $\begin{pmatrix} a+c & d-b \\ d+b & a-c \end{pmatrix}$; (iv) $-2+13j$

9. (ii) (b) $\begin{pmatrix} 10 & 0 \\ 0 & 15 \end{pmatrix}$;

(d) C is an ellipse, centred at O, with major axis along $x + 2y = 0$; maximum chord is $2\sqrt{10}$ long.

Investigation (p. 245)
(i) 6 reflections. (a) 4 axes, angles $2\pi/3$ or $4\pi/3 = 8$ rotations
  (b) 3 axes, angle $\pi = 3$ rotations. $8 + 3 + 1(\text{identity}) = 12$
(ii) (a) 3 axes joining midpoints of faces, angles $\pi/2$ or π or $3\pi/2 = 9$ rotations
  (b) 6 axes joining midpoints of opposite edges, angle $\pi = 6$ rotations
  (c) 4 axes joining opposite vertices, angles $2\pi/3$ or $4\pi/3 = 8$ rotations.

$$9 + 6 + 8 + 1 = 24.$$

(iii) The orientation of the cube is determined by the position of the 4 space diagonals, which are permuted by the rotations.
(v) (a) 6 axes joining midpoints of faces, angles $2\pi/5$, $4\pi/5$, $6\pi/5$, $8\pi/5 = 24$ rotations
  (b) 15 axes joining midpoints of opposite edges, angle $\pi = 15$ rotations
  (c) 10 axes joining opposite vertices, angles $2\pi/3$, $4\pi/3 = 20$ rotations.

$$24 + 15 + 20 + 1 = 60.$$

Index

Index